TOP SECRET

LIGHTCRAFT
FLIGHT HANDBOOK
LTI-20

A Riveting Preview of 2025 Space Command's
LTI-20 Lightcraft

Leik N. Myrabo
John S. Lewis

TOP SECRET

2025 Space Command Mission: The LTI-20 lightcraft must transport a 12 person crew to the far side of the planet, loiter in the atmosphere on-station for 2 weeks undetected, and return to the continental United States without refueling in the conventional sense – a "Mission Impossible" for all but beamed energy propulsion.

DEDICATION
This book is dedicated to the memory of the late Arthur Kantrowitz, without whose vision, beamed energy propulsion and lightcraft technology would still be decades away.

All rights reserved under article two of the Berne Copyright Convention (1971).
We acknowledge the financial support of the Government of Canada through the Book Publishing Industry Development Program for our publishing activities.

Published by Apogee Books, an imprint of Collector's Guide Publishing Inc.,
Burlington, Ontario, Canada, L7R 2B5
http://www.cgpublishing.com
http://www.apogeebooks.com
Printed and bound in Canada

Lightcraft Flight Handbook *LTI-20*
ISBN: 978-1-926592-03-9 – ISSN: 1496-6921
©2009 Apogee Books

TOP SECRET

LIGHTCRAFT
FLIGHT HANDBOOK
LTI-20

A Riveting Preview of 2025 Space Command's
LTI-20 Lightcraft

Leik N. Myrabo
John S. Lewis

TOP SECRET

2025 Space Command Mission: The LTI-20 lightcraft must transport a 12 person crew to the far side of the planet, loiter in the atmosphere on-station for 2 weeks undetected, and return to the continental United States without refueling in the conventional sense – a "Mission Impossible" for all but beamed energy propulsion.

~~TOP SECRET~~

T.O. 1LTI-20A-1

USSC SERIES **LTI-20A** AND **LTI-20B** (AND LATER LIGHTCRAFT)	THIS PUBLICATION INCORPORATES OPERATIONAL SUPPLEMENT T.O. 1LTI-2A-1S-170, DATED 12 FEB. 2025. SEE TECHNICAL ORDER INDEX T.O. 01-01-05 AND SUPPLEMENTS THERETO FOR CURRENT STATUS OF LIGHTCRAFT TECHNICAL MANUALS, SAFETY AND OPERATIONAL SUPPLEMENTS, AND FLIGHT CREW CHECKLIST. COMMANDERS ARE RESPONSIBLE FOR BRINGING THIS PUBLICATION TO THE ATTENTION OF ALL AFFECTED LIGHTCRAFT PERSONNEL.

LIGHTCRAFT
FLIGHT HANDBOOK
LTI-20

Leik N. Myrabo
John S. Lewis

Contributing Editors: Dan A. Gross, Kenneth A. Myrabo, John M. Likakis, Marilyn R.P. Morgan
Prepared By: Space Defense Automated Printing Service (SDAPS).

DISTRIBUTION STATEMENT: Distribution authorized to U.S. Government agencies and their contractors for Administrative or Operational Use, 1 June 2025. Other requests for this document shall be referred to Space Command Hdqtrs.
DESTRUCTION NOTICE: Destroy by any method that will prevent disclosure of contents or reconstruction of the document.

PUBLISHED UNDER AUTHORITY OF THE SECRETARY OF THE U.S. SPACE COMMAND.

DOD/SPACE COMMAND 17 Oct. 2025 REPRINT BASIC AND ALL CHANGES HAVE BEEN MERGED TO MAKE THIS A COMPLETE PUBLICATION.	Q **1 JULY 2020** CHANGE 5 – 1 JUNE 2025 (DECLASSIFIED 17 OCT. 2025)

~~TOP SECRET~~

PREFACE

Re: Declassification of LTI-20 in Official Float-Out Ceremonies

The 21st Century's most revolutionary spacecraft, just recently unveiled, is the ultra-lightweight, hyper-energetic LTI-20 Lightcraft. The official float-out ceremonies on December 1, 2025 were hosted at the corporate headquarters of Lightcraft Technologies International (press photo below), an event that was extensively covered on *WorldNet Virtual* (WNV). Because of the high level of secrecy placed on Beamed Energy Propulsion (BEP) technology, few realize that this "hypersonic airship" first entered flight status three years ago. The LTI-20B, which is designed to carry a crew of 12 in its primary role of planetary defense against extraterrestrial threats, has also proven to be an effective agent in the war against international terrorism because of its ultra-stealth capabilities.

LTI-20 Unveiling at Lightcraft Technologies International on 1 Dec. 2025.

The **LTI-20 Lightcraft Flight Handbook**, intended originally for training of military Lightcraft crews and other affected personnel, contains no DOD classified material (downgraded 17 October 2025). Certain aspects of the BEP engine and launch system, sensor arrays (Chapter 11), and planetary defense missions (Chapter 21) are omitted from this document for obvious reasons. Note that, because of the requirement of nuclear-powered, directed energy sources for military launches, civilian transport versions of the LTI-20 are not presently practical. Simply put, the commercial space power-beaming infrastructure for BEP has yet to be built.

Nevertheless, NASA is to be applauded for successfully deploying the first operational prototype (i.e., SPS-01) of a 20 GW rechargeable power-beaming station into low Earth orbit (see Chapter 19) – a threshold event that portends a bright future for space solar

PREFACE

power in both BEP and terrestrial electric grid applications. I see this as a perfect role for government and industry collaboration: Building the energy-beaming infrastructure for "Highways of Light." And because one month ago, this NASA station easily tractored the LTI-20 into LEO, I predict that the price of civilian Lightcraft flights will fall by 100- to 1000-fold below that of chemical rockets in the coming decade, finally making spaceflight affordable to a large cross-section of the world's population. The long-awaited revolution in low cost, highly reliable space access is clearly on the horizon – a spin-off of military BEP / Lightcraft technology developed over the past two decades.

William Q. Storer, Ph.D.
Secretary of the U.S. Space Command
31 December 2025

AUTHOR'S FOREWORD

What is this book about? The *LTI-20 Lightcraft Flight Handbook* examines the potential of today's very real, advanced technology and physics to create a phenomenal revolution in low-cost, safe space access – with beam-propelled Lightcraft. The 20-meter diameter disc-shaped, microwave-powered LTI-20 is proposed for US Space Command's hyperenergetic "space superiority" missions in the year 2025. The LTI-20 is, in essence, a hypersonic balloon (i.e., pressure airship structure) propelled by electromagnetic engines that are efficient, noiseless, and environmentally benign, with extraordinary performance levels that heretofore were thought to be in the realm of science fiction. The emerging *Lightcraft Revolution* will ultimately replace today's commercial jet airlines with hypersonic Beamed Energy Propulsion (BEP) craft, able to transport passengers to the far side of the planet in less than an hour – at competitive prices. Such craft will fly on microwave and laser energy beamed from satellite power stations, eliminating dependency upon limited fossil fuel supplies. Imagine spacecraft so energetic that a 3 G flight to orbit requires throttle settings barely above idle. In this alternative future, getting to space is finally "easy" – in sharp contrast to today's familiar rocket engines that must run at the ragged edge of melting down or blowing up. BEP flight technology will enable passengers to call for a Lightcraft flight, much like a cab is hailed today, and within minutes their flight will arrive at the local community "LightPort" to speed them to the destination. This revolutionary approach to space access must first establish an energy-beaming infrastructure upon which mass-produced Lightcraft can ride.

This document is a "pre-flight briefing handbook" that would be given to the crew of the LTI-20 for indoctrination during training or prior to a military mission. The level of treatment is highly detailed, covering all aspects of the ship and its technology: e.g., vehicle structures, beamed energy propulsion, human factors and crew support functions, flight operations, command and control systems, emergency systems, etc., all supplemented by a comprehensive index.

Why have we written a "technical manual" for a vehicle that does not yet exist? Our goal is to convey a technically accurate understanding of the structure and operation of this highly innovative and unfamiliar new technology in a visually engaging and interesting manner. For this purpose, and as an aid to understanding and visualization – and in "welcoming the reader aboard" – we have adopted the unusual approach of casting our story in the "flight manual" format.

What is the "story" behind this book? The Lightcraft Flight Handbook is the culmination of a two decade-long adventure at Rensselaer Polytechnic Institute (RPI), ranging from theory to flight experiments with scale models, to create a positive, ultra-energetic vision for the future of space flight. This manual constitutes a detailed "case study" (one of three at RPI) of BEP applied to Lightcraft designed for affordable and safe access to space.

Why the 2025 Space Command connection? Exemplary "Horizon Missions" for the Space Command's LTI-20 (and power-beaming infrastructure) include:
a) the war against terrorism,
b) the ability to visit remote and unprepared landing sites throughout the world,
c) planetary defense against both internal and extraterrestrial threats, and
d) Earth-launch of supplies and personnel for a lunar base or interplanetary missions.

History reveals that the military has often been the "bridge" to revolutionary flight technology for the commercial sector.

How "real" is this Lightcraft stuff? Instead of meaningless "techno-babble," this manual explains the logical extension of remarkable, tested capabilities afforded by Lightcraft technology. In total, the manual describes a very real alternative future for manned exploration of space enabled by BEP. The purpose of this book is to introduce a

AUTHOR'S FOREWORD

viable and exciting vision for the future of flight – one that could easily transform how we access low Earth orbit. Such BEP craft will enable safety levels equal to or exceeding today's jet airlines while reducing carbon dioxide emissions to near zero: they will fly on energy beamed from remote solar power stations in orbit, instead of limited fossil fuels. The technology to enable such BEP Lightcraft is within our grasp. As a society, we must decide at what point we cease to consume valuable fossil fuels in jet aircraft and rocket launch vehicles, cutting back on carbon dioxide emissions in the process. Creating this BEP transformation – "making it real" – is a matter of will, rather than "access to technology." We note, and deplore, that the 40 years that have passed by since man first walked on the moon have brought us no closer to making space flight accessible to the masses. Lightcraft technology can bring safe, affordable, and quiet access to space for us all, not just for an elite astronaut corps and billionaires.

To achieve maximum realism, extensive studies of the design of laser and microwave-propelled vehicles have been pursued in RPI's "TransAtmospheric Vehicle Design" (TAVD) course for the past 15 years. The present case study employed the Horizon Mission Methodology (HMM) developed by the late John Anderson, formerly of NASA Headquarters in Washington, D.C.

When did this adventure begin? It all started in 1991 with a phone call to Leik from Lee Valentine at the Space Studies Institute (SSI) in Princeton, NJ, offering a challenge and a proposal on behalf of the SSI. Basically, Lee affirmed that the SSI Board of Directors (of which John S. Lewis is a member) was quite impressed with Leik's laser propelled Lightcraft designs, but (and the conversation went something like this): *Frankly, we're still waiting for those gigawatt lasers that can propel your Lightcraft into orbit. Whereas in our estimate, gigawatt-level microwave and millimeter wave transmitters are feasible today – if enough sources are assembled into a phased array. Do you think it's feasible for a Lightcraft to fly into orbit on a powerful microwave beam?* Up to this moment, Leik had been exclusively pre-occupied with laser propulsion, but assured Lee that he would give serious thought to the microwave alternative. To make it tempting, SSI offered to team Leik with any scientist in the world in a six-month SSI-supported effort to solve a critical problem facing microwave Lightcraft: after a brief study, it had become crystal clear that microwave-powered Lightcraft must, in essence, be "hypersonic balloons," and that active control over their aerothermodynamics, on a massive scale, was perhaps **the** potential "show stopper." In short, a Directed Energy AirSpike (DEAS) was needed to streamline this blunt body (balloon) instead of a "massive" conical forebody.

Within months, SSI successfully forged a collaboration with Prof. Yuri Raizer (Institute for Problems in Mechanics, Russian Academy of Sciences, Moscow) to develop the theoretical foundations for DEAS airspike physics. The results of this ground-breaking research collaboration were published as an AIAA technical paper in June 1994. Exactly one year later, the first successful experimental demonstration of the airspike concept was performed in Rensselaer's Hypersonic Shock Tunnel (HST), simulated with an electric arc in Mach 10 flow. Aided by Prof. Henry Nagamatsu, Jack Marsh, Paulo Toro, Marco Minucci, Don Messitt, Ryan Bracken, Dean Meloney, Ernie Diaz, Joyel Kerl, Matt Filippelli, Keith Shanahan, Chris Hartley, Tom Portwood, Greg Mann, Chaz Misiewicz, and several others, increasingly complex airspike experiments were carried out in the HST facility, conclusively proving that airspikes do indeed work. Dean Head and Junghwa Seo investigated airspike performance along representative launch trajectories. Dr. Mikhail Shneyder (Princeton University) developed sophisticated computational fluid dynamics (CFD) simulations based on experimental airspike data. Airspike physics are now very well understood, as hypersonic laser-induced airspike experiments are underway at the Henry T. Nagamatsu Laboratory of Aerothermodynamics and Hypersonics at IEAv-CTA in Sao Jose dos Campos, Brazil – by Col. Marco Minucci, Dr. Paulo Toro, Dr. Antonio Carlos Oliveira, J. Brosler Jr., and Israel Salvador.

ACKNOWLEDGEMENTS

Over two hundred RPI students have contributed to the microwave Lightcraft Case Study in collaboration with their instructor (Prof. Leik Myrabo). The study began with the 1992 Fall semester *Transatmospheric Vehicle Design (TAVD)* course to create a single-place, 10-m diameter microwave Lightcraft. Team members included: Joseph Carroll, Hung-Sheng Chern, John Dec, Kevin Donofrio, Leonard Goldschmidt, Joel Limmer, Paul Messer, Martin Rahn, Velimir Randic, Gregory Ruderman, Cary Schenkenberger, James Sroczynski, Warren Trent, Christopher Wood, and Leonard Yowell. This first of its kind investigation of hypersonic pressure-airships, propelled by microwave power identified the requirement for DEAS to drastically cut aerodynamic drag and heat transfer.

Student teams in two different Spring 1993 courses wrestled with ancillary Lightcraft technology that ultimately influenced the contents of this LTI-20 Lightcraft Flight Handbook: 1) *Introduction to Space Technology* (special term project: Advanced Crew Escape Pod – ACEP): Team members included: Eric Achman, Christopher Bahner, Edward Barney, Robert Becker, Karen Craver, Thomas-James Curry, James Dairaghi, Kristopher Daniel, Kamalkishore Dasari, Mark Denley, Erin Edwards, Brenda Falwell, Anthony Flroy, Rachael Flynn, James Gateau, Jason Hylan, Arwen Isaac, Ja'Affar Ismail, John Kim, Chad Lagace, Hefen McCormick, Jeeshan Naqvi, Frederick Oko, Daniel Orchard-Hayes, Charles Player, Martin Polinski, Senter Reinhardt, Anthony Sage, George Sandall, Edward Schernau, Phoumchay Soratana, Kurt Stresau, Melanie Torres, Christopher Vorhees, Kentaro Watanabe, and Michael Wilson. 2) *Theory of Propulsion* (special term project: Microwave Lightcraft Propulsion & Performance): Charles Alex, Chanaka Amarasinghe, Chris Aulbach, Kenneth Bates, Coreen Bombardier, Kenneth Bouchard, John Capizzi, Joseph Carroll, Hung-Sheng Chern, Eric Cohen, Denise Combs, Gregory De LaFuente, John Dec, Anthony DeGeorge, Mark Doll, Dwight Drumtra, Christos Ekonomidis, Brandon Engle, David Fennell, Sandra Fruehauf, Ria Galanos, James Hingst, William Jellig, Corinne Johnson, Joel Limmer, Michael Lobo, Robert Lotz, Sean McKenna, Keith Orr, Michael Paguette, Chris Pirrera, Conlee Quortrup, Velimir Randic, Monica Renaud, Paul Rieder, Gregory Ruderman, Akhil Sahgal, Carey Schenkenberger, Yoshihiko Shiraishi, Craig Squier, Richard Tesoriero, Andrew Tong, Christopher Tunney, Donna Vulin Patrick Yagle, Mika Yamamoto, David Youker, and Leonard Yowell.

The Fall semester 1993 *TAVD* course carried the 10-m microwave Lightcraft investigation to a deeper level than the 1992 class, evolving more realistic propulsion and structural concepts: Team members included: Randall Biggers, Andrew Charles, Dean Head, Steven Hong, William Jellig, Mun-Hong Lim, Alexander Nalevanko, and Jungwa Seo. The Fall 1994, the *TAVD* class pursued a larger, 15-m diameter microwave Lightcraft with a 5-person crew designed for the ultra-fast lunar shuttle mission: Toshitaka Asai, Daniel Bellisario, Khulekani Dlamini, Eric Gaus, Benjamin Gottlieb, Robert Griffiths, Debra Ocejo, Daniel Orchard-Hayes, David Tremonti, and Rob Wall.

The Spring 1995 *Introduction to Space Technology* course included a special term project on Lightcraft human factors. Team members included: Fritz Bueker, Louise Courtois, Travis Dahl, Tri Dinh, Stephen DePascale, Tim Gallus, Christopher Hardy, Phillip Hoffman, Erin Johnson, Ki Lee, David Lewis, Dean Meloney, Peter Mitton, Jim O'Sullivan, Travis Pahl, Dustan Skidmore, Joshua Smith, Annette Strassberger, Sam Raney, David Tremonti, Ryo Tsukada, Ryan Tyler, Rob Wall, and Nathan Warner. Many aspects of this "imagineering" effort found their way into the LTI-20 Lightcraft Flight Handbook.

Take Note: As the principal objective of the Fall 1995 *Spacecraft Design Studio* course, a team of six highly motivated students – Joe Almeida, Tim Gallus, Dean Malone, Sam Raney, Annette Strassberger, and Russell Mohammed (artist) – created the first 3-D CAD model of the 20-meter microwave Lightcraft (i.e., the subject of this manual). This ground-breaking design effort produced the detailed exterior and interior Lightcraft architecture, propulsion system, pressurization concept, biconic escape pod geometry, entrance ramps, Maglev landers – nearly all the requisite systems.

The Spring 1996 *Introduction to Space Technology*, as a special term project, conjured up strategies and logistics for reconnaissance / surveillance / rescue missions by the 20 m microwave Lightcraft (i.e., 2025 Space Command context): Kobie Boykins, Justin DiVirgilio, Ronen Elkoby, Chris Gribbin,

ACKNOWLEDGEMENTS

Carleton Hanna, Wai Wai Lim, Randy Longtin, Fabien Nicaise, Douglas Parker, Stephen Pickering, Thomas Salisbury, Theodore Santaguida, Trevor Seaman, Keith Shanahan, Mark Skarulis, Xerxes Vania, Masamiro Watanabe, and Tom Winegar.

The comprehensive "imagineering" of a profusely illustrated "official manual" for the 2025 Space Command's 20-m Lightcraft was tackled by a team of 20 students as the principal objective of the Fall 1996 TransAtmospheric Vehicle Design (TAVD) course. By semester's end a 55% complete, fledgling rough draft of the LTI-20 Lightcraft Flight Handbook had been assembled by: Joe Almeida, Arthur Chen, Edward Clements, Karen Cramer, Shane Dover, Ronen Elkoby, Wayne Fu, Tim Gallus, Barry Kusumo, David Lewison, Dean Meloney, Jung Oh, Chris Parlato, Chris Perfetto, Sam Raney, David Simon, Annette Strassberger, Natilie Turner, Ryan Tyler, and Chris Tyrell. Their documentation efforts were expertly facilitated by Marilyn R.P. Morgan ("writingintensive" coach and editor). Douglas Parker created a credible 3D CAD model of the biconic escape pod based largely on von Braun's 1950's capsule, merged with a Grumman biconic aerobraking concept. Other contributions came from David Johnson, Christopher Bautista, Jeff Mcjurk, Chris Tyrell, and Russel Mohammed (graphics). The talented computer graphics team of Senter Reinhardt, Barry Kusumo, David Lewison, and Hernan Orellana created high-quality artistic renderings of the LTI-20 (using CAD models and Adobe Photoshop), flight animation sequences, and an RPI Lightcraft website. Kevin Hunt guided the website construction efforts. Please note: Artwork and illustrations in **The LTI-20 Lightcraft Flight Handbook** that were originally created or improved by the above-mentioned Rensselaer student design teams (and subsequent collaborations) are credited "Courtesy of RPI"; most, but not all, have been altered and recast into a consistent format for inclusion in the final manuscript.

We thank our technical advisors and consultants: Adrian Alden (microwave rectifying antennas), the late John Anderson (Horizon Mission Methodology), James Benford (microwave sources), William Brown (microwave rectennas), Brice Cassenti (structures and heat transfer), Richard Dickinson (microwave power transmission), Tom Dickerson (mission analysis), Arthur Kantrowitz (laser propulsion), Peter Glaser (solar power satellites), Dan Gross (superconducting magnetic energy storage), Abe Hertzberg (liquid droplet radiators), Brook Knowles (endoatmospheric ion propulsion; now deceased), Tom Mattick (liquid droplet radiators), Kenneth Myrabo (liquid / gaseous ventilation), Yuri Raizer (laser plasma physics, directed-energy airspikes), John Rather (highenergy laser systems analysis), Richard Rosa (magnetohydrodynamics), Mikhail Shneider (CFD, directed-energy airspikes), Euan Sommerscales (rectenna cooling), and Lt. Col. Matthew O. Warren (flight / crew operations) – with special thanks to our departed advisors: John Anderson (Horizon Mission Methodology), William Brown (microwave rectennas), Abe Hertzberg (liquid droplet radiators), Arthur Kantrowitz (laser propulsion), Brook Knowles (endoatmospheric ion propulsion), and Richard Rosa (magnetohydrodynamics).

We also thank Tim McElyea (Media Fusion), Bob Sauls (Frassanito & Associates), Tom Casey and Tom Nypaver (Home Run Pictures), Senter Reinhardt (Intensive Images Inc.), Chuck Lindgren and Chung Kim (RPI) for their impressive artistic renderings and computer animations of the LTI-20 and power-beaming stations.

Finally, we'd like to thank John Cole, Jack Lehner, John Mankins, Ivan Becky and Roger Luidens (NASA), Neville Marzwell and Henry Harris (Jet Propulsion Laboratory), Mitat Birkan (AFOSR), Franklin Mead, Jr. (AFRL), Lenard Caveny (SDIO), Stanley Sadin and Jack Sevier (USRA) for their support of Lightcraft research at RPI.

Now, after a decade-long incubation (stasis?) and considerable effort, we present the "official" **LTI-20 Lightcraft Flight Handbook** – incorporating all relevant technological advances from the intervening years. The *fledgling first draft* has been substantially re-edited and expanded to include a credible story for the microwave power-beaming infrastructure, construction chronology, and historical origins. Formerly missing or incomplete sections and chapters have been meticulously "filled-in." As affirmed on the title page, **"Basic and all changes have been merged to make this a complete publication."** We hope you find this manual to be fascinating reading.

Leik N. Myrabo
John S. Lewis

CONTENTS

Preface 5
Authors' Foreword 7
Acknowledgements 9

1.0 Introduction to LTI-20 Lightcraft 15
1.1 Lightcraft Fleet Acquisition 15
1.2 Mission Objectives 15
1.3 Design Lineage 18
1.4 Lightcraft Construction Chronology 18
1.5 Power-Beaming Infrastructure Deployment 19
1.6 Lightcraft Systems Development Milestones 21

2.0 Lightcraft Structure 26
2.1 Main Lightcraft Structure 26
2.2 Secondary Structure 27
2.3 Hull Layers and Pressurization 27
2.4 Intelligent Active-Structure Systems 29
2.5 Rectenna Arrays, Figure Controls, and Coolant System 32
2.6 Superconducting Magnets 32
2.7 Photovoltaic Arrays 33
2.8 Emergency Repair and Maintenance 33

3.0 Command Systems 34
3.1 Main Bridge 34
3.2 Crew Assignments, Roles, and Duties 35
3.3 Bridge Operations 36
3.4 Flight Operation from Escape Pods 40
3.5 Basic Control and Terminal Use 41
3.6 Flight Control 42
3.7 Mission Operations 43
3.8 Guidance and Navigation 44

4.0 Computer Systems 46
4.1 Computer Systems Architecture 46
4.2 Escape Pod Computers 46
4.3 Human / Machine Interface Options 46
4.4 Personal Access Displays 48

5.0 MHD Propulsion System 49
5.1 MHD Propulsion Physics 50
5.2 MHD Propulsion Conversion 52
5.3 Beamed Power Reception 58
5.4 Air Plasma Generation 58
5.5 Airspike Production 59
5.6 Thermal Management System 60
5.7 Vertical vs. Lateral Flight Modes 63
5.8 Emergency Procedures 65

6.0 Pulsed Detonation Engine (PDE) Mode 67
6.1 PDE Propulsion Theory and Application 67

CONTENTS

6.2	Propulsion System Components	69
6.3	Magnetic Nozzles and Thrust Vectoring	69
6.4	Microwave Power Reception	71
6.5	PDE Operations and Safety	72
6.6	Performance Specifications	72
6.7	Emergency Procedures	73
7.0	Ion Propulsion System	74
7.1	Theory and Application	74
7.2	Essential Physical Processes	77
7.3	Ion Propulsion System Components	79
7.4	Laser and Solar Power Considerations	80
7.5	Performance Specifications	81
7.6	Evasive Maneuvers	84
7.7	Emergency Procedures	85
8.0	Utilities and Auxiliary Systems	87
8.1	Major Utilities Networks	87
8.2	Additional Utilities Systems	88
8.3	Exterior Connect Hardpoints	88
8.4	In-Space Reaction Control System	89
8.5	Airspike System Auxiliary Functions	89
8.6	Magnetic Tractor / Repulsor Field	89
9.0	Communications	90
9.1	Intra-Ship Communications	90
9.2	Ship-to-Ground Communications	90
9.3	Ship-to-Ship Communications	90
9.4	Personal Communicators	91
9.5	Long-Range Transceiver	92
10.0	Maglev Ship Boarding System	93
10.1	Maglev System Introduction	93
10.2	Maglev System Operation	93
10.3	Limitations of Maglev System	96
10.4	Maglev Landers in Mated Mode	97
10.5	Emergency Maglev Pickup Operations	97
11.0	Remote Sensing Systems (CLASSIFIED – Section Removed)	99
11.1	Sensor Systems	99
11.2	Long-Range Sensors	99
11.3	Navigational Systems	99
11.4	Lateral Sensor Arrays	99
11.5	God's-Eye View Imaging Holograph	99
12.0	Tactical Systems	100
12.1	Tactical Doctrine	100
12.2	High-Power Laser Mirror	100
12.3	Pulsed Laser Weapons System	100
12.4	Microwave "Active Denial" System	100
12.5	Space Power Station Engagement	101
12.6	Personal Stun Weapons	101
12.7	Lightcraft Auto-Destruct Systems	102

CONTENTS

13.0 Environmental Systems 103
13.1 Environmental Control System 103
13.2 HeliOx Life Support System 105
13.3 Artificial Gravity Generation 105
13.4 Emergency Environmental System 106
13.5 Waste Management System 107

14.0 Crew Support Systems 109
14.1 Crew Support 109
14.2 Medical Systems 109
14.3 Crew Quarters System 110
14.4 Food Rehydration Unit 110
14.5 High-Quality VR Environment 110
14.6 Life Support Options 112
14.7 Ultra-G Personal Protection System 112
14.8 Lightcraft Boarding Options 114
14.9 Escape Pod Architecture 115

15.0 Auxiliary Lightcraft Systems 121
15.1 Maglev Lander Operations 122
15.2 Magnetic Docking Bays 122
15.3 Maglev Lander Configurations 123
15.4 Extravehicular Activity in Space 124
15.5 Space Plasma Shield Operations 125

16.0 Flight Operations 127
16.1 Introduction to Flight Operations 127
16.2 Mission Types 127
16.3 Flight Modes and Maneuvers 128
16.4 Takeoff and Landing 135
16.5 PDE Pickup of Water-Filled Maglev Lander 136
16.6 Water Collection from Cumulus Clouds 137
16.7 Procedure for Rescue of Downed Lightcraft 138
16.8 Abduction of Ground Transport Vehicles 139
16.9 Fleet Maneuvers / Operations 139
16.10 Stealth Procedures 141

17.0 Flight Dynamics 142
17.1 Introduction to Flight Dynamics 142
17.2 Subsonic Flight Regime 143
17.3 Supersonic PDE Flight Regime 146
17.4 Hypersonic Flight Regime 149
17.5 Reentry and Aerobraking Methods 150

18.0 Emergency Operations 153
18.1 Introduction to Emergency Operations 153
18.2 Lightcraft Hull as Protective Gas Bag 153
18.3 Biconic Escape Pod Ejection 154
18.4 Crew Retrieval Methods 155
18.5 Fire Suppression 158
18.6 Emergency Use of Maglev Landers 159
18.7 Emergency Airspike Support 159

CONTENTS

19.0	Power Beaming Infrastructure	162
19.1	SMES for Power Beaming Stations	162
19.2	Millimeter Wave Sources	164
19.3	PDS Basing Considerations	164
19.4	Microwave Transmission Through Atmosphere	164
19.5	Ground-Based Power Stations	166
19.6	Orbital Power Relays	170
19.7	Space Based Power-Beaming Stations	172
19.8	Nuclear Orbital Power Station: PDS-03	173
19.9	Solar Orbital Power Station: SPS-01	182
20.0	Human Factors and G-Tolerance	195
20.1	Introduction to Human Factors	195
20.2	Improving Human G-tolerance	196
20.3	Improving High-G Ventilation	198
20.4	Improving Communication	202
20.5	Super-Human Survivability Levels	205
20.6	Advancing BioElectronics Technology	206
20.7	Space Activity Suit Origin	206
20.8	Biconic Escape Pod Origins	206
21.0	Planetary Defense Missions (CLASSIFIED — Section Removed)	210
21.1	Monitoring Cislunar Space	210
21.2	Orbital Debris Mapping and Removal	210
21.3	Ballistic- and Cruise Missile Defense	210
21.4	NEO Asteroid Imaging	210
21.5	BEP Launch of Deep Space Probes	210
21.6	NEO Trajectory Modification Tests	210
21.7	Intercept, Escort, and Retrieval of Unidentified Spacecraft	210
APPENDIX:	Historical Origins of LTI-15 Concept	211
A1	Lawrence Passenger Spaceship	211
A2	Nuclear Propulsion of Spacecraft	213
A3	Artificial Gravity in Space	215
A4	Reentry Vehicle Options from Apollo Era	215
A5	Plasma Shields Against Solar Flares	218
A6	Auxiliary Functions for Onboard Electromagnets	221
A7	LTI-15 Propulsion Systems Integration	222
A8	Design Evolution of LTI-15 Concept	232
A9	Ultra-Fast Lunar Shuttle Mission	237
A10	Microwave Power Transmission Limits	244
	References	250
	Acronyms	261
	Index	263
	About the Authors	280
	Authors' Afterword	282
	Additional Picture / Illustration Credits	284

1.0 INTRODUCTION TO LTI-20 LIGHTCRAFT

The LTI-20 Lightcraft is a revolutionary atmospheric and exoatmospheric vehicle whose flight principles, structure, and propulsion systems are radical departures from familiar operational systems, and whose unique performance permits a wide range of novel missions. This manual is intended for use in the flight training of crew members who have had no previous experience with this type of vehicle and its systems.

1.1 Lightcraft Fleet Acquisitions

The LTI-20 Lightcraft is a product of Lightcraft Technologies International, built under contract 2019-USSC-8112-A with the U.S. Space Command. That contract was awarded for development and construction of a fleet of hyper-energetic, 6- to 12-person transatmospheric vehicles designed for both deep reconnaissance and aerospace superiority roles. The need for rapid engagement at global distances in response to brush-fire wars, terrorist activity, and insurgencies led to the requirement that the new vehicle be capable of operation in the atmosphere or in space, with global range and top speeds exceeding low-Earth orbit (LEO) velocity. Elementary energy considerations rule out the use of chemical propellants by a large margin, driving the design in the direction of externally supplied beamed power. The current availability of solar and nuclear (military) power satellites in low-Earth orbit, combined with their ability to deliver powerful laser or microwave power beams, makes global Lightcraft missions possible.

Definition: A lightcraft is any vehicle (i.e., flight platform) that is externally powered by an intense beam of electromagnetic radiation, usually in the laser or microwave spectrums. Microwave Lightcraft must be extremely light in structure, generally featuring semi-rigid inflatable structures. They may experience extremely high accelerations and may achieve velocities high enough to exit the atmosphere, enter orbit, or escape from Earth's gravity altogether.

1.2 Mission Objectives

Pursuant to Space Command contract 2019-USSC-8112-A the following objectives were established for the LTI-20 Lightcraft:

1.2.1 To depart from the continental United States (CONUS) with a crew complement of 12; fly halfway around the world in 45 minutes; land at unprepared and potentially hostile sites; perform covert surveillance, intelligence-gathering, or rescue functions such as the recovery of a downed aeronaut or astronaut; then return to the original CONUS base without refueling in the usual sense.

1.2.2 To exploit appropriate natural resources at the destination, such as water from lakes, streams, or reservoirs, to accomplish the return flight. (In covert missions, conventional propellants would not be available from the point of entry into hostile airspace.)

1.2.3 To demonstrate an all-weather capability and to function undetected at the chosen destination under all conditions of illumination, operating autonomously for days or weeks if necessary to assure mission success.

To assure achievement of these essential mission objectives, LTI recommended to Space Command that the LTI-20 Lightcraft meet or exceed the following design specifications given in 1.2.4 through 1.2.9.

1.2.4 Propulsion

1.2.4.1 To demonstrate a vertical takeoff and landing capability on unprepared surfaces in remote areas, operating with a stealthy propulsion system designed to loiter at its destination for days to weeks.

1.2.4.2 To equip the Lightcraft with two thruster modes designed for subsonic maneuvering and station-keeping in the dense lower atmosphere:

  a) an ion-propulsion system for subsonic cruise, and
  b) a pulsed detonation engine (PDE). For

stealthy operation, the PDE mode must operate at frequencies slightly above the human audible range limit of 20 kHz.

1.2.4.3 To incorporate gigawatt-class, ultra high power density rectifying antennas (rectennas) that can also serve as concentrating reflectors needed for the PDE thrusters. Rectennas are actively-cooled, solid-state devices that convert beamed microwave energy into DC electric current. To do so requires that the high-power side of the vehicle's lenticular hull must be transparent to microwaves.

1.2.4.4 To include a low-power, megawatt-class thin-film photovoltaic array to provide power for the ion engine. This array must be integrated with and completely cover the low-power side of the lenticular hull. The photocell array can draw power from either solar energy, with about 25% efficiency, or from beamed laser energy, with at least 65% efficiency, so that it can function both day and night.

1.2.4.5 To accelerate away from a potential ground or air-borne threat with the PDE thrusters at a minimum of 20 times the acceleration of Earth's gravity (20 G, or about 200 m/s^2) and a maximum of 220 to 360 G. This (i.e., >20 G) is faster than the human eye can follow, causing the illusion of instantaneous disappearance (the so-called "hyperjump" maneuver).

1.2.4.6 To exploit magnetohydrodynamic (MHD) slipstream accelerator technology to accelerate the Lightcraft to orbital speeds while still within the upper atmosphere.

1.2.4.7 To integrate a special expendable-coolant system for the Lightcraft's accelerator-class MHD engines that will efficiently remove waste heat from the rectennas by heating on-board de-ionized water and ejecting it as steam. This thermal management system is essential for the transatmospheric boost back to CONUS.

1.2.5 Structures

1.2.5.1 To make the Lightcraft hull an ultra-lightweight pressure vessel in which 95% of the primary structure is loaded in tension to exploit the highest performance silicon-carbide (SiC) thin films available.

1.2.5.2 To use a perimeter toroidal pressure vessel with an overall diameter of 20 meters and a tube diameter of 2 meters, inflated to a maximum pressure of 10 atmospheres, as the structural backbone of the Lightcraft, to which a pair of primary superconducting magnets are attached. The rest of the hull and all interior structures and components are built onto this toroidal primary skeleton.

1.2.5.3 To incorporate an actively cooled (with gaseous HeliOx), thin-film double hull for thermal protection during transatmospheric boost and during reentry from space, when magnetic aerobraking can be employed.

1.2.5.4 To accommodate within the pressurized airship-type structural shell of the Lightcraft a gas-bag impact protection system for the crew provided by the hull itself.

1.2.5.5 To instrument the thin-film Lightcraft structure with numerous sensors and actuators, monitored and controlled by the Flight Management System (FMS) computer (i.e., extensive use of smart materials).

1.2.5.6 To use the structural silicon-carbide thin films as a substrate material for adding other Lightcraft functions, including vapor-deposited solid state components such as rectennas, the low-power photovoltaic array, and electro-luminescent interior surfaces to provide lighting for the crew.

1.2.6 Mission

1.2.6.1 To operate independently throughout cislunar space for extended periods of time while conducting a wide range of covert missions including surveillance, intelligence gathering, interdiction, and rescue.

1.2.6.2 To demonstrate a sufficiently high degree of artificial intelligence in Lightcraft functions that the crew may perform a top-secret mission on the ground while the 20-m Lightcraft hovers autonomously overhead in a remote-controlled autopilot mode.

1.2.6.3 To incorporate two small magnetic levitation (maglev) landers, each capable of transporting 6 occupants to and from a hovering mother craft. The magnetic docking system must accommodate an expedited subsonic flyby and pickup maneuver from treetop level to avoid detection by hostile surface radar.

1.2.6.4 To integrate multiple functions into the maglev landers, including transport for 2400 kg of water for use as coolant for Lightcraft transatmospheric boost, or for use as an emergency escape pod for 6 people, including reentry from space.

1.2.7 Crew Environment

1.2.7.1 To provide a customized 2.15 m long escape pod for each crew member. Each pod must also provide acceleration protection for its occupant, conferring superhuman capabilities during ultra-high acceleration and deceleration maneuvers up to levels as high as 220 to 360 G. Liquid immersion in a centimeter-thick liquid layer and partial liquid ventilation with highly oxygenated perfluorocarbons can in combination provide near-perfect support for the human body under extreme G loads.

1.2.7.2 To inflate the vehicle with a breathable Helium-Oxygen (HeliOx) mixture at a pressure of at maximum 1.5 atm (0.5 atm. gauge) at sea level altitudes, including an oxygen partial pressure of 160 mm Hg, thereby providing a normal breathable atmosphere while the crew is on the observation deck and not confined to their escape pods.

1.2.7.3 To accommodate maglev flying belts for the crew to soar up to and board a hovering Lightcraft for rapid departure from a dangerous area.

1.2.7.4 To provide for extended operations in orbit by including a space radiation protection system, effectively a kind of "plasma shield" against 200 Mev solar proton storms.

1.2.7.5 To provide artificial gravity in space by spinning the vehicle after arriving in orbit, enabling the crew to walk on the curved outer walls at $1/6$ g (lunar gravity equivalent).

1.2.8 Tactical

1.2.8.1 To equip the LTI-20 Lightcraft for extreme accelerations which provide maximum flexibility and survivability in covert operations. Kinetic energy munitions, including rockets, bombs, and bullets, are impractical because of the excessive mass penalties associated with carrying them and their launch systems. Beam weapons, made attractive by external sources of power, are the only feasible option for the LTI-20. Weapons would be fired only in the most extreme circumstances because the mission profile normally entails stealthy entrance to and exit from hostile airspace without leaving evidence of the visit.

1.2.8.2 To endow the Lightcraft with all the tactical firepower commensurate with its aerospace superiority role, including on-board megawatt-class lasers and microwave weapons, the ability to relay gigawatt-level (GW) space laser power off the maglev lander's lower surface as a "fighting mirror," and finally the ability to call down direct space power at 10 GW levels for threat negation as a last resort. Basic mission philosophy is "walk softly, but carry a big stick."

1.2.8.3 To electrically link all on-board magnets into a Superconducting Magnetic Energy Storage (SMES) "battery" with capacity of 900 megajoules (MJ), available to power a megawatt-class pulsed laser or 100 MW-class microwave weapon in self-defense. The SMES unit can also power the ion thrusters in a "silent running" mode for hour-long periods independent of external power supplies of beamed energy.

1.2.8.4 To equip the Lightcraft to levitate and electomagnetically transport 2400-kg conventional steel-frame automobiles off roadways by employing the on-board superconducting magnets in its propulsion system.

1.2.9 Design Life

Given the limitations of exotic thin-film structures, the LTI-20 prototype is not expected to demonstrate a design life beyond 5 years. During the life of the prototype, significant advances in Lightcraft technology are anticipated. Production LTI-20s will incorporate these advances to enable a design life expectancy of about 15 years. Minor refurbishment and upgrades to production LTI-20 systems are anticipated at approximately 5-year intervals.

1.3 Design Lineage

The first Lightcraft model capable of carrying a single human passenger was the LTI-X10 experimental prototype. Construction of the PDS-01 power station (6 GW, 35 GHz) at the High Energy Microwave Systems Test Facility (HEMSTF) located at White Sands Missile Range (WSMR), New Mexico was completed in time for suborbital flight demonstrations of the LTI-X10 (Figure 1.3.1) late in its flight test program. Models -X11 through -X13 were design concepts that never led to full-scale flight prototypes, although several flight tests of much larger but empty LTI-X13 hulls were carried out to verify the inflation systems and the design of the radial pressure cells. The LTI-X14, which built upon the -X13 hull with new propulsion and computer systems, was capable of carrying a crew of four, but generally flew unmanned or with a crew of two, the balance of the payload weight being assigned to test instrumentation. Two flight articles and one static test mockup were built. The only test flight that resulted in catastrophic failure and loss of the vehicle was fortunately one of the unmanned LTI-X14 flights.

The LTI-X15 vehicle was, in essence, a stripped, slightly modified -X14 hull with completely new sensor arrays, a functional ion propulsion system, and an advanced, lightweight piezoelectric actuator system for structural control. This vehicle flew unmanned several times, and with a crew of two for several test flights. On its final flight, with much of the test instrumentation removed or replaced by ultra-lightweight operational electronics, a full crew complement of five was carried. All LTI-X15 flights were powered under the USSC's mountain-top PDS-02 facility (15GW, 94 GHz); power was routed to the LTI-X15 by a 550-m diameter relay station in LEO.

The LTI-20 construction contract was signed during flight testing of the LTI-X14. The improved LTI-X16 hull concept, already under in-house development at LTI, was chosen for use in the LTI-20. The unmanned LTI-X17 propulsion test bed, funded under an earlier NASA contract, also fed directly into the LTI-20 program. Both the LTI-X18 and -X19 were conceptual designs that were never built. The ground-based PDS-02 (and LEO relay) provided microwave power for the LTI-X20's entire flight testing program, right up until the global (launch and return) excursions began which required a transition to the orbiting PDS-03. Until NASA's SPS-01 became operational in mid 2025, PDS-03 was the only option for such flights.

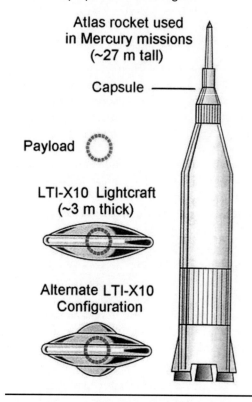

Figure 1.3.1: LTI-X10 size relative to historic Atlas-Mercury rocket.

1.4 Lightcraft Construction Chronology

2012 LTI-X10 contract signed with Space Command 2013 USSC power-beaming facility (PDS-01) is operational (6 GW, 35 GHz).

2014 First LTI-X10 aeroshell inflation, stress, and tethered flight tests. First unmanned test flights with automated "ion breeze" propulsion and control system.

1.0 INTRODUCTION TO LTI-20 LIGHTCRAFT

2015 LTI-X10 first unmanned supersonic test flights with automated PDE thrusters and control system. PDS-01 is upgraded to 9 GW just in time for supersonic flight test program.

2016 LTI-X10 piloted suborbital flight tests with power up-linked from PDS-01, then down-linked from new 550 m diameter passive relay satellite in LEO. LTI-X15 contract signed with Space Command.

2017 PDS-02 facility operational (15 GW, 94 GHz).

2018 First subsonic unmanned flight tests of LTI-X15 prototype. LTI-X15 avionics and propulsion systems fully flight qualified. LTI-X15 First free flights with crew of 2. LTI-X15 exoatmospheric flights with crew of 5.

2019 LTI-X20 contract (No. 2019-USSC-8112-A) signed with Space Command.
Second LEO microwave relay satellite (250 m) is fully operational (USSC). Extensive high performance flight testing of LTI-X15 on PDS-02.

2020 LTI-X15 modifications completed. First flights into LEO, and return.
LTI-X16 hull inflation and fatigue tests (for LTI-20 program).

2021 X17 propulsion test bed flown in tethered mode (in preparation for LTI-20 program).

2022 PDS-03 facility is fully operational in LEO: 20 GW, 140 GHz (USSC).

2023 LTI-20A high performance, exoatmospheric flights with crew of 6.

2024 LTI-20B global flights with crew of 12, and return (without refueling).
NASA's SPS-01 power-beaming station is fully operational.

2025 First suborbital and orbital boosts of LTI-20A using NASA SPS-01.
Official LTI-20 "float-out" (declassification ceremony) at LTI headquarters.

1.5 Power-beaming Infrastructure Deployment

See Chapter 19 for specific details on the USSC and NASA microwave power-beaming facilities and infrastructure for "highways of light."

1.5.1 Space Command PDS Construction Chronology

2011 Construction begins on PDS-01 power beaming facility at WSMR. 2012 Construction begins on 550 m diameter microwave relay satellite (MRS-01) in LEO.

2013 USSC microwave power-beaming facility (6 GW, 35 GHz) – designated PDS-01 – is now operational at the High Energy Microwave Systems Test Facility (HEMSTF) at WSMR, NM.

2014 Construction begins on PDS-02 power-beaming facility on secure 3-km mountain site in southwest (sun belt) USA.

2015 PDS-01 station power is upgraded to 9 GW at 35 GHz.

2016 LEO microwave relay satellite (MRS-01) is now operational. (Circular orbit = 476 km altitude; Inclination = 28.5°; Daily repeating orbit over WSMR.)

2017 USSC's PDS-02 facility is fully operational (15 GW, 94 GHz).

2018 Construction begins on PDS-03.

2020 PDS-02 is upgraded to 20 GW and 140 GHz. 2022 PDS-03 orbital power-beaming facility is fully operational (20 GW, 140 GHz) – enabling global Lightcraft flights, both deployment and return. (Circular orbit = 476 km altitude; Inclination = 28.5°)

2023 Six "Monocle" laser relay mirrors are launched into equally-spaced 3000 to 5000 km orbits. LEO orbital debris removal begins. Construction begins on 2nd LEO microwave relay satellite.

2024 Low earth orbital debris cleanup program is complete (courtesy of FELs aboard PDS-03).

2025 Second LEO microwave relay satellite (250 m diameter) is now operational, anticipating retirement of 12 year old MRS-01 in the coming decade. (Circular orbit = 476 km altitude; Inclination = 28.5°; Daily repeating orbit over PDS-02.)

1.5.2 NASA SPS Construction Chronology

2018 Construction begins on SPS-01 solar power satellite in LEO.

Chapter 1

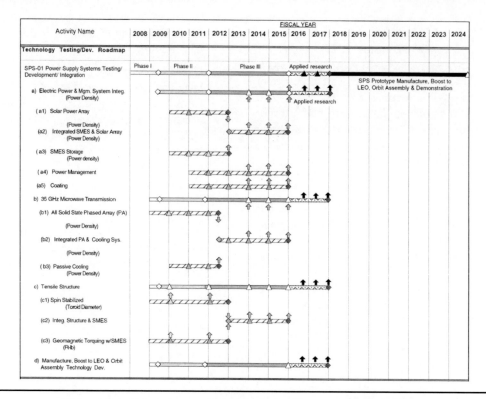

Figure 1.5.1: SPS-01 System Development Milestones. *(After J. Lehner.)*

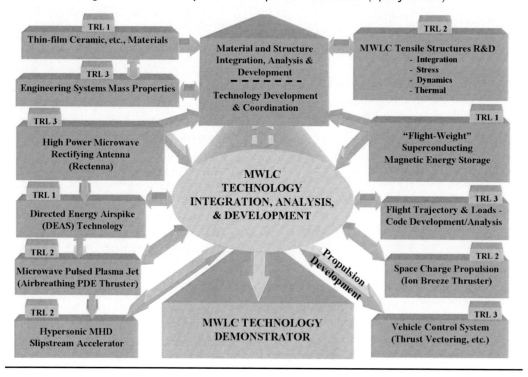

Figure 1.6.1: MWLC Technology Integration Challenges. *(After J. Lehner.)*

2020 "Phase A" construction of SPS-01 Solar Power Satellite is complete; 1/16 of complete system.

2024 SPS-01 fully operational (20 GW at 35 GHz) in LEO. Circular orbit = 476 km altitude; Inclination = 28.5°; Daily repeating orbit over Cape Canaveral, FL.

Figure 1.5.1 presents the development milestones for SPS-01, a 1 km diameter rotating satellite solar power station in LEO, rated at 20 gigawatts of beam power at 35 GHz.

1.6 Lightcraft Systems Development Milestones

The Microwave Lightcraft (MWLC) project was unofficially launched as a "black" program in mid-2008 (FY 2009 funds) with little fanfare. Figure 1.6.1 summarizes the technology integration challenges facing development of the MWLC demonstrator at that "project inception" stage. The first of its kind, and radically unlike any such previous attempt in history, microwave lightcraft demanded engine and structures technology so tightly interwoven that only a massively parallel, systems analysis and integration / engineering effort could hope to bring the first technology demonstrator into reality. At the start, the Technology Readiness Level (TRL – see Figure 1.6.2) of ALL contributing technologies were judged to be only at the TRL 1 to TRL 3 stage.

The *MWLC Technology Systems Development Plan* for the embryonic, first four years of the program is given in Figure 1.6.3. Summarized in Figures 1.6.4 and 1.6.5 are the task structures for the *Propulsion System Technology Development Plan*, and *Vehicle Structure / Materials Technology Development Plan*, respectively. Specific programmatic details on the sub-task structure for **Phase I** (i.e., FY 2008-2009 period) of that *Vehicle Structure / Materials Technology Development Plan* are presented in Figure 1.6.6.

FY 2008 activities centered on:

1) Examination of the baseline MWLC concept propulsion system, and structures and materials (including a SOA technology survey) for technical understanding and initial opinion on feasibility;
2) Code development and preliminary analysis of baseline concept propulsion, structures, thermal, materials, and aerodynamic characteristics and loading;
3) Propulsion testing and analysis R&D; and,
4) Continual assessment of development progress (e.g., including test results, design issues, concept feasibility, etc.) to identify "roadblocks" and immediate remedies.

The FY 2009-2010 activities addressed:

1) Evolution of propulsion code and technology R&D (test article enhancements, analysis, etc.);
2) Development of a lenticular test article, test plan, and begin structure tests;
3) Code refinements / validation and analysis of the baseline concept structural, thermal, materials, and aerodynamics characteristics / loading; and
4) Reporting on development progress which again included test results, design issues, concept feasibility, technology "roadblocks," and recommendations.

At this point, the MWLC concept was determined acceptably feasible, so the following prototype design activities were initiated in FY 2011:

1) Develop preliminary design and "more applicable" test articles, and perform analyses of new designs;
2) Initiate tests for validation and development of 35 GHz microwave rectenna technology;
3) Conduct propulsion R&D testing, and wind tunnel experiments for concept validation and evolution; and,

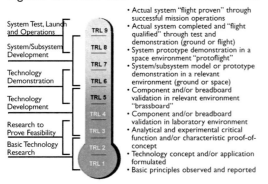

Figure 1.6.2: Technology Readiness Levels: TRL 1 to TRL 9. *(Courtesy of NASA.)*

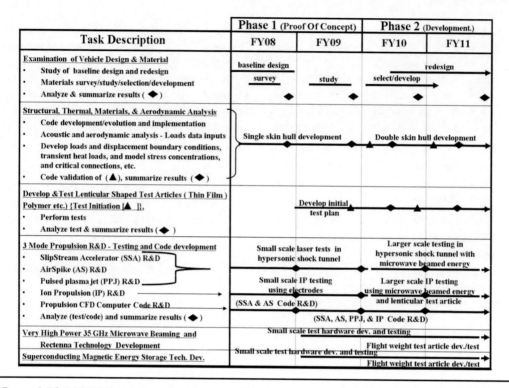

Figure 1.6.3: MWLC Technology Development Plan: Embryonic Phase – First 4 years. *(After J. Lehner.)*

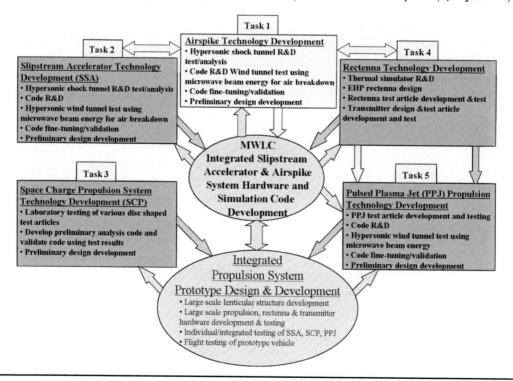

Figure 1.6.4: MWLC Propulsion Systems Technology Plan. *(After J. Lehner.)*

1.0 INTRODUCTION TO LTI-20 LIGHTCRAFT

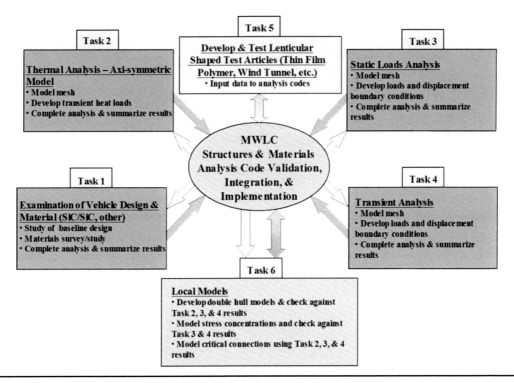

Figure 1.6.5: MWLC Structure and Materials Technology Plan. *(After J. Lehner.)*

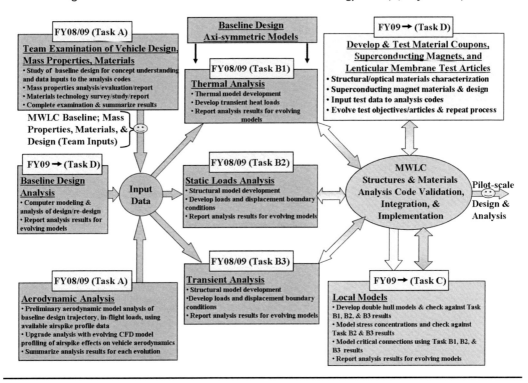

Figure 1.6.6: MWLC Structures Development Plan: Initial Phase – First 2 years. *(After J. Lehner.)*

Chapter 1

4) Document design "issues," identify redesign data bases, assess redesign feasibilities, and examine / analyze test results of relevance to the baseline MWLC prototype concept.

Figures 1.6.7 and 1.6.8 give the *MWLC Systems Development Milestones* and project overview for the complete 17-year history of the microwave Lightcraft program. This chronology details the major program events and advances that ultimately led to the LTI-20A's first high-performance exoatmospheric flights in 2020. Note that thin-film polyimide disc structure investigations began in 2009 with the fabrication of a 2 m diameter test article, followed by a rapid succession of increasingly larger 2-, 3-, and 6.4 m models before the first 10 m inflatable was tested. The objective was to identify deficiencies (i.e., in structural stability / integrity under rigorous dynamics testing) and refine the MWLC structure concept with inexpensive / disposable polyimides prior to constructing the exotic and expensive thin-film SiC/SiC discs. In the propulsion arena, note that hypersonic shock tunnel tests on airspike, pulsed detonation thruster, and MHD slipstream accelerator technology were initially conducted with "available" high-power pulsed infrared lasers (10.6 µm, TEA-CO_2 sources) to create the requisite air plasmas. Subsequent experiments transitioned into the microwave regime, exploiting 35 GHz pulsed gyrotron sources.

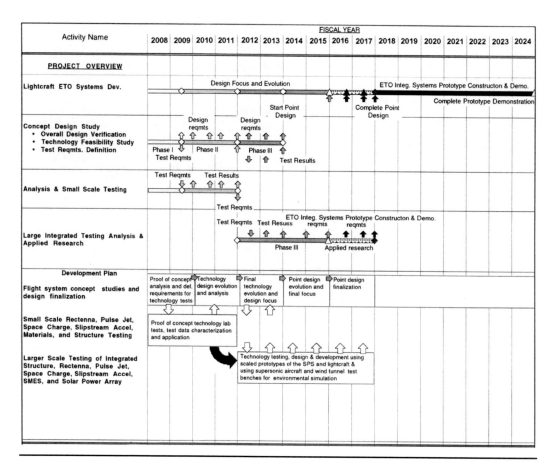

Figure 1.6.7: MWLC Systems Development Milestones: Project Overview. *(After J. Lehner.)*

1.0 INTRODUCTION TO LTI-20

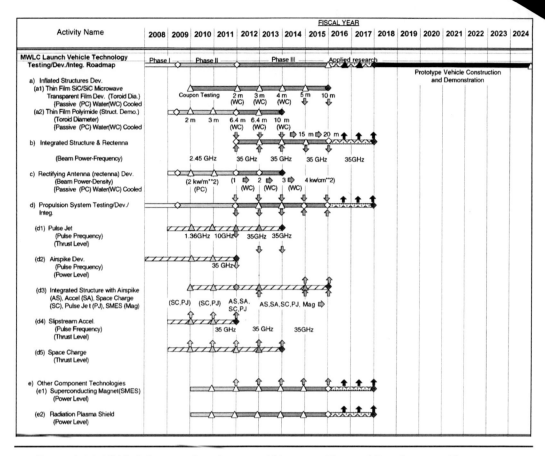

Figure 1.6.8: MWLC Systems Development Milestones: Testing / Development / Integration Roadmap. *(After J. Lehner.)*

STRUCTURE

The structural system, comprising approximately 1200 kg of SiC, is the main backbone of the LTI-20 Lightcraft. All components of the Lightcraft play an integral role in providing sound structural support and safety to the crewmembers and payload.

2.1 Main Skeletal Structure

The LTI-20 Lightcraft is a stiff, thin-walled inflated vehicle capable of attaining supersonic to hypersonic speeds in the atmosphere as well as high-velocity space flight. The primary framework of the Lightcraft is built around a toroidal pressure vessel at the rim of the lenticular Lightcraft. The main hull section, shaped like a shallow dome, is supported by the pressurized toroid. The pressure vessel and the main framework of the vehicle are fabricated from an interlocking series of silicon-carbide (SiC) films and frames of various sizes and shapes.

The toroid is pressurized to 7.5 (sea level) to 10 (in space) atmospheres to maintain the lenticular Lightcraft geometry against severe propulsion system and acceleration / loads. The hull, like the pressure vessel, is made of SiC, a high-temperature isotropic thin-film composite ceramic structure reinforced with nano-fibers. The film used in the Lightcraft skin is transparent to microwave radiation and visible light; it has a normal peak operating stress of 3.0 GPa, ultimate tensile strength of 3.7 Gpa, and a density of 2.05 g/cm^3 (Table 2.1.1). The toroidal pressure vessel uses film that is 0.37 mm thick, whereas the film used in hull construction varies in thickness depending on expected loads (Table 2.1.2).

Table 2.1.1: Lightcraft skin characteristics.

Property	Value
Density (kg/m^3)	2050
Elastic Modulus (GPa)	255
Poisson's Ratio	0.25
Ultimate Tensile Stress (GPa)	3.70

The main spacecraft hull must be pressurized to 1.5 atmospheres (0.5 atm. gage) at sea level.

Table 2.1.2: Hull film thickness vs. load.

Component	Thickness (mm)
Magnet's Vacuum Tube Wall	0.540
Torus Strut Membrane	0.059
Torus Outer Hull Facesheets	0.373
Torus Inner Membrane	0.327
Main Hull Facesheets	0.280
Inner Hull Membrane	0.280
Rectenna Facesheets	0.100
Main Strut Membrane	0.074
Payload Bulkhead Membrane	0.100

During transatmospheric flights to space, this hull gage pressure of 0.5 atmospheres is held constant by high-capacity pumps that scavenge, pressurize, and cool excess hull HeliOx, then inject it into the toroid; this ultimately increases the toroid pressure from 7.5 to 10 atmospheres. Because of this pressurization, all of the primary load-bearing structural members in the vehicle are loaded in tension. Other key components in the framework of the Lightcraft are the two high-power parabolic rectennas that dominate nearly half the vehicle.

Table 2.1.3: Mass breakdown for primary structure and HeliOx pressurant.

Component	Mass (kg)
Expendible Coolant (H$_2$O for LEO boost)	2400
MWLC SiC/SiC Structure	1288
Torus HeliOx Pressurant	220.5
Hull Inflatant Gas (HeliOx)	92.5
Superconducting Magnets (*Confidential*)	<1200
Payload (crew and escape pods)	800 - 1200

The rectenna assemblies are suspended from the pressurized hull by catenary SiC curtains. The central parabolic rectenna is also supported by ultra-light I-beam trusses that are the only Lightcraft members subject to compressive loads. Panels are connected from the outer hull to the rectennas and lie in the radial-axial plane. This orientation serves to minimize rectenna deflections that might result from changing stress distributions in the Lightcraft's framework caused by rapid accelerations. Table 2.1.3 gives the mass breakdown for the LTI-20 primary structure and HeliOx pressurant.

This basic mechanical framework assures physical integrity of the pneumatically-inflated vehicle during all phases of operation. Active vibration dampers are attached to critical parts of the structure. These dampers can detect vibration and react by generating counter-vibrations, opposite in phase to the detected vibrations, thus canceling out potentially detrimental effects due to vibrational stressing of the Lightcraft hull and contents. Many system components, including the microwave rectennas and photovoltaic power array, are built into the hull structure.

2.2 Secondary Structure

Suspended from the primary pressurized spaceframe by means of tension curtains is a secondary structure that forms an axisymmetric envelope. This envelope is divided into 12 equal gores, each of which may be individually sealed and pressurized. These gores are used as cabin space for crew members and as storage space for payload or mission-related equipment.

Other vehicle components, including the escape pods and the engine systems, are attached to this secondary framework. The tensile structures supporting different components have thicknesses proportional to the masses of the components they support. Since these members experience only tensile stresses, buckling is not a dominant factor.

2.3 Hull Layers and Pressurization

The exterior shell of the LTI-20 Lightcraft is a double hull composed of SiC films 0.28 mm thick. Silicon-carbide fibers run perpendicularly between the outer and inner layers maintain a constant "core" depth of 1 cm between the two layers. This open-cell structure (see Figure 2.3.1) permits forced convection of gaseous HeliOx coolant between the hull layers, assuring that the external skin temperature remains within its

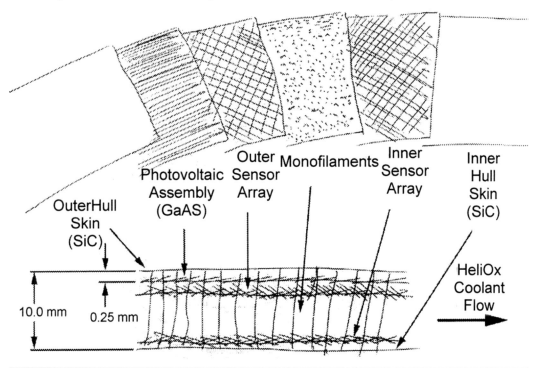

Figure 2.3.1: Double hull sandwich construction showing active HeliOx coolant system of photovoltaic hull side. *(Courtesy of RPI.)*

Chapter 2

thermo-structural limits during the brief trans-atmospheric boost and reentry maneuvers.

The double hull is designed to prevent penetration of the inner hull layer by micrometeoroids. The outermost skin is sufficiently robust to fragment such micrometeoroids before they reach the inner hull, thus protecting the pressurized volume and its contents. Sensors on the outer hull monitor the temperature, stability, and integrity of the skin and notify the main computer of developing problems. The main computer uses these inputs to maintain the proper pressure and flow rate of the gaseous HeliOx coolant.

The hull structure, as shown in Figure 2.3.2, differs slightly between the two lenticular halves of the Lightcraft. The "high-power" side of the vehicle containing the MIRV ejection portals and parabolic rectenna arrays does not have a layer of photovoltaic (PV) cells.

On the side of the LTI-20 that is powered by solar and beamed infrared laser light, a PV array covers the entire half of the vehicle. The thin-film PV array is deposited directly on the inner surface of the outermost skin layer of the hull. This placement assures that the PV array receives the maximum possible flux of incoming light while still realizing some protection from micrometeoroid impact. The central computer's sensor grid on the outer hull is bonded to the inner side of the PV array on the visible-light-powered side of the Lightcraft, but is bonded directly to the outermost hull skin layer on the rectenna side of the Lightcraft. The inner hull sensor grid is bonded to the outward-facing surface of the inner hull skin. See Figure 2.3.1 for the detailed layering of the hull components.

The outer hull is compartmentalized into 48 individual "pie slice" cells running radially outward. The individual cells are separated by SiC curtains 0.25 mm thick. The inflation pressure of each cell is monitored and regulated at collection points on the rim. As the temperature of the HeliOx coolant flowing between the inner and outer hull layers rises, whether by absorption of incoming light by the PV arrays or by atmospheric friction during high-speed atmospheric flight, the coolant flow rate is increased to compensate for the rise in hull temperature. Coolant collection takes place at the outermost rim where the main toroidal tube meets the outer hull and along the inner rim of the maglev lander bays. After filtration to remove particulate and molecular contaminants, the HeliOx coolant cycles from these collection points to other sections of the hull or to the coolant heat-transfer system. This system of separate cells ensures proper circulation over the entire hull area, facilitates repairs and maintenance, and allows for continued operation even with small punctures in one or more cells.

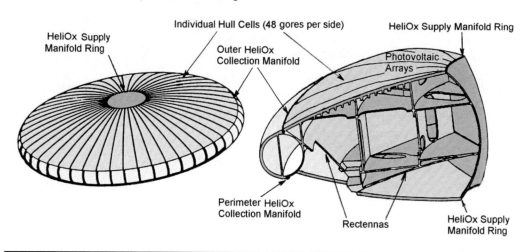

Figure 2.3.2: Hull gore structure with active HeliOx coolant system. *(Courtesy of RPI.)*

A constant pressure of 1.5 atmospheres of HeliOx mixture fills all the main compartments during normal operation at low altitudes; in space, the pressure is reduced to 0.5 atmosphere. The main toroidal tube is maintained at a pressure of 7.5 (sea level) or 10 (space) atmospheres at all times, except during emergencies.

Each of the 6 innermost compartments of the LTI-20, along with the 12 sub-compartment crew quarters, can be sealed off and pressurized independently from the rest of the interior. The annular walkway area is divided into 6 main segments, with pressurization doors on either side of the main landing ramps (see Figure 2.3.3). The chambers directly adjacent to the main landing ramps serve as air-locks when these ramps are in use. Monitoring and gas storage equipment, located around the center section of the Lightcraft, are responsible for maintaining pressure in the hull and crew compartments.

Each Multiple Independent Reentry Vehicle (MIRV) escape pod loads into its own launch tube which, with the MIRV, serves as an entry air-lock. The launch tube is pressurized or depressurized only after closing the hatches located at both ends.

2.4 Intelligent Active-Structure Systems

Multiple sensor systems are integrated directly into the structural framework of the LTI-20 Lightcraft. Using input from these systems, the computer network can assess both the status of the Lightcraft's structure and the external conditions around the Lightcraft. By analyzing the information gathered through the various sensor systems, the computer network can determine the Lightcraft's mission worthiness and relay this information to the crew.

Temperature sensors are placed throughout the LTI-20 because much of the craft's interior and exterior experiences severe temperature changes. Large amounts of waste heat are generated internally by the rectenna arrays and externally by high-speed flight through the atmosphere.

Temperature sensors are also widely distributed in the hull to monitor its temperatures. An additional temperature sensor array is co-located with the rectenna arrays to assure that

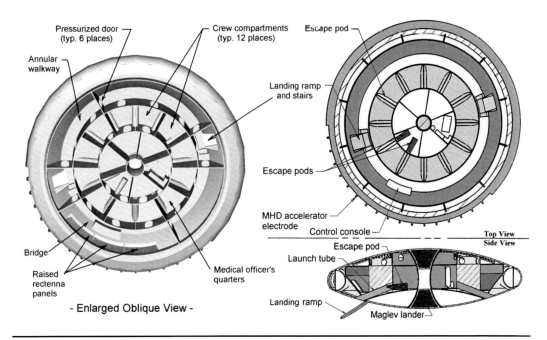

Figure 2.3.3: Interior functional layout of pressurized compartments within lenticular envelope.
(Courtesy of RPI.)

waste heat from the rectennas is dissipated without heating the interior of the LTI-20 beyond safe limits. Other temperature sensors monitor the 9 primary superconducting magnets to warn the computer system of possible magnet overheating, which could result in abrupt loss of their superconductive properties. A thermal "runaway" of the magnets could result in catastrophic failure.

Radiation sensors are included in the external hull sensor arrays to enable the computer network to monitor the entire electromagnetic spectrum impacting the surface of the Lightcraft. During flight through space the computer network can verify whether the Plasma Shield is effectively repelling solar proton storms during flares. If, for example, proton irradiation fluxes are found to be intolerably high, the computer network signals red alert status, warning the crew to enter their shielded escape pods until the crisis has passed.

The LTI-20 is a pressurized structure that depends on specific pressure levels to maintain structural integrity. Pressure changes in the vessel's interior, supported by other diagnostics, can alert the LTI-20's computer network to imminent structural damage. Pressure sensors are located throughout the LTI-20's hull, the main toroidal pressure vessel, and all pressure cells. These sensors ensure that proper pressure levels are being maintained at all times. Pressure is also monitored in the hull layers to ensure proper gaseous coolant circulation. Coolant pressures are also heavily monitored around the parabolic rectenna arrays to assure proper operation of the expendable (open-cycle) water coolant system. The sensors in the toroidal tube allow the computer to diagnose the stability of both the main structural backbone, under 15 atmospheres pressure, and the coolant surrounding the superconducting magnets.

Sensors that monitor static electric and magnetic field levels are concentrated in the hull sensor arrays, supplemented by others distributed throughout the interior of the vehicle. Hull sensor arrays are used during the ion propulsion mode to regulate the electric and magnetic fields present on the hull. During magnetohydrodynamic (MHD) and pulsed detonation (PDE) flight propulsion modes, sensors monitor both the magnetic fields used to vector the ionized air around the Lightcraft and the electric fields across the hull electrodes. Interior magnetic sensors are used to monitor the magnetic environment the crew experiences as they operate the LTI-20.

The mechanical strength of the LTI-20's spaceframe is augmented by the structural integrity system (SIS). This system provides an extensive network of piezoelectric actuators that compensate for propulsive and other structural loads that could compromise the configuration of the spaceframe. The rectenna arrays are especially at risk because the alignment of the antenna elements must be maintained to a precision of 0.5 mm.

Many frame members throughout the Lightcraft have piezoelectric films such as barium titanate chemically bonded to them. Under extreme acceleration loads these members may experience deformations sufficient to alter the load distributions and the flight characteristics of the external aeroshell. The computer network monitors the deformation of all critical structural components and sends a commensurate electrical current through the barium titanate matrix to correct such changes in figure.

The SIS can also be used to tense the muscles of critical structural members of the LTI-20 in anticipation of extreme flight maneuvers. By prestressing the superstructure, forces arising from extreme accelerations or changes in direction cause a minimum of disturbance to the framework geometry. Prestressing minimizes fatigue in the structure by reducing fluctuations of stress loads throughout the vehicle.

Hull structural sensor arrays include a network of fiber optics distributed throughout the framework of the LTI-20. As shown in Figure 2.4.1, this web network of fiber optics runs through the hull. By sending light through these fibers and watching for light pattern changes or interruptions, and by correlating these data with pressure and strain monitors, the central computer can detect and assess structural

LIGHTCRAFT STRUCTURE

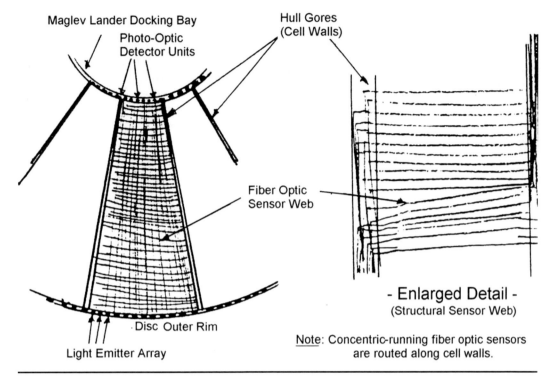

Figure 2.4.1: Fiber optic sensor array for hull structure. *(Courtesy of RPI.)*

breaches in the hull. The layout of the fiber-optic network is designed so that the central computer can determine precisely where hull integrity has been breached, whereas pressure and strain sensors can only determine which pressure cell has been compromised.

Structural fiber-optic sensors are also chemically bonded to the local load-bearing members of the frame. The material properties of these fibers are matched to those of the structural materials to which they are bonded, assuring that both fail simultaneously under ultimate loads. These sensors, supplemented by strain sensors, alert the computer network to local fractures in structural members before catastrophic failure of a structural member. This system gives the computer network sufficient time to alter load distributions and avert serious component failure.

Strain sensors are attached to primary load-bearing members and to the hull of the LTI-20. Hull-strain sensor arrays tell the computer network whether the stress loads being experienced by the hull layers are within safe operational limits. If stresses are found to exceed these limits, the main computer will attempt to reduce stress levels by changing the hull cell pressure distribution, altering the vehicle's profile through the use of the SIS, or diverting the Lightcraft's flight path. From these data the computer network assesses the load distribution throughout the vehicle, ensuring that the structure is not compromised.

Acoustic and vibrational sensors and actuators are located at key areas around the Lightcraft. These sensors detect harmful levels of vibration that could damage or impair structural members, on-board equipment, or crew. After detecting excessive vibrational levels or an approach to resonance, an array of intelligent actuators sends counter-pulses to damp such oscillations. Similarly, in the Lightcraft interior other acoustic actuators cancel out audible and ultrasonic sound waves by emitting sound waves of the same amplitude and frequency, but perfectly out of phase with the original wave. This system is built into the structure of the LTI-20 so as to isolate the inner structure of the vehicle from the superconductors positioned throughout the ship. This cancellation of vibration prevents damage to the ship that might otherwise be caused by the vibratory fluctuations that occur in the superconductors when they are operating in certain frequency ranges.

2.5 Rectenna Arrays, Figure Controls, and Coolant System

The rectenna array is the main source of high electrical power for the Lightcraft. Two rectennas, one at the center and another (annular) at the rim, provide the Lightcraft with the power required to operate the MHD propulsion and auxiliary electrical systems by capturing beamed microwave energy. These rectennas are supported from the underside by twelve ultralight trusses made from SiC-SiC ceramic matrix composites.

The rectennas must be able to adjust their focal length during flight to keep the desired shape under a high acceleration load. A complex truss work of adjustable-length struts made of active "smart" materials braces the underside of the rectenna array. Tension in the ultralight struts is adjusted actively during flight to keep the rectennas at the desired shape and focus. Another network of piezoelectric actuators connected to the underside of the rectenna array allows for high-precision adjustment of the rectenna focus. These materials, which change shape or length under the influence of an electric current, virtually eliminate all mechanical devices from the rectenna control system, with large weight savings.

A water storage system is incorporated into the structure of the Lightcraft. A tube with a diameter of 22 cm runs beneath the toroidal pressure vessel. This area is where the 2400 kg of purified water that is used for the rectenna cooling and temperature regulation systems is stored. The stored water is also used for drinking and food preparation. The tube has an elaborate internal baffle system that prevents the sloshing of water during high acceleration, which might otherwise throw the Lightcraft off balance during routine velocity or attitude changes.

The water tank is connected to pipes that feed water from the water collection and purification modules.

2.6 Superconducting Magnets

An array of "flightweight" superconducting magnets serves as an electrical power storage system for the Lightcraft, and the resultant high magnetic fields enable maglev belt and lander systems.

A total of 9 superconducting magnet coils run circumferentially around the Lightcraft's inner and outer rims. Two primary magnets have a mass of approximately 1200 kg and follow along the top and bottom of the toroidal pressure vessel. These, the largest magnets found on the Lightcraft, provide most of the capacity of the superconducting magnetic energy storage unit, about 900 MJ. Each magnet is comprised of a 20 cm diameter, hollow cylindrical cable made from metal matrix superconductors (MMSC) with integral coolant passageways for circulating the liquid-helium coolant to prevent magnets from warming above the superconductive transition temperature. As in Figure 2.6.1, each superconducting cable is suspended within a 30-cm diameter toroidal vacuum tube, braced by a radial mesh of high-strength insulating fibers that are loaded in tension. These non-conducting fibers precisely position the cable within a structural vacuum tube; the thermal insulation system is so efficient that liquid helium evaporation from all nine magnets totals

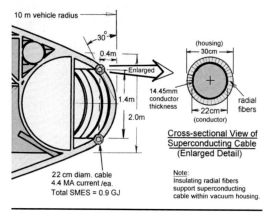

Figure 2.6.1: Cross-sectional view of primary superconducting magnets at rim. *(Courtesy of RPI.)*

only milliliters per day. The MMSC magnet material contains high strength fibers that run circumferentially around the cable to prevent its destruction by extreme self-induced magnetic pressures. Hence the structural load path starts with forces generated within the magnet material itself that are transferred through the insulating fibers to the 30 cm structural vacuum tube that is intimately tied into the Lightcraft's primary structure, the perimeter toroidal tube. In the event that a superconducting magnet goes normal, it is protected by an external electrical circuit that dumps most of the circulating power into the rectennas (see Figure 5.2.3 in Section 5.2).

The secondary superconducting magnets lie near the top and bottom of the lenticular domes where the "donut hole" is located. The other 5 magnets, which line the inside of the "donut hole," are also used to store power. These smaller magnets provide the "close range" magnetic lift for docking the maglev landers and personnel lifted by maglev belts. The large primary magnets do the bulk of the lifting at large ranges: i.e., 5 to 20 meters.

2.7 Photovoltaic Arrays

The "low power" face of the Lightcraft (i.e., opposite the rectennas) contains the photovoltaic arrays (Figure 2.3.2). Twelve panels made of gallium indium phosphide cover the entire surface. These panels provide the Lightcraft with a source of power when the Lightcraft is on the ground or in the ion propulsion mode. During these times microwave beamed energy cannot be used because the "high power" face of the Lightcraft containing the rectenna array is facing the ground. The photovoltaic array produces 100 kW of power at 25% efficiency under solar radiation, and several megawatts at 65% efficiency when illuminated with a 650 nm laser beam. During high-acceleration flight, or during normal operation when microwave energy is being used to power the Lightcraft, the photovoltaic arrays are turned off.

2.8 Emergency Repair and Maintenance

To prevent dangerous electrical discharges from outer hull surface irregularities, the entire surface of the Lightcraft must be kept mirror-smooth and free of foreign debris at all times during flight. To maintain the craft's structural integrity it is also imperative that all 12 hull sections remain fully inflated. Because of these requirements, certain methods must be employed to maintain the outer hull of the Lightcraft.

Several methods can be employed to keep the outer surface of the hull free of any dirt particles. The first involves cycling the SMES units to clean the surface. Electrical current can be pulsed through the magnets at sub-audible (20 Hz or lower) to ultrasonic frequencies (>20 kHz). The rapidly oscillating magnetic field shakes off any dirt particles clinging to the hull. This procedure effectively removes all foreign particles from the surface of the Lightcraft, allowing a uniform electric charge distribution across the hull during the low-power ion propulsion mode. This magnetic ultrasonic cleaning works much like a conventional ultrasonic cleaner used on fine jewelry and other delicate materials, in that it can clean the surface without damaging it.

The second cleaning method involves using the liquid-helium coolant that normally flows through the main toroidal magnets. A small quantity of helium can be sprayed into the gaseous HeliOx coolant system to cool the outer skin rapidly. The large temperature difference between the Lightcraft's hull and the ambient atmosphere supersaturates the air closest to the ship, resulting in a film of water vapor that condenses at the surface and runs off, taking dirt and debris with it.

3.0 COMMAND SYSTEMS

Command and management of the LTI-20 Lightcraft is a joint but hierarchical effort between pilot and machine. Numerous on-board intelligent computers are all interlinked and ultimately connected to a central Lightcraft computer, creating a highly advanced and autonomous Flight Management System (FMS). The normal 6-person (LTI-20A) to 12-person (LTI-20B) crew complement of the Lightcraft contributes the human aspect of its intelligence. Although the FMS computer is sufficiently intelligent that any two highly qualified bridge officers (i.e., pilot and co-pilot) can fly the LTI20 with absolute precision, typical Space Command missions ranging from 2 days to 2 weeks, demand the full crew complement.

The first two sections of this chapter (3.1 and 3.2) deal with the main bridge of the Lightcraft and the crew operations performed there. Section 3.3 addresses how the craft is to operate when the crew cannot be on the bridge, but must retain control of the Lightcraft from the safety of their escape pods. Section 3.4 examines the controls and terminals used to control the Lightcraft. Section 3.5 discusses control of the vehicle; Section 3.6 covers mission operations; and Section 3.7 deals with guidance and navigation.

3.1 Main Bridge

The primary area from which the Lightcraft is controlled, and from which communication with both ground and orbital stations is maintained, is the "main bridge" on the Lightcraft's observation deck or "flight deck."

The bridge consists of that volume inside the craft that is adjacent to the exterior viewing window (see Figure 3.1.1). Despite the large amount of visual and other information obtained directly from the FMS, there is a need for a flight deck from which the exterior surroundings can be viewed directly. At low speeds, 3 segments of the annular rectenna can be retracted to allow the crew on the flight

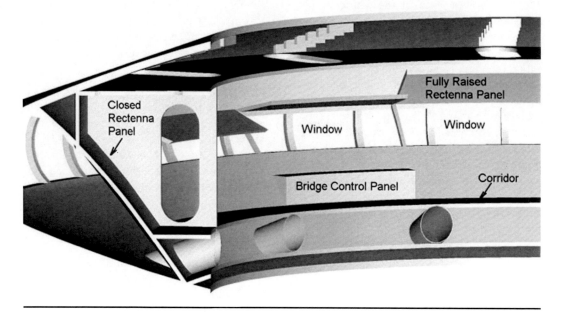

Figure 3.1.1: Bridge showing three rectenna array panels being raised to permit direct viewing of landing site through toroid. *(Courtesy of RPI.)*

deck to view the outside through the transparent toroidal tube. This fail-safe feature enables controlled landing and docking maneuvers in the event of complete failure of the FMS remote viewing systems.

Controls and communications are available on the bridge through the use of the FMS console placed there for that purpose. Both the control panel and the chairs where crew members on the bridge sit must rotate between two positions: when the Lightcraft is in the atmosphere the direction of gravity is downward, and the chairs and console are in their normal position. When the vehicle is slowly rotating in space, however, the apparent direction of gravity is radially outward from the spin axis, toward the outside rim of the craft. Thus the chairs and console must swivel in concert so that the crew can be seated in this new configuration.

The bridge is not always in use; indeed, full utilization of the bridge is possible only when the crew can be outside their escape pods and maglev landers. Under rotating and hence gyroscopically stabilized flight conditions, when the Lightcraft is either hovering or maneuvering in the dense atmosphere at low speeds, the crew is despun within the centrally docked maglev landers. During high-acceleration maneuvers, such as the transatmospheric trip to orbit, the main bridge would be empty of crew. Any needed flight control input could be carried out from within the escape pods with the aid of the FMS.

3.2 Crew Assignments, Roles, and Duties

The standard Lightcraft crew complement consists of 6 members in the LTI-20A, and 12 in the LTI-20B (see Table 3.2.1). The standard USSC crew selection criteria for the LTI-20A permits personnel masses up to 100-kg (inclusive of 1.5 kg for space-mobility suit); the LTI-20A escape pods are reconfigurable for statures from 163-cm to 196-cm.

LTI-20B missions restrict crewmember masses to the range of 40 to 60 kg (i.e., *LCJ-class* only), with a mandatory flight-average of 50 kg (inclusive of the 1.36 kg suit); LTI-20B escape pods accommodate statures from 130-cm to 162-cm. "LightCraft Jockeys" (known in their ranks as LCJ's) are, of course, world renowned for their athletic prowess, high G-tolerance, and technical skills in maglev-belt and other USSC special-ops. (That Space Command has been, and continues to be a "low-key" sponsor of national equestrian competitions, is not a coincidence.)

Ultra-energetic Space Command missions demand that the total crew complement mass (inclusive of escape pods) does not exceed 1200 kg for both the LTI-20A and LTI-20B.

(*Note: Certain mission-specific LTI-20B rescue / escape pods can reconfigure for small children: 15 to 30 kg; 107 to 126-cm; 5th to 95th percentile mass-for-stature.*)

Every crewmember has specific functions to fulfill on a given mission. That function is implied by the crewmember's title and professional

Table 3.2.1 Space command crew complement for LTI-20 lightcraft.

	LTI-20A	LTI-20B
Mission Leading Officer (MLO)	1 (pilot)	1
Second-in-Command Officer (SCO)	1 (pilot)	1
Pilot (helmsperson)	—	2
Navigation Officer / (pilot crew-member)	1	1
Load Master/ Scanner (LMS)	1	1
Flight Engineer/ (pilot crew-member)	—	1
Communications/Tactical Officer (CTO)	1	1
Mission Specialist	1	3
Flight Medical Officer (FMO)	—	1

background (pilot, doctor, etc.). However, all crew members must have detailed knowledge of the essential Lightcraft systems. On global missions, each of the three (mandatory) pilots will fly the craft on 8-hour rotating shifts; Naval crews retain their traditional 4-hour "on," 4hour "off" watches. All crewmembers except the flight medical officer and certain mission specialists serve on the bridge as well as in their own specialties. Among these crew members, only specific personnel (3-normal, 2-minimum) are on the bridge at any time. For the abbreviated 6-person crew of the LTI-20A, the highly trained MLO and SCO serve as pilot and co-pilot, respectively, and the navigator must be a pilot crewmember; the flight engineer's functions are adequately provided by the FMS, and a mission specialist often displaces the FMO.

3.3 Bridge Operations

The only times when the bridge is in use for atmospheric flight are during extended loiter use of the ion propulsion system or brief intervals (i.e., low-g maneuvering) with the pulsed detonation engine (PDE) thrusters, and while rotating in space. These are the times when the Lightcraft is not in high-performance acceleration or deceleration mode. Only at these times can the 6 - 12 person crew leave their escape pods to assume their appropriate roles on the "flight deck" (bridge). During subsonic cruise periods with ion thrusters – wherein the Lightcraft is spin stabilized in "frisbee mode" (FM) – bridge operations are normally transferred to the despun maglev landers.

Only certain bridge stations are in use at the same time. The choice of which crew members are on the bridge is determined by the level of security under which the Lightcraft is operating.

3.3.1 Security Levels

In military operations, the two security levels are L2 (normal operation) and L1 (*battle stations*). These security levels and the tasks of the crew members at each station require further description.

When a Lightcraft is operating at L2 it is flying undetected (neither being chased nor under attack) and maneuvering at low performance.

Most of the crew are therefore not at their stations. During L2 operations, only 3 of the 6 total stations on the bridge are staffed. On the LTI-20B, these stations are pilot, communications / tactical, and command (bridge). Each station is staffed by a 3-person rotation, with each serving 8-hour watches in each 24 hours. The pilot watch is filled in turn by the 2 pilots and the navigation officer (a pilot crewmember). The CTO and two mission specialists take shifts at the communications / tactical station. The bridge watch, entailing command of the bridge, is staffed by the SCO, flight engineer, and load master / scanner, in rotation.

The decision to change security level is made by the officer on bridge watch or the mission leader.

When at security level L1 on the LTI-20B, a total of 6 positions on the bridge are staffed and operational. These 6 positions are pilot, communications / tactical, engineering, navigation, second-in-command, and mission leader. During transition from L2 to L1 a shift of personnel occurs. All positions must be staffed by their appropriate personnel. Specifically, the officer on L2 helm duty is relieved by a pilot, the communications / tactical station is taken over by the CTO, the officer on L2 bridge watch is relieved by the MLO and SCO, and the navigator and flight engineer assume their stations. (Note: *Under L1, these bridge duties can be precisely performed whether designated bridge crewmembers are serving on the flight deck, in one of the two de-spun maglev landers, or within their individual escape pods, linked by VirtuNet in a virtual "bridge" environment.*) This bridge configuration continues until the security level is dropped back to L2. At that time, only the pilot watch remains on station.

3.3.2 Crew Station Duties

Although it is clear that different personnel occupy different stations at security levels L1 and L2, the responsibilities of the crew are specific to the station occupied. These duties are outlined below.

3.3.2.1 Mission Leading Station

The mission leading station is, of course, staffed by the mission leading officer (MLO) only, and only during L1 operation. At other times, under L2 status, the MLO does not serve on any watches, and remains free to tour the Lightcraft checking up on the other crew members, equipment, and mission-specific communications. The MLO is often working in his/her crew quarters. When the security level rises to L1 the MLO goes immediately to the bridge. All mission-specific information from bridge personnel and the FMS computer is sent to the mission leader by way of the personal communicator console or by direct verbal communication. Final decisions, judgment calls, and orders are issued by the MLO and relayed to the bridge officer, who acts accordingly. The MLO's interaction with the FMS computer with regard to Lightcraft status is mostly by query and response. The MLO's principal duties on the bridge are to assure that the mission plan is followed, to respond to threats and other changes that might impact mission success, and to make command decisions to assure the safety of the Lightcraft and its crew. On the LTI-20A, the MLO is also the pilot in command.

3.3.2.2 SCO Station

The second-in-command officer (SCO) station on the bridge is occupied only during L1 operation. The principal role of the SCO is to monitor command decisions and suggest different points of view and alternative courses of action to the MLO. The SCO uses the FMS computer console to monitor actions taken by any bridge officer and, under instruction from the MLO, to assure that orders are fully executed. The SCO cannot issue an order to engage the enemy or fire weapons, which is the sole responsibility of the MLO. The SCO assumes full control and full MLO command authority when the MLO is incapacitated. On the LTI-20A, the SCO is also the co-pilot.

3.3.2.3 Navigation Station

The navigation station, which is staffed only by the navigation officer (a pilot crewmember), is occupied only during L1 operation. The navigation officer must be a specialist in orbital mechanics, hypersonic aerothermodynamics, and atmospheric reentry physics, and is responsible for planning and monitoring the flight path. Flight path planning begins before liftoff with preplanning of the overall flight path profile (one active, two alternates) for the entire mission using the FMS computer. In flight, the navigation officer does real-time flight path planning, including low-performance cruise, hovering, and response to emergencies. The navigation officer normally plans all flight paths interactively with the FMS computer which provides (among other functions) global autonomous guidance using selectable Global Positioning System (GPS), Inertial Navigation System (INS) or integrated GPS-INS navigation. The FMS can readily download (by encrypted laser DataLink) highly detailed maps of any part of the world onto the navigation computer console, import mission-specific data, compute optimum flight paths and times to destination, provide instantaneous speeds and headings along such paths, and explore emergency options. These data permit the navigation officer to predict and monitor any flight path of the Lightcraft during high-performance modes (e.g., hyperjump, orbital boost, and reentry) when the FMS computer is navigating. The FMS computer continuously monitors the real-time flight path, as determined by inertial and external sensors, and compares it to the predicted flight plan. Should discrepancies arise, the navigation officer can enter corrections into the FMS computer and alert the MLO/SCO. The navigation officer's role is to plan and monitor the Lightcraft's trajectory and assure its consistency with overall mission objectives and crew safety.

3.3.2.4 Helm Station

Piloting of the Lightcraft and monitoring of the flight path during FMS autopilot control are the responsibility of the helm officer. The operations of the helm station during L1 and L2 are the same irrespective of who is on watch.

Piloting of the Lightcraft can only be done in conjunction with the FMS computer, aided by the navigation officer who plans the flight path and relays preplanned information to the helm officer during L2. The helm officer accepts input data on course, speed, and orientation relayed from the navigation officer and allows the FMS autopilot to guide the Lightcraft. Ultra-energetic maneuvers, under L2 conditions, are "laid in" (i.e., preprogrammed) to the FMS computer, and executed at the discretion of the pilot. In the event of any subsequent course deviation from the preplanned trajectory, the pilot may intervene and attempt a real-time correction of the discrepancy or, after completion of the maneuver, relay the information to the MLO for new orders. In all high-performance flight modes, the FMS autopilot controls the Lightcraft, and it is the duty of the helm officer to monitor the flight and input course corrections as needed. Under emergency conditions in low-performance flight, under L1 conditions, the helm officer can take control from the FMS autopilot at any instant and pilot the Lightcraft manually using the *FailSafe NeuralNet* – a "backup" fly-by-light (i.e., fiber-optic) control system with triple redundancy.

Every helm officer has received the most rigorous Lightcraft pilot training that USSC has to offer, emphasizing ultra-stealth and hyper-energetic missions. Of course, that program still begins with instruction on electrodynamic airships with 5 MeV endoatmospheric ion drives (EID), before transitioning to more energetic BEP systems.

3.3.2.5 Tactical / Communications Station

The tactical / communications station also has the same functions during L1 and L2 operation, regardless of who is on watch. The duty of the officer at this station is twofold, as the title implies.

The tactical role involves checking for threats at all times, using the FMS computer and long-range sensors at various levels to watch for enemy presence. Based on threat assessment by the FMS computer, more detailed inspection of the enemy may be requested by the tactical officer, or the decision may be passed up to the commanding officer (the bridge watch officer, SCO, or MLO) pending further action. A more detailed threat assessment requires using scanned information along with data stored in computer memory to determine the firepower and flight performance capability of the enemy.

Under orders from the commanding officer, the tactical officer selects and fires the requested weapons, usually on-board and remote directed energy weapons (both laser and microwave). Thus the tactical duties involve using the FMS computer to request, receive, and analyze data relevant to preserving the safety and tactical advantage of the Lightcraft.

The communications duties include keeping in constant contact with the Space Command headquarters and with any crew members deployed in hostile territory. The communications officer must also be able to open communication with the enemy through an onboard universal translator. Computers serve the communications station by constantly seeking the clearest and most secure communications channel. In the event of interference or jamming, communication mode or frequency changes must be swiftly implemented. Of course, the LTI-20's advanced *PulseTimeDomain* communications are virtually undetectable and jam-proof, as is the laser *DataLink* to Space Command headquarters.

3.3.2.6 Engineering Station

The engineering station is under the most intensive use during L1 operation, and is staffed by a flight engineer who may also be a pilot crewmember. In flight, the FMS computer constantly monitors and records (i.e., so-called "*black box*" function) time-resolved Lightcraft performance levels along the flight trajectory, inclusive of all mechanical, electrical, and environmental systems performance, and available power. To ensure maximum Lightcraft performance, the flight engineer must occasionally run diagnostic programs on the health of all Lightcraft systems, often favoring the three propulsion modes:

a) subsonic, low-energy ion propulsion,
b) ultra-energetic PDE thrusters, and
c) hypersonic MHD slipstream accelerator and airspike.

For example, on the ion-propulsion (i.e., EID) engines the engineer can scrutinize the real-time or most recent hull charging history by accessing stored FMS data to reveal the recent performance of all 24 relativistic electron beam accelerators (E-beam guns, 5 MeV). In EID flight, the Lightcraft hull is normally charged up to the breakdown limits of the surrounding atmosphere (1-3 MW/m at sea level altitude), so the EID diagnostic program can readily reveal a "dirty" hull that needs cleaning, faulty E-gun metal foils or improper orientation of specific E-guns, and the like. Other FMS diagnostic programs track the PDE thrusters, MHD slipstream accelerator and Airspike performance, revealing how well the Lightcraft has maintained alignment with the microwave power beam, whether the pulses were repeated at the right frequency with the proper pulse waveform and pulse energy, etc. Critical atmospheric propagation issues are revealed by spatially-resolved data on received power density distributions striking the Lightcraft hull and are also tracked by the FMS computer. Microwave beam-spreading effects caused by atmospheric turbulence, thermal blooming, and jitter can be determined from this FMS recorded data, in addition to extinction estimates. All such diagnostic programs are run by the engineer at the engineering station, relying on FMS computer data. The engineer has the responsibility to keep all structural, electronics, life support, and propulsion systems of the Lightcraft performing at their maximum attainable efficiency so that optimum use of available power may be achieved, consistent with acceptable safety levels.

3.3.2.7 Load Master / Scanner Station

The Load Master / Scanner (LMS) works closely with the flight engineer(s) in a team effort to assure safe operating procedures, during ramp extension / retraction, kneeling / unkneeling the craft, loading and unloading cargo and personnel, as well as guaranteeing the proper load distribution, weight and balance. When the craft is being serviced (i.e., recharged and "refueled" with H_2O) at a USSC base, the LMS assures proper removal of the external power cable and related gear. It is the responsibility of the LMS as the Safety Officer to secure the craft (with tie-downs) on highly uneven terrain, when it is resting on the tri-pod landing gear. When escape pods are used to quickly deploy or extract crew for special ops, the LMS is to issue "clear" commands for opening and closing the pod doors. The LMS also monitors maglev belt and maglev lander operations to assure safe deployment and retrieval of special ops crewmembers. The LMS is the "eyes and ears" outside the craft during landing and near-ground operations, monitoring crew entrance / exit through Lightcraft pressure bulkhead doors to the landing ramps, or to the maglev landers. LMS duties include inspection for electrical arcs, shorts and fires during engine startup, including potential microwave / laser ignited fires on the ground, and to assure that personnel are not in harms way from "friendly" microwave and laser discharges. As a regular in-flight duty, the LMS conducts a visual inspection of the Lightcraft critical areas, looking for leaks, spurious electrical discharges, smoke, etc. (i.e., Scanner role) on an hourly basis. One major LMS duty is to assure that the 1200 kg load limit (i.e., ~1.2 tonnes maximum load) of the LTI-20 is strictly respected, and that the load is uniformly distributed about the Lightcraft for proper balance. For example, under a normal 20-, 30-, or 50 G acceleration, this load becomes 24, 36 or 60 tonnes (respectively), but at momentary 100- to 200 G levels, it reaches 120 to 240 tonnes – i.e., matching or exceeding the load limit of large USSC jet transports. The LTI-20 is essentially a hypersonic balloon / airship with a limited design load capacity in order to safely perform USSC hyper-energetic missions.

3.3.2.8 Flight Medical Officer Station

The flight medical officer is a specialist in both high-pressure HeliOx, and partial liquid ventilation systems, insertion / removal / diagnosis of standard USSC issued micro-electronic implants (specifically communications and life-sign monitoring functions), treating burns from incidental exposure to microwave and laser radiation, and diagnosing / treating high-G related injuries. In addition to the traditional general practice training, the FMO skills include specialized procedures for treating injuries sustained from transient exposure to charged particle beams and high magnetic fields, as well as administering laser surgery and trauma stabilization (and emergency treatments) of special ops related injuries.

3.3.3 Flight Management Computer System

The intelligent computer network interacts with and links all bridge stations into a highly integrated Flight Management System (FMS). The computer system also has its own autonomous functions over which it has full oversight, essentially those with data rates so high and response times so short that real-time human intervention is impossible. The FMS constantly monitors ship maintenance by means of mechanical, thermal, optical, and electronic sensors positioned all over the Lightcraft. When structural or functional deviations from the norm are detected the FMS orders compensating actions for the purpose of keeping the Lightcraft in equilibrium. If emergency conditions are detected, the FMS notifies the appropriate officer so that corrective measures can be taken.

Overall, most commands originating from the bridge under both L2 and L1 conditions are generated by human intelligence. The FMS computer's only autonomous authority is general monitoring of Lightcraft integrity and maintenance. The FMS also mediates extensive exchange of data between different crew stations.

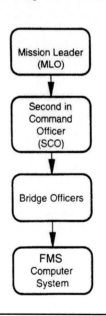

Figure 3.3.1: Chain of command. *(Courtesy of RPI.)*

The general chain of command, illustrated in Figure 3.3.1, has the mission leader (MLO) at the top, the second-in-command (SCO) second, the other bridge officers third, and the FMS computer fourth.

The flow of information in the Lightcraft is quite different in nature, as depicted in Figure 3.3.2. Direct two-way communication takes place between commanding officers and bridge officers, between commanding officers and FMS computer, and between bridge officers and FMS.

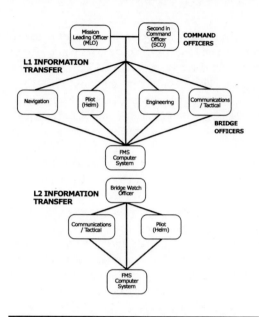

Figure 3.3.2: Information transfer during bridge operations (L1 and L2). *(Courtesy of RPI.)*

Each level can send and receive information directly or through a mediating party, such as using the helm officer as a mediator between the FMS computer and the commanding officer.

3.4 Flight Operation from Escape Pods

During high-performance flight using the Pulsed Detonation Engine (PDE) or magnetohydrodynamic (MHD) acceleration, all 12 crew members are in their escape pods, and command must therefore be shifted from the bridge to the escape pods. In the pods, the crew are protected by their ultra-high-G suits and surrounded by a 1-cm-thick liquid layer. They will either be breathing pressurized HeliOx, or oxygenated perfluorocarbon liquids, instead of air. Although the crew cannot talk in the normal manner, they can still communicate with the FMS computer and with each other using virtual reality information systems.

In the PDE mode, and especially in the MHD propulsion mode, most of the flight operations are preprogrammed into the FMS computer so as to enable it to fly the Lightcraft autonomously. Speed, orientation, and airspike maintenance are monitored by the FMS continuously during flight. The FMS also monitors sensor data and feedback systems. Another autonomous computer function is to monitor conditions relevant to the health and safety of the crew, including biomedical and environmental instrumentation. In the event of hull damage, major system failure, or any other imminent threat to the crew, the FMS assumes full command and all 12 pods are ejected in a fraction of a second. However, if a remediable problem such as a fire or minor hull breech occurs, the FMS automatically extinguishes the fire or seals off the damaged section. It then automatically compensates for lost performance and, when appropriate, modifies the internal environment to keep it within acceptable limits. Note that sufficient time may exist for a crew member to evacuate a damaged compartment before that section is sealed off and the fire suppression system is activated.

During both PDE and MHD propulsion modes, when the FMS assumes full command of Lightcraft flight operations, the crew remains in the command loop. All important flight information, including speed, orientation, altitude, location, and sensor feedback, is displayed via the FMS computer network to all 12 crew members. The crew members receive all information essential to carrying out their specific assignments. The crew are kept constantly aware of Lightcraft status by means of their FMS computer interfaces, comprising both consoles and personal communicators. They are therefore able to monitor for warnings from the FMS and for discrepancies between planned or projected data and actual performance. In addition to these computer interactions, all crew members are cross-linked by direct communication channels, so that any crew member can communicate with or check on any other member in their respective escape pods.

Once the FMS computer is programmed, it can assume command of the Lightcraft during high-performance PDE and MHD propulsion modes, subject to monitoring by the crew. The command structure in this case is essentially the inverse of that used during bridge operations. Figure 3.4.1 schematically illustrates information transfer during high-performance flight modes.

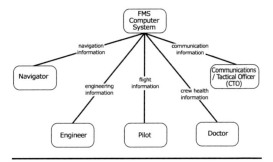

Figure 3.4.1: Information transfer during high performance modes. *(Courtesy of RPI.)*

3.5 Basic Control and Terminal Use

The LTI-20 Lightcraft and FMS computer network are designed to be efficient as well as user-friendly. The FMS control panels aboard the Lightcraft, which act as the interface between crew, passengers, and various onboard systems, can be reconfigured to suit their users" preferences.

3.5.1 Basic Systems Numerous subsystems on board the Lightcraft can be monitored and adjusted from FMS control panels throughout the vehicle. These service-providing systems are mainly those concerned with the comfort and necessities of the users. The environmental, entertainment, food storage, hatch operation, and other systems are all accessible from these panels.

3.5.2 Environmental System

The FMS computer-managed environmental system is responsible for maintaining life-support, atmospheric quality, and waste management. These functions can easily be adjusted by the individual occupants. For example, personal comfort control within the crew quarters, which is one of the most frequently used functions of the environmental system, permits each user to input a desired temperature and humidity. Automatic controls on each crew member's body suit permit the user to also vary the conditions to match the selected settings.

3.5.3 Entertainment Systems

The LTI-20 is equipped with a number of entertainment utilities to help with the passage of long travel times and extended missions. The user may elect to watch a movie, play video games, listen to music, take virtual-reality "tours," or enjoy other activities available from the panels. Linkage of the entertainment system with the virtual reality system permits complete, effortless immersion in whatever entertainment mode is selected. This is particularly useful during week-long "on-station" loiter periods for special ops, or during 5 ½ hour to 3 day+ treks between Earth and the USSC lunar base.

3.5.4 Hatch Operation System

This system controls the operation of all hatches within the vehicle, including the boarding ramp illustrated in Figure 3.5.1. Although the hatches are normally governed automatically by the FMS computer, a manual override is incorporated into the control panel to safeguard the users.

3.5.5 Food Storage System

Since mass and space are at a premium aboard the Lightcraft, all food products are dehydrated and stored in a central storage area. In response to user requests entered through a terminal, the FMS computer automatically rehydrates and, where appropriate, heats the food items and serves them.

3.5.6 Panel and Terminal Layout

Control panels are located throughout the Lightcraft at convenient locations, including inside the maglev landers and escape pods. The control terminals, as illustrated in Figure 3.5.2, have the same basic layout for all subsystems. Upon activation a panel displays a menu that presents the user with the list of options available. From this menu the user selects the desired subsystem and makes the appropriate changes. Any changes that affect all crew members, such as life-support alterations, require an authorization code. This precaution is intended to prevent an accidental alteration of the function of any essential system component or system by any single person.

3.6 Flight Control

The flight control station shown in Figure 3.6.1 is primarily responsible for the piloting and navigation functions of the LTI-20. This station is located on the main bridge and is normally staffed by the flight officers (i.e., pilot, copilot, navigator or flight engineer).

Figure 3.5.1: Boarding ramp / hatch deployed, with escape pods serving as landing gear. *(Courtesy of RPI.)*

Figure 3.5.2: Control panel and terminal layout. *(Courtesy of RPI.)*

Figure 3.6.1: Flight control station on the Bridge, with "heads-up" display. *(Courtesy of RPI.)*

3.6.1 Functions

The flight control system on the bridge has four major functions, all intimately intermeshed with the FMS computer. These functions are navigation and course plotting, supervision of automatic flight systems, manual flight operations, and monitoring of overall Lightcraft status. "Manual" control can be assumed by the pilot – using the *FailSafe NeuralNet* – a backup fly-by-light (i.e., fiber-optic) control system with triple redundancy.

3.6.2 Navigation and Course Plotting

The flight control system (FMS computer) automatically displays data from the navigational sensors (three inertial Nav units with dual G.P.S.), overlaying them on current position and course projections. This function allows the crew to confirm the heading and course of the Lightcraft visually. The flight officer (pilot) and/or mission commander can then use this information to guide the Lightcraft toward its destination.

3.6.3 Supervision of Automatic Flight System

During hypersonic flight, and under most propulsion modes, the LTI-20 is under the control of the FMS autopilot system. Although the flight control system is designed with quadruple redundancy (i.e., four separate fiber-optic cables) to avoid catastrophic failures, manual supervision is still required. To aid the flight officer in this duty, the FMS computer reports any abnormality as a warning message. Upon complete FMS computer failure (which has never occurred), the pilot disengages the autopilot, engages the *FailSafe NeuralNet* system (which uses 3 of the 4 fiber-optic cables), and assumes "manual" control. The transfer is instantaneous.

3.6.4 Manual Flight Operations

Under subsonic (and some supersonic) fight modes the flight officer has the option of exercising manual control. Under manual control the pilot may use some combination of voice commands, the joystick, or the advanced hand-pad, illustrated in Figure 3.6.2, to interface with the FMS computer and execute maneuvers or course corrections in real-time. The hand-pad control is a fly-by-light input device to the FMS

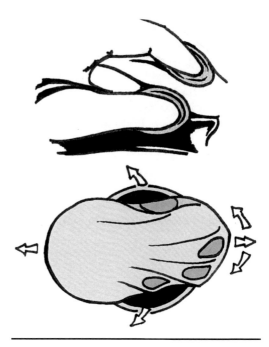

Figure 3.6.2: Hand-pad for manual control of flight operations. *(Courtesy of RPI.)*

computer, much akin to the traditional joystick interface, which some USSC pilots still prefer.

3.6.5 Overall Lightcraft Status

The final function of the flight control station is to provide the crew with accurate and timely updates on the status and health of all Lightcraft systems and subsystems, including life support, structures, electronics, flow control, SMES units, etc. Hull conditions, ion thruster systems, the PDE propulsion system, the MHD slipstream acceleration system and its airspike are all monitored automatically by the FMS computer and displayed on the flight control console. Any abnormality is brought to the attention of the crew by means of a warning message.

3.7 Mission Operations

The "mission ops" system is responsible for providing additional support for operations management and for monitoring the progress of secondary mission functions. Secondary objectives are those that are intended to be accomplished once the primary mission goal has been met, and which must not be allowed to interfere with accomplishment of the primary mission objectives.

The principal purpose of mission ops is to *execute the mission* by coordinating processes and resource allocation in accordance with the guidelines specified by USSC operations management. At least two contingency flight paths are normally built into the master flight / mission plan. For example, in a reconnaissance / surveillance mission, operations management may specify the priorities of several targets. Within these guidelines, the mission ops system selects the sensor suites appropriate for each target and the Lightcraft's time over target. The mission ops system is also responsible for the actual collection and storage of sensor data. Mission ops also assumes the secondary function of monitoring station-keeping during such a mission to help assure a stable platform for the sensors. Although real-time management of these functions is handled by the FMS computer, a human operator is required to monitor performance and to insure that none of the mission priorities are compromised.

The mission ops system is responsible for resolving low-level priority conflicts, but will refer any conflicts involving primary mission objectives to operations management. Given a specific Lightcraft system failure, the FMS computer would automatically select a contingency flight plan. But if the survival of either the Lightcraft or its occupants is endangered, mission ops will refer the conflict back to USSC operations management.

3.8 Guidance and Navigation
The FMS guidance and navigation system is crucially important to the mission of a Lightcraft, entailing as it does several different modes of propulsion, high-speed flight, extreme acceleration, and trans-atmospheric capabilities. The guidance and navigation system must detect variations in the Lightcraft's flight velocity, orientation, and trajectory in order to maneuver safely from one location to another. The position of the Lightcraft relative to outside objects must be known with a precision of 0.5 m so that collisions may be avoided.

3.8.1 Guidance
The FMS computer provides global autonomous guidance using selectable Global Positioning System (GPS), Inertial Navigation System (INS) or integrated GPS-INS navigation. Precise, continuous sensing of the Lightcraft's flight orientation by the FMS computer is necessary so that trajectory corrections can be made quickly enough to avoid hazards. If the Lightcraft is pitched, rolled, or laterally displaced by a strong wind gust, quick, automatic adjustments are required to correct the trajectory.

Accurate guidance of the Lightcraft ensures that it remain precisely aligned with the microwave or laser power beam. Misalignment causes a serious drop in propulsive power and may endanger those standing beneath the Lightcraft by spilling power to the ground.

3.8.2 Navigation
The Lightcraft must have up-to-date information on its position and the location of external objects near it, both in space and in atmospheric flight.

A primary responsibility of the FMS navigation system during atmospheric flight is to monitor nearby ground contours and elevations, including the locations and shapes of obstacles such as mountains and trees, both for collision avoidance and for providing cover for low-altitude and surface operations. The navigation system also contains, and updates as needed, a database of cities, bridges, transmitter towers, roads, and other features that may be significant navigation hazards, or that may be useful as landmarks. The craft must also know its position with respect to both friendly and hostile vehicles and personnel. This information will allow the Lightcraft to avoid collision with, or detection by, vehicles and personnel in its vicinity, and to take effective evasive action when necessary.

In space, FMS navigation of the Lightcraft takes on new complexities. The Lightcraft must retain the information necessary for its operations in the lower atmosphere, but must also be aware of the positions and motions of a number of other objects such as satellites, space stations, and cataloged orbital debris. Orbital debris must be detected with the LIDAR (laser imaging, detection, and ranging) system and targeted by onboard megawatt-class pulsed lasers, either to fragment, vaporize, or divert them so that they do not collide with the Lightcraft (Section 18.0). The masses and projected orbital paths (ephemerides) of the Moon, planets, and other Solar System bodies are, of course, known to the FMS computer as important navigational reference points and their gravitational accelerations on the Lightcraft are automatically accounted for.

3.8.3 Sensors

The data required for Lightcraft guidance and navigation are secured by numerous onboard and off-board sensors. The FMS computer automatically assigns a high priority to monitoring and interpreting sensor data. The sensor data streams are double-checked and cross-correlated before being relayed to the crew or being used to make necessary adjustments to the Lightcraft's course.

4.0 COMPUTER SYSTEMS

The computer systems aboard the LTI-20 are of extreme importance under all circumstances of Lightcraft operation. The data rates delivered by the extensive sensor array and the response speed necessary to monitor and control the propulsion and navigation systems rule out direct human control. To handle its many essential functions, the LTI-20 computer system consists of a massively parallel central computer with 1024 identical Terahertz-class milspec radiation-hardened processors and a number of dedicated separate processors located in each piece of equipment, ultra-G space suit, maglev lander, and escape pod, all linked by high-data-rate optical interfaces.

4.1 Computer Systems Architecture

The central computer, which is functionally a part of the bridge, is physically delocalized into eight major processing clusters, two redundant solid-state file servers, and two dedicated virtual-reality drivers, each with an isolated power supply and its own backup power. Each processor cluster in this Virtual Computer Network has direct communication with every sensor array and every major piece of equipment. Each bridge station is fully reconfigurable in software, and all computer stations are identical in hardware. All displays are accessible via virtual reality goggles that constitute an essential part of the ultra-G space suit. Alternatively, any display can be presented on thin-film display panels at each bridge station and at other locations throughout the Lightcraft. The capability for virtual-reality display and the use of optical data links permits extreme flexibility in accessing the system. Voice recognition, retinal and finger print scans are used by the FMS computer during low-performance flight to identify crew members and set the virtual reality displays to individual custom settings. In the interest of interchangeability, such human-computer interfaces (HCIs) employ window-based displays that can be altered (within limits) to suit the user, while still sufficiently conforming to "standardized" USSC formats.

4.2 Escape Pod Computers

The escape pod computers are designed so that each provides a fraction of the on-board data storage capability. While docked, these pod computers also contribute their processor power to the central computer system. Operating autonomously, the escape pod computers run the rudimentary sensor array of the pod, assist in basic navigation and communications functions, monitor the life-support package, and interact with global navigation and search-and-rescue (SAR) systems via EGPS, GLONASS2, and SARSAT links. Output interfaces are tailored to mesh with the virtual reality functions built into the ultra-g suit of the pod occupant. In emergencies when the individual escape pods and their occupants forcibly separate from the Lightcraft, the pod computers provide all flight management functions necessary for autonomous reentry guidance (if needed), drogue chute and main parachute deployment and steering to precision preprogrammed landing coordinates.

4.3 Human-Machine Interface Options

Personal access to the FMS computer and its associated communications, navigation, and systems-control functions can be achieved through flight-deck consoles, terminals located in each area of the Lightcraft, or virtual reality goggles. The flight-deck and distributed-terminal interfaces are capable of voice recognition and speech synthesis for ease of operation. Under low-performance flight (L2) conditions most interactions with the FMS computer system will be through reconfigurable control panels on the bridge and elsewhere, based on thin-film display screens. At all times, speech-synthesis functions are used to call attention to warning messages and anomalous sensor readings. Normal command interaction with the FMS computer during high-performance flight occurs through simultaneous use of the hand-pad control (or joystick) and body-position feedback sensors built into the ultra-g suit, supplemented by tracking of eye movements by the virtual-reality goggles.

Under extreme flight (and partial liquid ventilation) conditions, when speech may be impossible, or when security issues prohibit speaking aloud, radio-frequency and microwave links to communications implants in the heads of the crew members and subvocalization sensors are employed. For access to highly-classified material, the virtual reality goggles are the normal crew-machine interface.

During space flight all crewmembers are required to wear virtual-reality goggles in order to access the LTI-20 Virtual Computer Network, which provides access to the FMS computer, other crew members, and the outside environment via both sensors and communications channels.

The design of the LTI-20 Virtual Computer Network is closely constrained by human factors considerations; as always, form follows function. The interface is exceptionally user-friendly. Figure 4.3.1 shows the first-level interface to the system, which contains 12 icons representing different menus of functions. The crew can select one of these second-level menus by gazing at the corresponding icon. The virtual reality goggles monitor eye movement and provide feedback to the crew. The LTI-20s High Quality Virtual Reality Environment is explained in more detail in Section 14.5.

The view window, located on the left side of the display panel in Figure 4.3.1, gives access to the detailed contents of all submenus. Each icon selects a particular function:

1. *Intra/Inter Communications* enables crew members to communicate with each other whether inside or outside the Lightcraft.
2. *Mission Briefing* provides top-level secure information on the current mission, subject to password, retinal identification, and need-to-know.
3. *Lightcraft Diagnostics* accesses sensor data on, and complete analysis of, the physical structure and outside environment of the Lightcraft.
4. *Medical* enables crew members to access medical information and support, including direct contact with the flight medical officer.

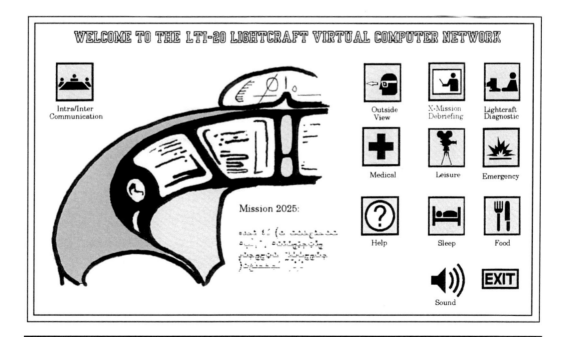

Figure 4.3.1: FMS virtual computer network interface. *(Courtesy of RPI.)*

5. *Leisure* gives access to an extensive library of movies, games, music, and educational materials as well as locally available external media.
6. *Outside View* provides framed, panoramic, holographic, or annotated virtual-reality views of the surroundings of the Lightcraft.
7. *Emergency* enables emergency evacuation via the escape pod.
8. *Sleep* activates the screen saver of the display panel and the energy-conserving mode of the virtual reality goggles. Some personalized applications of this mode may be used to assist a crew member in falling asleep, with timed soothing music and visual displays, defaulting to computer sleep mode when biomedical sensors report the subject is asleep. The most popular such application is the routine *Count_Sheep*.
9. *Food* provides access to food and drink, and to the consumables inventory list.
10. *Help* gives access to all on-line archives, manuals, and diagnostic programs.
11. *Sound* enables "room" sound through electrostatic speakers in the control console.
12. *Exit* deactivates the virtual reality goggles and the suit's position and proprioceptor sensors while leaving the biomedical sensors in the ultra-g suit on-line.

4.4 Personal Access Displays

Although each terminal may be personalized and customized, security considerations and the need to minimize confusion in oral communications may require the frequent use of the personal access displays in the virtual-reality goggles even under low-performance flight conditions. The same feedback mechanism that tracks eye movements in the goggles also performs iris scans to verify the identity of the wearer of the goggles. The visual "desktop" presented in the goggles is highly customizable, with personal icons and special-purpose displays.

5.0 MHD PROPULSION SYSTEM

The magnetohydrodynamic (MHD) slipstream accelerator is used for propulsion at speeds greater than Mach 2. The MHD propulsion system uses electric power to accelerate an electrically conductive "atmospheric fluid." This acceleration is accomplished by means of powerful on-board magnetic fields (2 Tesla) which exert a controlled force on the external slipstream air, which is partially ionized. The non-conducting slipstream air entering the MHD accelerator is first "seeded" by an injection of intense electron beams (from the perimeter array of E-guns) to make it conductive; under certain atmospheric conditions, this working fluid must be further energized by a focused burst of microwave power. In the process, a conducting "paddle" is created, which is maintained in a conducting state by an electric current driven by a potential applied to electrodes around the Lightcraft rim.

The MHD propulsion system is energized by beamed microwave power that is converted into electric power by the Lightcraft. This conversion of beamed power into electric power is accomplished by two 35 GHz rectifying antennas (rectennas) with an efficiency of 85%. The antennas are configured to make good use of the 5- to 7% of the microwave beam power that is normally reflected and lost. The central rectenna has a parabolic contour that reflects this "waste" microwave power into the airspike; the annular rectenna has an off-axis parabolic contour that reflects the "waste" power into the slipstream air plasma. To avoid overheating under extreme power loads, the rectenna arrays are actively cooled with highly purified water, fed at very high pressure through microchannel heat exchangers that exploit the phase change to steam. Upon emerging from the rectennas, this relatively cool "vapor" is run through the MHD generator electrodes and finally vented overboard to provide "film cooling" over the vehicle rim, which is exposed to the highest heat transfer load behind the airspike.

The MHD accelerator starts when the microwave power station illuminates the Lightcraft with its power beam. The Lightcraft then rides that beam, using it as a source of propulsive power. The propulsion system is capable of accelerating the LTI-20 either axially along the beam, or transversely, at right angles to the beam direction (Figure 5.0.1). The latter is also referred to as the lateral flight or "Frisbee mode" (FM). The Lightcraft is designed to accelerate through the atmosphere (i.e., its propulsive fluid) anywhere around the Earth that a microwave power beam can be directed. In the MHD mode, the beam must be aligned with the symmetry axis of the Lightcraft with a precision of a few degrees. If such close alignment is not achieved the useful power from the beam is greatly diminished, leading to possible propulsion emergencies.

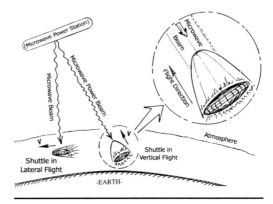

Figure 5.0.1: Vertical and lateral flight propulsion modes.

The principal advantages of the air-breathing MHD propulsion system are high propulsive efficiency during hypersonic transatmospheric flight, compatibility with a high-temperature plasma environment, annihilation of sonic booms, and use of completely "green" liquid consumables (i.e., water).

The overall efficiency of the MHD propulsion system, measured as the fraction of the onboard electrical power delivered to the accelerated slipstream, is about 80%. Since the annular MHD engine employs "action-at-a-distance" forces to electromagnetically accelerate the entire volume of partially ionized air captured by the airspike inlet, it delivers high efficiency without imposing severe local stresses on the skin or structural framework. In transatmospheric flight the Lightcraft hull would quickly be driven to temperatures that can greatly exceed 2700 C (i.e., the temperature limit for silicon carbide), so hot that ceramic hull materials and an active hull cooling system are mandatory elements in maintaining structural integrity.

5.1 MHD Propulsion Physics

A plasma is a partially ionized gas that differs from an ordinary gas in that electric and magnetic fields strongly affect its behavior. A charged particle (ion) in a magnetic field experiences a force that is proportional to the charge on the particle, the strength of the magnetic field, and the crosswise component of the particle's velocity relative to that field. This force, called the Lorentz force, acts on a charged particle moving in a uniform magnetic field to cause it to circle around the magnetic lines of force. The particle's motion parallel to the magnetic field lines is free and unobstructed, so that it experiences no force in that direction. The speed of ions in a plasma is at the very least governed by the plasma temperature, with the mean speed of the ions increasing as the square root of the Kelvin scale temperature.

Figure 5.1.1 is an illustration of a microwave-boosted LTI-15 Lightcraft in the axial MHD slipstream accelerator mode with the airspike energized. The lower maglev lander is omitted to reveal hull geometry necessary for the space radiation shield (Section 15.5). Note the enlarged right inset – a cutaway view of the Lightcraft rim – showing the high-pressure HeliOx toroid (the structural "backbone" of the vehicle), the two primary superconducting magnets, and MHD accelerator electrodes. Three orthogonal vectors indicate the directions of the applied electric current (from the rectenna array), the 2-Tesla magnetic field, and

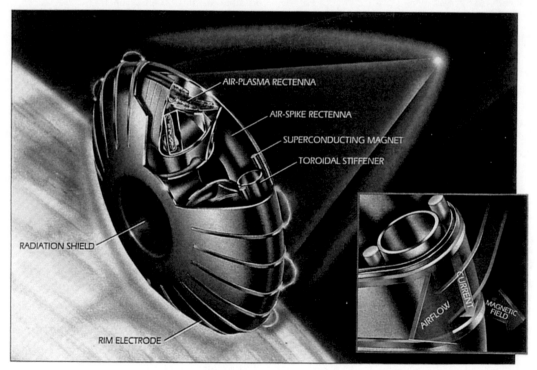

Figure 5.1.1: LTI-15 lightcraft in axial MHD slipstream accelerator mode with airspike energized. *(Courtesy of P. Dimare; Courtesy of NASA.)*

MHD PROPULSION SYSTEM

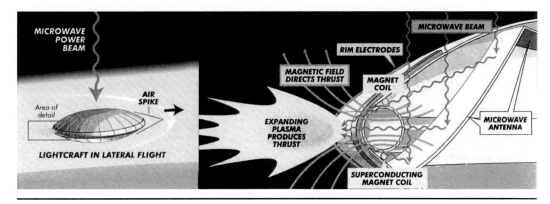

Figure 5.1.2: Pulsed detonation engine in lateral propulsion mode with linear aerospike engaged. *(Courtesy of Christian Science Monitor.)*

the direction of accelerated air plasma flow from the applied Lorentz forces. The two superconducting magnets are running in a conventional "bucking" configuration (i.e., carrying electric currents in opposing directions) and hence repel each other to generate the radial magnetic field indicated in Figure 5.1.1. In the larger drawing, note also the diffuse electric discharge that arcs between alternate pairs of electrodes around the vehicle rim. It is important to emphasize that the airspike delivers compressed inlet air (i.e., not an electrically conductive air plasma) to the MHD accelerator entrance station (the upper Lightcraft rim), and this air must again be ionized by the E-guns and a focused burst of microwave power before it can be accelerated with Lorentz forces. This "ionization" power is pulsed out through the Lightcraft's rim (between each set of electrodes), creating a plasma "paddle" that sweeps the largely neutral air ahead of it like a snowplow. By using a series of such paddles, the ionization power spent by the MHD engine can be minimized. Note that the MHD slipstream accelerator mode is fundamentally different than the PDE thruster mode (Section 6.0), for which the applied electric current (see inset of Figure 5.1.1) is missing and unnecessary (i.e., unless a pitch, roll, or spin torque is desired). In a non-uniform magnetic field, such as that generated at the rim of a Lightcraft by onboard "bucking" magnets, the microwave-heated, pulsed air plasma expands freely parallel to the magnetic field lines, eventually converting all of its internal energy into kinetic energy (energy of motion) along the field lines. The escaping ionized air stream exerts a reaction force on the magnetic field, which is firmly anchored to the Lightcraft. Thus the repetitively-pulsed PDE thruster propels the Lightcraft in the direction opposite to the motion of the air plasma exhaust (see right side of Figure 5.1.2) accelerating out its "magnetic nozzle." *The PDE is formally classified as a BEP "virtual" jet engine because the inlet air is simply scavenged from the surrounding atmosphere (i.e., there is no designated physical inlet); crossing the magnetic field lines is no hindrance for the working fluid since it is not yet ionized.*

In axial MHD propelled flight the gaseous airspike forebody of the Lightcraft (Section 5.5) acts as a hypersonic air inlet for the repetitively-pulsed Lorentz force engine (Figure 5.1.3). The use of air as the ejected fluid

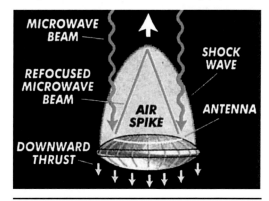

Figure 5.1.3: Axial MHD slipstream accelerator mode with airspike on.

exploits the presence of the atmosphere to great advantage (i.e., momentum exchange), removing the necessity of carrying propellants aboard the Lightcraft. Note that a conventional rocket requires both a source of energy and an exhaust mass flow to eject: chemical rockets carry combustible materials and oxidizers to provide both energy and ejected mass, and therefore must be very large (and heavy) to carry the required propellant load. The strength of the Lightcraft is twofold: it carries no propellants, since it derives its working fluid from the atmosphere, and it receives the power needed for its propulsion from an outside source such as beamed microwave or laser power. Since a Lightcraft under MHD propulsion can develop acceleration all the way to the top of the atmosphere, it can easily attain orbital or escape velocity.

The MHD propulsion system can be started at speeds as low as Mach 2 or 3. For an MHD engine to work successfully, the incident microwave beam must reach the Lightcraft rectenna arrays with minimal power loss after traversing the sensible atmosphere. The beam transmission system must be able to work under very diverse, sometimes hostile, environmental conditions. Certain microwave frequencies, corresponding to wavelengths shorter than 2 mm, are severely attenuated by absorption and scattering in cloud droplets. A power beam weakened in this manner may be insufficient to propel a Lightcraft, or even cause the beam to miss its target in part or completely. Thus the Lightcraft would suffer a failure of its propulsion system.

Another factor that must be considered is the electrical breakdown behavior of the atmosphere at different heights and wavelengths (Figure 5.1.4). Note that the Lightcraft's 35 GHz power beam reaches a limit of about 4 kW/cm^2 at 30 km altitude.

5.2 MHD Propulsion Conversion

Air breathing MHD propulsion converters are ideally suited for the hypersonic accelerator role in transatmospheric flights to orbit or even 2X escape velocity. To transport useful payloads, MHD-propelled flight demands gigawatts of

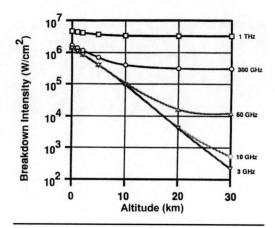

Figure 5.1.4: Microwave breakdown threshold of ambient air vs. frequency and altitude. *(Courtesy of RPI.)*

onboard electrical power, a requirement that cannot be satisfied by low-energy-density chemical fuels. Only beamed energy coupled with ultra-high power density converters (e.g., rectennas) can meet such stringent mass limitations for flights into orbit. MHD propulsion systems are critically dependent upon five principal ingredients: directed-energy airspike (DEAS) inlets, flight-weight superconducting magnets and action-at-a-distance forces, pneumatically-inflated tensile structures, aggressive thermal management, efficient "plasma-paddle" production, highly versatile "snow plow" Lorentz force thrusters, and bow shock wave annihilation.

The DEAS inlet is a massless alternative to the pointed external compression inlets of experimental hypersonic spaceplane concepts at the turn of the millennium. The airspike enables precise control over the bow shock wave's position at the entrance to the MHD engine. Furthermore the airspike gives the Lightcraft control over its own aerothermodynamics, greatly reducing the aerodynamic drag (and heat transfer) on the Lightcraft forebody; in turn this reduces the thrust required from an MHD engine to demonstrate a given acceleration performance.

The LTI-20's axisymmetric geometry is in large measure defined by the simple superconducting "bucking" magnet design needed to generate 2-

Tesla fields around its rim. By employing action-at-a-distance forces in its MHD thruster, the air plasma working fluid does not have to be fully enclosed by the engine; hence the Lightcraft has only to cool its outer hull, which, coupled with the reduced forebody heating load due to the airspike, minimizes the overall thermal management problems faced in trans-atmospheric flights.

The nearly exclusive use of tensile structures for bracing superconducting magnets and all major Lightcraft structural elements places MHD propulsion system mass fractions into the hyper-energetic range: 20- to 50 g, and beyond.

Because electromagnetic (Lorentz) forces accelerate the air working fluid, the MHD propulsion system can demonstrate high efficiencies over a very wide range of flight Mach numbers from 2 to 25 or more. The use of pulsed ionized "plasma paddles" reduces the ionization power penalties to a minimum when linked to a "snow plow" type of MHD accelerator design. Because of the extreme energetics of MHD, the accelerator "channel" for the LTI-20 is just under 2 meters in length. Given sufficient power, the MHD engine is easily able to annihilate the bow shock wave aft of the vehicle rim (i.e., the engine exhaust station), thereby enabling completely silent transatmospheric flights into orbit.

In summary, reliance on beamed power permits the mass of on-board power conversion equipment for MHD engines to be minimized. A Lightcraft can achieve hyper-energetic levels of performance with far less environmental noise and mass than a conventional space planes. The low noise level, meaning the absence of intense sonic booms, is achieved largely by using beamed power to control the Lightcraft's aerothermo-dynamics, pre-compress inlet air for the MHD engine, and to accelerate the processed slipstream mass flow while being careful to minimize the pressure discontinuities that produce shock waves in the Lightcraft's wake.

5.2.1 MHD System Components

The MHD accelerator system comprises the directed-energy airspike hypersonic inlet system, 2 rectenna arrays, 9 superconducting magnets, rim electrodes, power-switching circuitry, and an open-cycle rectenna cooling system. Note that once they are fully charged, the superconducting magnets require no electrical power to fulfill their function. Propulsion power is delivered from the rectenna arrays to the rim electrodes, and thence to the ionized slipstream, which exerts a reactive force on the magnets to propel the vehicle. The system is protected from overheating by the high-pressure, open-cycle water cooling system that ejects steam overboard.

5.2.2 Microwave Beam Reception

As shown in Figure 5.1.4, the maximum power density for a 35 GHz power beam is about 4 kW/cm^2. The beam targets a circular area 18 meters in diameter, for a total incident power of 10.2 GW.

The microwave power beam is transmitted from a LEO power-beaming station, where the microwave power is generated either from a large array of solar cells (SPS-01) or nuclear power source (PDS-03). In contrast, the mountain-top PDS-02 station must first transmit its beam to space (i.e., "uplink"), and then reflect the beam off a LEO microwave relay station to accomplish the downlink that can successfully "tractor" the LTI-20 into orbit (see Chapter 19). When a space-based power station such as the SPS-01 or PDS-03 employs a LEO relay station, this transmission leg is referred to as a "cross-link." The "downlink" microwave power beam is pointed at the Lightcraft, which monitors the incoming beam position by means of a computer-controlled feedback system which continuously transmits guidance data (by means of a beacon signal) to the microwave power-beaming station or relay satellite. For this beam control system (BCS) to function at maximum efficiency in the MHD or PDE propulsion mode, the axis of the Lightcraft must point to within a few degrees of the power beam's centerline. Note that "rechargeable" SPS-01 and PDS-03 satellites have the capability to store up electrical power for 1 or 2 orbits and transmit power levels well in excess of (e.g., >2x) the LTI-20's instantaneous needs.

5.2.3 Rectenna Arrays

Rectifying antenna arrays, commonly known as "rectennas," have been designed to provide direct current (DC) power for the propulsive engines of the microwave Lightcraft. These very high density sub-arrays (solid-state electronics) convert the incident 35 GHz microwave power (~4 kW/cm^2) with an efficiency of 85%. Each sub-array (see Figure 5.2.1) is an ensemble of tiny dipoles and Schottky diode elements deposited upon and within an advanced semiconductor substrate. The substrate is anchored on a low-loss SiC wafer carrier (structural web) equipped with microchannel heat exchangers to reject 1 kW/cm^2 (maximum) of waste heat into pressurized water coolant.

As shown in Figure 5.2.2, the concentric rectenna arrays receive the microwave power beams from the satellite power-beaming station or relay. The nested 35 GHz antennas are 18 m in diameter and are located inside the Lightcraft hull. The rectenna panel thickness is 2.14 mm, with a reflecting backplane positioned about ¼ wavelength behind the front surface. Depending upon instantaneous engine demands, the arrays can be programmed to reflect 10% to 100% of the incident microwave power beam.

The unique rectenna design enables the periodic reflection and focusing of the

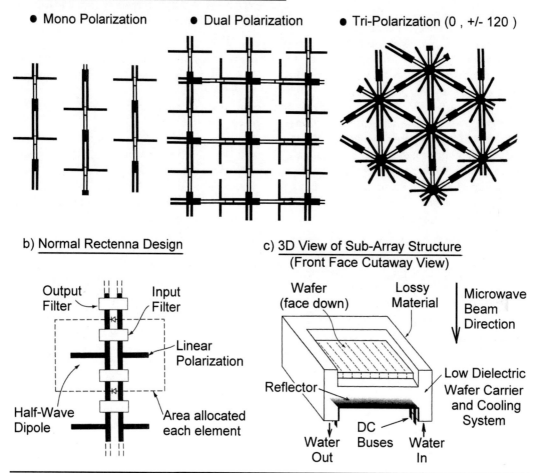

Figure 5.2.1: Tripolarization rectenna sub-array with microchannel cooling system (H$_2$O coolant).
(Figures b and c courtesy of A. Alden.)

MHD PROPULSION SYSTEM

Figure 5.2.2: Vehicle cross-section showing two concentric rectenna arrays (central on-axis, and off-axis parabolic). (Courtesy of RPI.)

microwave beam to maintain a Directed Energy AirSpike (DEAS) in front of the vehicle, in addition to other important functions. The central rectenna is designed to transmit onboard stored electrical energy (from the superconducting magnets) to briefly maintain this airspike in the event of a loss of microwave beam power. The outer (lampshade-shaped) rectenna permits focusing of the microwave beam at the periphery of the Lightcraft disk to cause electrical breakdown of the air. That air plasma is concentrated near the rim electrodes and perimeter superconducting electromagnets.

Throughout the microwave Lightcraft development program, high efficiency rectennas have been developed for increasingly high frequencies: first 35 GHz, followed by 94 GHz and 140 GHz. The state of the art has now advanced to permit rectennas that can function equally well for any two well-spaced frequencies. In fact, the latest incarnation, prototyped just two years ago, permits dual-frequency operation at 35 GHz and 140 GHz, prompting the USSC to place an order for a fleet of thirty LTI-20s with this upgrade. Although 220 GHz (another good atmospheric frequency "window") rectennas are aggressively being pursued, their efficiencies are still too low for LTI-20 use.

5.2.4 Superconducting Magnets

A main pair of superconducting magnets, having a combined mass of ~1200 kg, is located at the rim of the Lightcraft (Figure 5.2.3). Their 4 MA electric current normally runs in opposing directions such that the two coils repel or

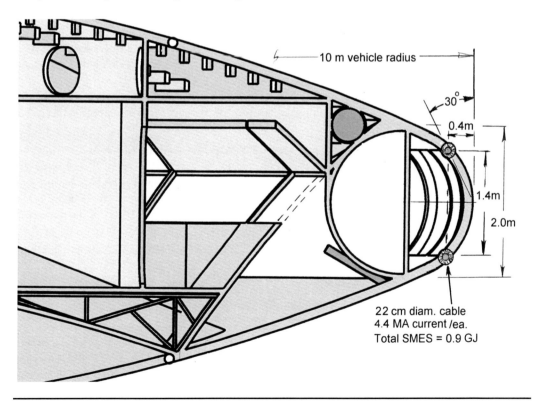

Figure 5.2.3: Rim location of twin primary superconducting magnets.

"buck" each other. Both coils are attached to the perimeter toroidal pressure vessel (measuring 2 m x 20 m) that serves as a structural backbone for the inflatable craft. The primary function of these superconducting coils, which are separated by 1.4 m (Figure 5.2.3) is to generate the 2-Tesla rim magnetic field required by the MHD slipstream accelerator and PDE engine modes.

The 20 cm (cross-sectional diameter) coils are manufactured from advanced Metal Matrix Superconductor Composites (MMSC) that employ imbedded carbon-fiber filaments (ultimate strength of 5.5 GPa; modulus of 827 GPa) with a superconducting $MgCNi_3$ high current density film surface layer. The carbon-fiber-$MgCNi_3$ superconductors are, in turn, imbedded in a beryllium matrix – a high electrical and thermal conductivity stabilizer metal of superior intrinsic strength and low density. These MMSC materials provide a combined performance function, both structural and electrical. The carbon-fiber filaments easily carry the magnet's "hoop" (i.e., explosive) stresses, whereas the beryllium matrix sustains the "magnetic pinch" (implosive) pressures. The LTI-20's MMSC electromagnets are cooled to 4.2 K with liquid helium, and sustain a design current density of 550 kA/cm^2 in an 8-Tesla magnetic field.

The perimeter superconducting magnets generate a powerful "magnetic nozzle" that can vector the expanding air plasma and constrain it to expand in a specific direction much like a jet engine with a vectorable nozzle. By adjusting the current in the two coils, the magnetic field can be directed to eject the hot plasma downward, creating a reaction lifting force that propels the Lightcraft upward. Alternately, a pitch or roll moment may be applied to the craft by lifting on one side and driving the other side down.

Other pairs of smaller electromagnets (Figure 5.2.4) are electrically linked with the two rim electromagnets to constitute a Superconducting Magnetic Energy Storage (SMES) system for the craft. This SMES system provides many power-related functions for the Lightcraft, including the coil protection circuitry shown in Figure 5.2.5. If

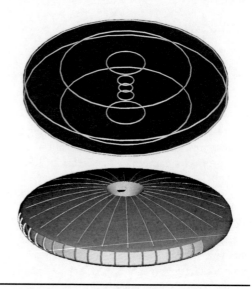

Figure 5.2.4: Location of 9 superconducting magnets comprising SMES unit. *(Courtesy of RPI.)*

an electromagnet overheats and approaches the superconductivity threshold temperature (i.e., an emergency condition), its energy may be diverted to the other magnets or hastily dumped out through the rectenna arrays (as a widely defocused microwave beam). Alternatively, under normal operations, the PDE engine can be activated with this SMES "battery" power to propel the Lightcraft through the atmosphere in a variety of propulsion modes, or to provide short-term (a few seconds) airspike support as described earlier.

5.2.5 Rim MHD Electrodes

A set of 48 electrodes, each 20 cm wide, is distributed around the Lightcraft's rim with an azimuthal spacing of 130 cm (see Figures 5.2.6 and 5.1.2). They effectively divide the rim (i.e., outer surface of the perimeter toroid) into 48 microwave-transparent windows measuring 110 cm square. Operating in either the "reflect" or "transmit" mode, the outer (lampshade-shaped) rectenna array can focus microwave power out these windows to trigger electrical air breakdown at whatever pre-programmed rim locations are selected. As a result, the air immediately surrounding the Lightcraft rim becomes partially ionized at desired locations by this pulsed microwave beam; the momentary air plasmas are then accelerated through the

MHD PROPULSION SYSTEM

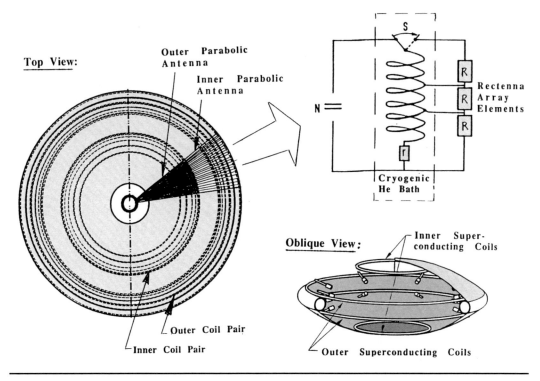

Figure 5.2.5: Coil protection circuitry linking rectenna arrays to SMES magnets.

action of pulsed Lorentz forces by driving pulsed electrical currents at right angles to the applied 2-Tesla magnetic field. In the MHD slipstream accelerator mode, Lorentz forces are rapidly commutated around the vehicle rim such that no two adjacent electrode pairs are activated simultaneously. If all electrode pairs were to be accidentally "switched on" at the same moment, the MHD accelerator would simply "short out," producing no thrust at all.

In essence the rim electrodes, superconducting "bucking" magnets, and MHD pulsed power system create an "unwrapped" linear electric motor that exchanges momentum with the hypersonic slipstream air. The temperature of these actively-cooled electrodes (made from refractory metals) is accurately monitored and maintained within thermo-structural limits by open-cycle steam coolant ejected from the rectenna arrays.

After flowing through the electrodes the steam coolant is ejected uniformly around the vehicle perimeter, just upstream of the "windows," to

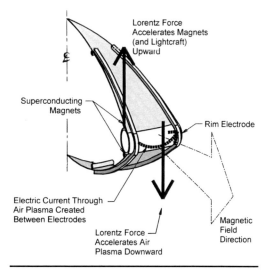

Figure 5.2.6: Active electrodes and main magnet coils for MHD slipstream accelerator; electric current, magnetic field, and Lorentz force vectors shown for axial flight mode. *(Courtesy of RPI.)*

"bathe" the rim with a curtain of water vapor. At hypersonic speeds, the vehicle rim is exposed to the highest heat transfer loads, which are greatly reduced by this "film" cooling mechanism.

5.3 Beamed Power Reception

Beamed power is received by either or both of the two nested rectenna arrays on the Lightcraft: the inner paraboloid (central), and the outer off-axis (lampshade-shaped) paraboloid – see Figure 5.2.2. These rectennas are positioned inside the double-hull, which is composed of two thin silicon-carbide (SiC) ceramic films in a sandwich-type structure that is transparent to microwaves. The function of the outer rectenna is to absorb the microwave beam and convert this beamed power into electrical power, or to reflect the power beam in a controlled manner and concentrate it just outside the rim of the Lightcraft into the slipstream air, thus preparing it to be accelerated by the MHD engine. The central rectenna is also used to generate electric power or to concentrate microwave power into the air above and near the spacecraft to partially ionize and heat it for the airspike support function. Power absorbed by the rectenna arrays and converted into electrical power is fed to the 48 rim electrodes of the MHD slipstream accelerator.

The high specific power of the water-cooled rectenna arrays – about 60 MWe per kg – is handled by the very large number density of solid-state electronic sub-array elements. The Lightcraft rectenna converts the 10 GW power beam (average incident intensity of 4 kW/cm^2) into electric power with an efficiency of about 85%; about 5- to 7% of the beam power is always reflected by the array so the MHD engine is configured to capture and make use of this energy. Power losses due to the incompleteness of electronic conversion in the sub-array elements appear as waste heat, which must be dissipated to protect the rectennas from overheating.

The peak operating temperature of SiC semiconductors is over 500° C. With a 10 GW power beam, the waste heat load is about 1 GW which is absorbed by pressurized water pumped through integral microchannel heat exchangers, and exiting as superheated steam. The water pressure within the microchannels is sufficiently high that boiling is entirely suppressed within the rectennas, even at extreme heat transfer rates approaching 1 kW/cm^2. After leaving the rectenna arrays, the water flashes into steam enroute to the MHD electrodes, cooling them before final ejection around the disc perimeter (see Section 5.2). Therefore, during high-power MHD propulsion such as boost into orbit, the waste heat is finally vented to the atmosphere. Due to the short duration of normal MHD accelerations (1 - 2 minutes), this open-cycle cooling system is far lighter and more flexible than a closed-cycle system, but does require replenishment of the water supply (~2400 kg) before every orbital boost. In fact, closed-cycle rectenna cooling systems are altogether infeasible for the LTI-20.

5.4 Air Plasma Generation

As mentioned in Section 5.2.1, only through the combined action of powerful 2-Tesla magnetic fields (produced by the two superconducting rim magnets) in unison with pulsed megampere electric discharges (driven through the slipstream, across the rim electrodes, at right angles to the applied magnetic field) can MHD engines successfully produce thrust with Lorentz forces by accelerating the slipstream air. However, MHD thrust production requires that the air working fluid first be transformed into an electrically conducting air plasma.

This air plasma generation process involves three basic steps:

1) Injection of relativistic "seed" electrons (1 - 5 MeV) into the slipstream air from any of the 24 pairs of perimeter electron guns;
2) Pulsed microwave-induced electrical breakdown of the air outside pre-selected microwave windows around the rim, to create "plasma paddles;" and
3) Inverse Bremmstrahlung absorption of microwave energy to maintain the ionization within each 1 meter square "paddle" (i.e., a two temperature plasma) that pushes unionized air ahead of it much like a snow plow.

Once in motion, the plasma paddle's electrical conductivity is maintained by resistive heating losses for the duration of the thrust pulse.

Step 1 requires the slipstream air to be first "seeded" with free electrons from the perimeter set of 24 E-guns; then the focused pulse of microwave power creates a high enough electric field strength to accelerate these electrons (in the presence of nearby heavy ions), initiating an electron avalanche / cascade sufficient to trigger electrical air breakdown leading to the creation of a plasma "paddle." For Steps 2 and 3, the microwave energy pulse can come from either of two sources:

a) beamed directly from the remote microwave power station, or
b) generated onboard by the outer rectenna array operating in the transmit mode using electric power from the SMES "battery."

With the latter approach, air plasmas can be triggered just outside the perimeter "windows" in very complex, pre-programmed patterns and sequences that promote optimum MHD engine propulsive efficiency.

The slipstream air swept into the high-field region close to the rim magnets is ionized intensely in this manner, and the high-energy air plasma is accelerated downward by MHD forces around the vehicle rim. The overall propulsive effect is loosely comparable to that of an external-combustion ramjet or scramjet engine, except that the increased kinetic energy in the exhaust is provided electromagnetically rather than by chemical combustion. The Lightcraft's rectenna arrays generate nearly 9000 MW of electric power to drive this MHD air-breathing propulsion system.

Upon reentry from orbit, the LTI-20 may apply MHD aerobraking to modify its descent trajectory by reversing the electric current in its MHD engine, much in the manner of turbofan engine with its thrust reverser activated. The craft's sophisticated magnetoaerodynamic systems can also permit the extraction of energy from the hypersonic slipstream for recharging the SMES "battery," as well as active control over vehicle aerodynamics that may or may not involve airspike maintenance. The limited flight envelope for "passive" magnetoaerodynamics frequently requires activation of the LTI-20's air plasma generation system, especially at the lower Mach numbers and altitudes of interest.

5.5 AirSpike Production

A fundamental feature of the LTI-20 propulsion system is the directed-energy airspike (DEAS) mentioned above. The DEAS concept, first experimentally proven in April 1995 at Rensselaer Polytechnic Institute, is applicable only at supersonic flight speeds. The airspike serves three basic functions: to reduce vehicle drag, to lower heat transfer into the Lightcraft forebody, and to pre-compress "inlet air" into the thin annular slipstream region indicated in Figure 5.5.1.

The airspike system employs the central parabolic rectenna array as a microwave reflector to concentrate the innermost portion of the microwave power beam toward a point ahead of the upward-moving Lightcraft. At the focal point, which is about one vehicle diameter away, the elevated beam intensity triggers electrical air breakdown and absorption of the rapidly pulsed microwave power, essentially creating a continuous series of explosions that produce a superheated, partially ionized air plasma. The resulting hot, high-pressure gases expand away from the focal point, and divert the incoming supersonic air out of the vehicle's path and into the annular MHD accelerator inlet. In the process, the airspike creates the parabolic detached shock wave shown in Figure 5.5.1.

In emergencies the central rectenna array, running on power extracted from the SMES "batteries," can transmit microwave power ahead of the Lightcraft in the direction of flight to create the airspike (as in Figure 5.5.1); the Lightcraft's aerodynamic shape is drastically modified by adding a virtual conical forebody comprised of low-density hot air. Similarly, in lateral flight ("Frisbee" mode) the forward-facing portion of the outer annular rectenna array can beam power to generate a wedge-shaped airspike that lowers drag and heat transfer to the hull.

Chapter 5

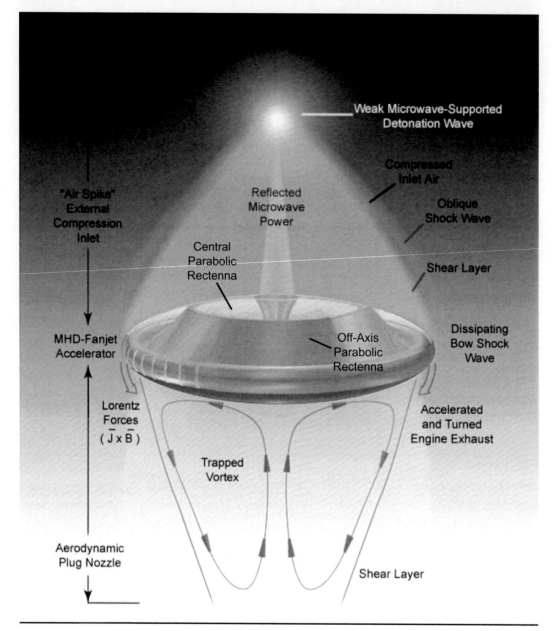

Figure 5.5.1: Directed energy airspike (DEAS) with MHD slipstream accelerator in axial flight mode. *(Courtesy of Media Fusion and NASA.)*

Depending upon the available microwave power and focus geometry, the airspike can reduce supersonic drag by as much as 10 to 30 times below that of the "unpowered" blunt body.

5.6 Thermal Management System

The LTI-20 employs an active thermal management system (TMS) to reject the waste heat generated in flight by the various propulsion modes. Heat rejection is essential at high supersonic and hypersonic speeds to prevent overheating of the hull, cryogenic refrigeration equipment, rectenna arrays, and other sensitive electronics, and to preserve safe internal cabin temperatures.

MHD PROPULSION SYSTEM

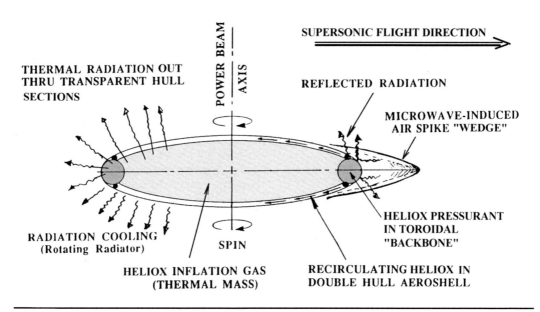

Figure 5.6.1: Low power thermal management system for hyperjump maneuver and short-term supersonic excursions in lateral flight mode.

The closed-cycle, *low-power TMS* is activated for subsonic flight and brief supersonic excursions – as with the "hyper-jump" maneuver. In lateral Frisbee-mode flight when the Lightcraft is slowly spinning, the hull becomes a rotating radiator with enhanced convective heat transfer. During brief supersonic dashes in this flight mode, the forward hull surface (which has high aero heat loads) is "time-shared" around the perimeter, as heated leading edges are continuously rotated rearward for cooling (Figure 5.6.1). With the lateral airspike activated in such maneuvers, heat transfer rates into the hull are minimal, warming the envelope pressurants by just few degrees centigrade. If hull temperatures approach structural limits, reserve liquid helium can be injected into the recirculating HeliOx hull coolant to quickly bring temperatures down (i.e., as long as hull inflation pressures stay within norms).

The open-cycle, *high-power TMS* is activated for ultra-energetic flights to the far side of the planet or into orbit, consuming up to ~2400 kg of ultra-pure water that is eventually ejected as steam into the atmosphere. On extended transplanetary missions, the LTI-20's on-board water purification system permits scavenging of water from the environment (e.g., lakes, streams, freshwater ponds, etc.) for the return flight to CONUS.

The TMS functions as three independent units, each responsible for a third of the internal volume of the Lightcraft, but with a reserve capacity for cooling other sections at need. This arrangement provides some redundancy for the Lightcraft in the event of equipment malfunction. Each TMS unit has both a HeliOx closed loop and a water-cooled open loop; the latter is activated for high-power TMS functions only.

5.6.1 Low-Power TMS Overview

The *low-power TMS* operates in the closed-cycle mode and uses the HeliOx gas mixture that inflates the craft to carry off excess heat from the rectenna arrays for subsequent rejection out the hull to the atmosphere (see Figure 5.6.1). The closed coolant loop captures waste heat from the rectennas by passing the HeliOx coolant through narrow, transverse-flow orifices across the rectenna elements. HeliOx coolant can handle rectenna temperatures as high as 500° C, but is used only when the rectenna arrays are operating in the low-power mode.

Chapter 5

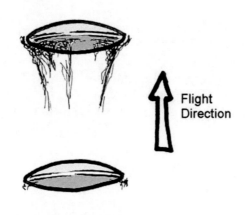

Figure 5.6.2: Perimeter steam curtain ejected by high-power TMS, for film cooling of rim. (Courtesy of RPI.)

The LTI-20's double hull acts like a pinned-fin type of heat exchanger through which recirculated HeliOx is pumped. The hull's CMC sandwich or core passageways are filled with high thermal conductivity fibers that pass waste heat through the external hull into the cooler slipstream air.

The low-power TMS may also be used in reverse (for short periods) to refrigerate the outside boundary layer air and condense atmospheric water vapor or form a cloud around the vehicle under high-humidity conditions. To quickly super-cool the HeliOx mixture for triggering such external condensation, a small quantity of on-board liquid helium may be injected into the recirculated HeliOx pressurant.

Each of the 3 TMS units consists of several subsystems, including an electric motor-driven HeliOx compressor, running on electric power from the rectennas and/or SMES units. In nominal operation each TMS unit cools a third of the Lightcraft; however, any one of these units could cool any part, or share the load in the event of failure of one of the units.

5.6.2 High-Power TMS Overview

The *high-power TMS* uses 2400 kg of pressurized, ultra-purified water flowing through the rectenna arrays within microchannel heat exchangers designed to absorb 1 kW/cm² of waste heat during ultra-energetic maneuvers. The water coolant is highly filtered and softened before use to prevent fowling of the microchannel passageways. Operating in the open-loop mode, water is pumped from the perimeter storage tank to the rectenna arrays through a network of pipes (intake manifolds) approximately 38 mm in diameter at water flow rates up to 150 kg/sec, a rate sufficient to cool the Lightcraft during 50 g boosts to 8 km/s in just 16 seconds. Running on rectenna or SMES derived power, the electric motor-driven pumps deliver water at elevated pressures to suppress boiling in the microchannels; upon emerging from the arrays it flashes into steam. Intricate exhaust manifolds direct this steam to the MHD electrodes before venting it around the Lightcraft's hull for film cooling of the rim (see Figure 5.6.2). This provides an insulating, relatively cool "curtain" to protect the rim against the searing temperatures of hypersonic flight, and also heavily absorbs any microwave power that strays beyond the rectenna arrays.

All three high-power TMS units contain the same components with identical functions. Pressurized water entering the rectenna arrays at 20° C exchanges heat with the high-power, integrated solid-state electronics and exits at a temperature of 500° C into the exhaust manifolds.

5.6.3 TMS Water Purification System

The LTI-20 is equipped with a three-stage Water Purification System (WPS) in order to replenish its coolant tanks (for intercontinental return flights) with water extracted from the natural environment: freshwater streams, lakes and ponds. The WPS is an ultra-lightweight version of conventional "make-up water" purification equipment used to replenish makeup water in ground-based Pressurized Water Reactors (PWR). The open-cycle water that cools Lightcraft rectennas must be ultra-pure in order to prevent fouling of the intricate microchannel passageways.

The first WPS stage employs a series of coarse and increasingly fine filters to extract and eject larger debris. Next the water is demineralized and deionized with the aid of disposable ion exchange columns, exploiting an approach not

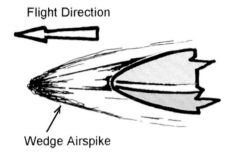 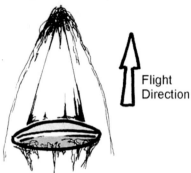

Figure 5.7.1: Lateral and axial flight modes for MHD slipstream accelerator with airspike engaged. *(Courtesy of RPI.)*

unlike that of Culligan® industrial systems (zeolite). The final stage uses reverse osmosis, followed by UV radiation to destroy any remaining microorganisms. Hence, the water vapor steam that is released into the environment during a high-performance Lightcraft boost is absolutely pure.

5.6.4 Human Factors and HeliOx TMS

For high-performance maneuvers such as the hyper-jump, suborbital boost, and orbital flight when the high-power TMS is activated, the crew is safely protected within the confines of their individual escape pods.

In sharp contrast, when the closed-cycle HeliOx TMS is engaged for extended subsonic excursions at low altitudes, the crew is deployed at the bridge and other stations throughout the vehicle. During these low power maneuvers, the crew may be exposed to slight (engine-driven) environmental variations in the HeliOx temperature, normally a few degrees Celsius, which is easily countered by their space activity suits. Of course, scrubbers continue to remove carbon dioxide along with other contaminants from the recirculated HeliOx, and humidity levels are continuously adjusted for human comfort.

5.7 Vertical vs. Lateral Flight Modes

The MHD slipstream accelerator is capable of propelling the LTI-20 in both the axial and transverse directions (Figure 5.7.1). The choice of flight mode is influenced by the position and orientation of the Lightcraft with respect to the microwave power source, the nature of the desired maneuvers, and of course, the MHD engine configuration (i.e., the selectable magnetic field geometry / orientation, active electrode sets, and electric current discharge direction through the air plasma working fluid).

The MHD propulsion system is normally used in the axial flight mode (see Figure 5.2.6), which is used exclusively for high-g acceleration into space. However, in the dense lower atmosphere, the MHD engine may also be used for lateral flight (Figure 5.7.2) until sufficient altitude is reached to permit full axial acceleration. A hyperjump (see Section 16.6) is usually done in the axial mode, performed in an oblique, vertical or horizontal direction. If a lateral hyperjump is mandated, the MHD engine must be engaged with a wedge-shaped airspike created ahead of the leading edge of the rim. The airspike reduces compressibility effects (drag divergence) over the lens-shaped vehicle. The goal is to change the high aerodynamic drag associated with the strong bow shock that normally forms over the blunt leading edge of the Lightcraft into a low-drag, detached oblique shock. The wedge-shaped airspike greatly reduces drag and heat transfer on the Lightcraft hull, increases propulsive efficiency, and gives the lateral flight mode much higher acceleration capability.

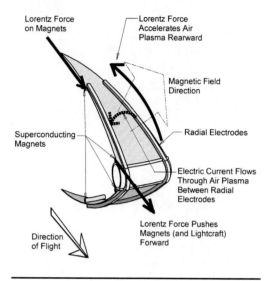

Figure 5.7.2: Orientation of applied magnetic field, electric current, and resultant Lorentz force for lateral MHD slipstream accelerator mode. *(Courtesy of RPI.)*

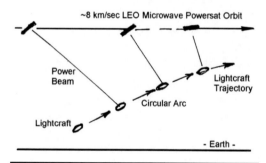

Figure 5.7.3: Extended circular arc flight trajectory option

The relative positioning of the microwave power beam with respect to the vehicle is important in governing the lateral flight mode. If the Lightcraft central axis of symmetry is not aimed directly at the power beam (i.e., within a few degrees alignment) the MHD propulsion system cannot properly function since the rectenna array may not be intercepting and converting enough of the power beam. Inaccurate positioning of the Lightcraft axis with the power beam allows part of the beam, possibly carrying several megawatts to gigawatts of power, to strike the ground. Hence, the flight management system (FMS) is programmed to avoid this situation at all costs since the beam intensity can be high enough to start fires and injure personnel on the ground.

It must be kept in mind that the power beam will almost never be vertical during the lateral flight mode. (Note that in the axial flight mode the beam is generally not vertical either, and often closer to horizontal.)

There are two principal options for lateral fight trajectories under a LEO (e.g., 476 km) microwave power satellite or relay station:

1) Continuous boost along extended circular-arcs, and
2) Boost / glide (e.g., "atmosphere skipping").

The first option is to travel extended circular-arc flight paths through the atmosphere using either PDE or MHD propulsion at supersonic or hypersonic speeds (Figure 5.7.3). For example, say the Lightcraft must fly in an easterly direction, expecting to be easily outpaced by the LEO power satellite passing overhead at ~8 km/s. With its axis of symmetry aligned to the beam, the vehicle would accelerate or "cruise" along a circular-arc path defined by the powersat location vs. time. Like a sailboat tacking in the wind, such trajectories can be greatly modified by varying aerodynamic lift produced from the craft's angle of attack to the relative wind. At any point along the trajectory, the Lightcraft may disconnect from the power beam and coast, engine-off. After a pause, the craft could then pitch over to reconnect with the same beam (or another), and initiate the next extended arc segment under boost.

The second lateral flight option is hypersonic "atmospheric skipping" in which the Lightcraft performs a consecutive series of hypersonic boost / glide maneuvers at the top of the atmosphere – a trajectory that resembles a stone skipping over water (Figure 5.7.4). This highly efficient MHD flight option permits nearly orbital flight speeds that greatly extend range, but places significant demands on the thermal management system. The boost / glide trajectory allows the craft and its propulsion system to cool down between boosts as it coasts along brief, quasi-ballistic arcs above the

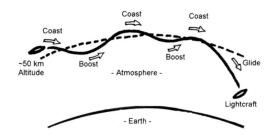

Figure 5.7.4: Hypersonic "atmosphere skipping" flight mode. *(Courtesy of RPI.)*

atmosphere. Upon recontacting the atmosphere again, the Lightcraft axis is realigned with the beam, the lateral MHD engine is fired up, and the cycle begins anew.

Running briefly on internal SMES power without the microwave beam, the PDE or MHD engine can vector the Lightcraft into alignment at the beginning of each new cycle, as necessary. The SMES unit is subsequently replenished during the MHD boost, using surplus rectenna power. Because of the intermittent nature of propulsion and the time spent coasting and adjusting vehicle orientation in this option, the average velocities range from Mach 10 to 20, significantly lower than orbital (i.e., Mach 25). Note that, in lateral flight, it is often possible to use multiple power beams. The Lightcraft is free to ride one power beam for a distance and then switch to another power beam as it becomes available.

5.8 Emergency Procedures

When accelerating at 20 - 200 gravities, or when traveling at high Mach numbers, flight control decisions must be made in milliseconds, which is too fast for human reactions. The LTI-20's FMS computer has the critical responsibility for identifying emergency conditions, and choosing and executing appropriate responses. In MHD propulsion mode, several types of emergencies are handled by the computer without recourse to the escape pods.

One potential emergency is the collapse of the airspike. If the power beam is interrupted for more than a few milliseconds while the Lightcraft is traveling at hypersonic speed, the airspike will collapse back onto the leading edge of the hull. The computer, however, is fast enough to sense such an event and react instantly, using the power reserves stored in the SMES system to reinflate the airspike. Restoration of the airspike is accomplished by using the rectenna in phased-array "transmit" mode to direct microwave energy into the airspike until the external microwave beam is reacquired.

If the computer senses that there is insufficient reserve energy in the SMES it will automatically initiate the escape pod ejection sequence, since a catastrophic hull collapse may be imminent.

Another potential danger is that of encountering a solid obstacle such as a projectile, missile, or bird at lower altitudes. The LTI-20 is surrounded by optical sensors that can detect objects as small as a few millimeters in diameter. Upon detection of such an obstacle the FMS computer redirects the Lightcraft flight path to avoid it while minimizing the deviation from the pre-planned flight path. The computer has the option of disengaging the MHD engine and switching to the PDE thrusters for active thrust vectoring, or to torque the vehicle into a high-performance pitch or roll maneuver. Depending on the flight Mach number, MHD thrusters may also be activated in a non-axisymmetric pattern to augment torquing of the vehicle. The PDE quickly "jumps" the LTI-20 out of danger and then back onto course, where normal MHD engine functions take over again.

In most cases, there is sufficient time for interactive diversion of the power beam: the round-trip "light-signal" time from the Lightcraft to the LEO powersat and back (assuming a maximum range of 1100 km) is about 7.3 microseconds, corresponding to only 2.4 cm of travel at a Lightcraft speed of Mach 10.

A second option for clearing smaller threats (e.g., debris, projectiles, etc.) from the flight path is to momentarily focus on-board or remote beam weapons on the obstacle to propel it away with ablative laser propulsion by rapidly evaporating its outer surface.

Another emergency that may arise is overheating of the Lightcraft rectenna or MHD engine. The rectenna array absorbs the microwave power beam with about 85% efficiency; about 5% is reflected, and about 10% appears as waste heat in the rectenna elements and elsewhere. Overheating may occur if the thermal management system malfunctions. The FMS computer then analyzes its sensor data and reacts before the situation becomes critical. If one of the TMS units fails, the central computer lowers the acceleration of the vehicle to compensate for the lower cooling capacity of the remaining TMS units. If all three TMS units fail, the computer disengages the MHD engine immediately.

If any emergency situation arises and the FMS computer decides that insufficient time or power is available to avoid disaster, it initiates crew ejection procedures. The computer seeks to minimize risks to human life. If there is an appreciable probability of catastrophic failure and no viable option remains for avoiding that failure, the computer will choose crew ejection via the escape pods. The crew must in any event be in their escape pods during high-performance flight, so that ejection can be essentially instantaneous. The escape pods are self-contained micro-spacecraft, capable of returning their passengers to safety from any altitude or Mach number, including reentry from orbit. The escape pod system is described in Section 18.3

Once the crew is safely ejected, the computer will attempt to save the Lightcraft.

6.0 PULSED DETONATION ENGINE (PDE) MODE

The pulsed detonation engine (PDE) is a high-thrust, high-performance propulsion system powered by beamed microwave power.

6.1 PDE Propulsion Theory and Application

This section introduces the basic theory of the pulsed detonation propulsion system and explains several of its applications.

6.1.1 Basic Theory

The PDE propulsion system is powered by beamed microwave power from a remote power station in space. The microwave beam, in order to propagate with little attenuation through the atmosphere, must have a frequency of 35, 94, 140 or 220 GHz (i.e., ideal "atmospheric windows"). The LTI-20 uses a 35 GHz microwave beam with a wavelength of 8.57 mm. This beamed energy first passes unattenuated through the SiC upper hull, then reflects off the outboard off-axis paraboloid rectenna and is focused by it into a ring just outside the rim of the Lightcraft's hull (Figure 6.1.1). The focal ring, which is about 1 m outside the rim of the hull, has such a high power density that electrical breakdown of the air occurs, making a ring of high-pressure plasma. This plasma cools as it expands radially outward from its source, guided by strong radial magnetic fields produced by superconducting magnets in the Lightcraft rim. The reaction force of the expanding plasma pushing against these fields accelerates the Lightcraft away from the plasma source.

The microwave power source is pulsed, allowing unheated air to rush in and refill the plasma source region between pulses. The pulses are applied at a rate of hundreds to thousands of pulses per second; i.e., fast enough to maintain a copious supply of plasma, but not so fast as to prevent fresh air from replenishing the source. To the crew, the effect is essentially one of continuous thrust. Often, especially to evade acoustic detection, the pulse rate is set outside the normal range of the human ear (~20 kHz or higher). Frequencies below the lower human hearing limit of about 20 Hz are less desirable for covert operations because they can be detected as vibrations, and because their effective range in air is very large.

Directional control of the PDE is achieved by using the two superconducting rim magnets to vector the engine thrust. The strength of the magnetic field is varied to force the expelled plasma either up or down to effect the desired flight direction or torquing maneuver of the Lightcraft. This thrust vectoring, combined with the ability to pulse separate sections of the hull perimeter selectively, allows complete three-dimensional freedom of movement.

While operating in the lateral flight PDE mode the Lightcraft can also use the off-axis paraboloidal rectenna to generate an airspike, just as in the MHD accelerator mode. An airspike, which reduces aerodynamic drag and heat transfer, enables the Lightcraft to accelerate to high speeds more rapidly. A lateral airspike creates a hot air "wedge" ahead of the Lightcraft when it is flying laterally (Figure 6.1.2).

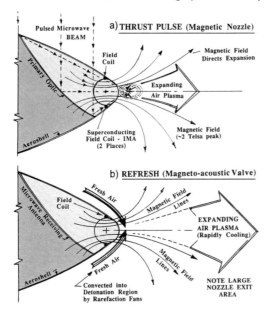

Figure 6.1.1: Pulsed detonation engine cycle.

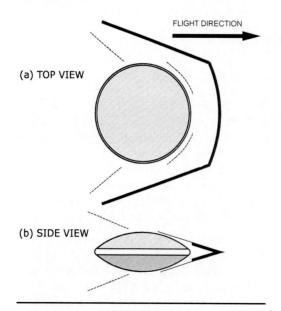

Figure 6.1.2: Supersonic / hypersonic flight mode with airspike engaged. *(Courtesy of RPI.)*

The focal point of the PDE's rectenna array is extended roughly 5- to 10 m in the direction of flight, depending on the flight Mach number. The focal point can be adjusted by using the outer (off-axis) rectenna ring as a phased array transmitter. Rapid pulsing at this focal point produces a series of quasi-cylindrical blast waves that coalesce to produce a "wedge" of hot air in front of the leading edge of the Lightcraft rim, effectively streamlining the blunt leading edge. The hot-air driven shock wave, expanded until it is in pressure balance with the surrounding supersonic slipstream, is far less dense than unshocked air, so that the Lightcraft is in effect flying through a much more tenuous gas, and experiencing much less drag, than it would if the airspike were not present. With the lateral airspike turned off, maximum flight speeds at sea level are limited to Mach 2, which creates a normal shock (very high drag) against the blunt lightcraft leading edge as shown in Figure 6.1.3.

6.1.2 PDE Applications

The PDE is the most versatile propulsion system of the three modes employed by the Lightcraft. It is capable of lateral speeds ranging from 0 to Mach 2 without the airspike engaged,

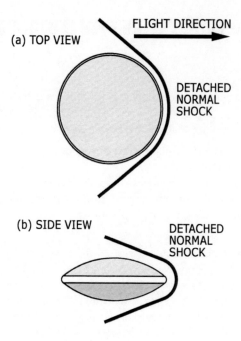

Figure 6.1.3: Lateral flight with airspike inoperative. *(Courtesy of RPI.)*

and up to Mach 6 with the airspike. The Lightcraft can function effectively under PDE propulsion from sea level up to about 20 km altitude. One of the most important functions served by PDE is as the propulsion system responsible for rapid acceleration, at rates up to 300 g, up to a speed of Mach 2. At 200 g, Mach 2 (~660 m/sec) is achieved in 336 msec. Beyond Mach 2 the MHD engine can function effectively and complete the boost into orbit. The ability of the PDE engine to operate from a standing start allows extremely effective evasive maneuvers such as "blink out" in hostile environments where the crew may not be situated for an immediate direct boost into orbit. A "blink out" is simply an acceleration too rapid for the human eye to follow (typically, an acceleration greater than 20 g). Such rapid acceleration gives the illusion of the craft simply vanishing with an occasional latent peal of thunder.

Although the Lightcraft can be flown at low altitudes using the ion propulsion system, the PDE engine is the propulsion system of choice when agility and speed are required, and top speeds above Mach 6 are not necessary.

6.2 Propulsion System Components

The design of the PDE, with internal (off-axis) paraboloidal rectennas, requires that the hull of the Lightcraft be transparent to a variety of microwave wavelengths employed by various Lightcraft systems. Without the transparent silicon-carbide (SiC) hull, the external energy source could not be absorbed, reflected, or focused by the rectennas. The PDE is relatively simple in structure, requiring only a few major components for its operation.

The first component is the off-axis paraboloidal rectenna, which in PDE mode does not extract any electrical energy from the incident power beam. The rectenna comprises solid-state receiver elements of SiC integrated microelectronics, such as diodes and transistors, that are disconnected from the power circuit so that 100% of the incident microwave power beam is reflected and focused out at the chosen air detonation distance. The rectenna can reflect a large range of microwave wavelengths used by the Lightcraft, from 1.0- to 10 mm wavelength, thus allowing the PDE to run in almost all weather conditions.

The second component crucial to the success of the PDE is the pair of superconducting magnet coils that encircle the outer edge of the Lightcraft hull (see Figure 6.1.1). These coils generate the magnetic fields that are used for thrust vectoring under PDE propulsion. The expanding plasma follows the steepest gradient in magnetic field intensity by exiting along open field lines as they depart from the hull. The pairs of magnetic coils can be adjusted in strength independently, thus providing both a magnetic nozzle to contain the escaping plasma and directional flexibility to vector the thrust. The two perimeter Lightcraft magnets, constituting the bulk of the SMES system, can store up to 900 MJ of energy at any given time. This stored power can be transmitted by the rectenna arrays for short time periods as microwave energy with a wavelength of 8.57 mm, providing propulsive power by creating and "detonating" air plasmas. This stored energy could also be used for a variety of other purposes, including autonomous PDE thrusting (without an external power beam). The SMES magnets provide insurance against brief interruptions in the supply of microwave power from orbit.

The final component necessary for efficient use of the PDE thrusters is the LTI-20's hull cooling system. The closed-loop HeliOx coolant system allows the hull to withstand the high temperatures produced behind the airspike at hypersonic flight speeds. Momentary pulsed air plasmas created by the PDE engine can exceed 3,000 K. The thermal management system (TMS) is explained in detail in Section 5.6. Open-loop cooling with venting of steam is used only when overheating is imminent.

6.3 Magnetic Nozzles and Thrust Vectoring

The maneuvering capabilities of the PDE are conferred by the use of the magnetic nozzle created by the rim electromagnets. Thrust vectoring is achieved by altering the electric current (and thus the strengths of the fields) of the magnet pairs.

The effect of equal field strengths in a magnet pair is illustrated in Figure 6.3.1. The symmetric magnetic field lines in the upper and lower magnets provide a radially vectored flow of air plasma, whose reaction force propels the Lightcraft in the direction opposite the expanding gas flow. The plasma is squeezed out between the two strong-field regions like rocket exhaust between the confining walls of a physical nozzle.

As the current is increased in one of the magnets, the direction of the open magnetic field lines is changed. A stronger current in the upper magnet diverts the magnetic nozzle downward, thus pushing the Lightcraft's rim upward, as illustrated in Figure 6.3.2. Similarly, increasing the strength of the lower field diverts the plasma flow upward and forces the Lightcraft's rim downward.

The direction of thrust vectoring is controlled not only by the magnetic nozzle orientation, but also by which parts of the Lightcraft rim are engaged in active PDE thrusting. For example, if

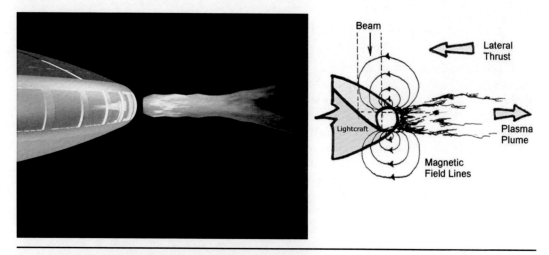

Figure 6.3.1: PDE exhaust with equal field strengths in magnet pair. *(Courtesy of RPI.)*

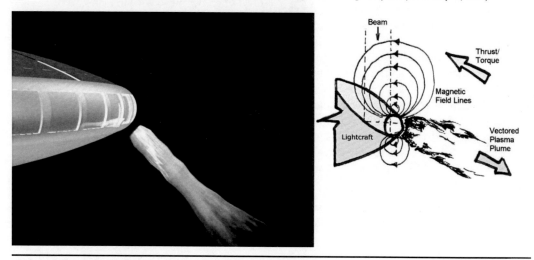

Figure 6.3.2: Vectored PDE exhaust with non-equal field strengths in magnet pair. *(Courtesy of RPI.)*

the rear-facing portion of the rim activates its PDE thrusters, the Lightcraft will move forward in a lateral flight mode (Figure 6.3.1). Lateral movement in any direction can be accomplished by firing the appropriate sector of PDE thrusters in the rim without the use of any mechanical or material control surfaces. Oddly enough, it's all done with (magnetic) mirrors.

To thrust vertically, the PDE is pulsed all the way around the rim with the upper magnet field strength maximized, thus forcing the magnetic nozzle downward to create an upward reactive thrust. Likewise, upward deflection of the nozzle all around the rim forces the Lightcraft downward.

Activation of the PDE over only part of the rim, either to fly upward or downward, will influence the pitch or roll of the Lightcraft. Pitch and roll can therefore be caused at will, as shown in Figure 6.3.3. Pitch changes can be prevented by the expedient of firing the PDE thrusters on opposite sides of the rim simultaneously in such a way as to cancel out the torque.

The steadiness of the PDE thrust is affected by the pulse repetition rate. During covert mission operations or at any time when stealth (in this case, low noise level) is necessary to ensure mission success, the pulses delivered to the LTI-20 must be in either the ultrasonic (above 20

PULSED DETONATION ENGINE (PDE) MODE

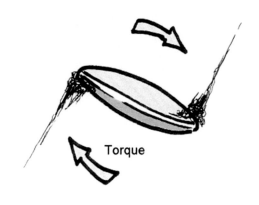

Figure 6.3.3: Pitch maneuver with PDE thrusters and "vectorable" magnetic nozzles. *(Courtesy of RPI.)*

kHz) or infrasonic (below 20 Hz) range. Pulsing at these frequencies keeps the Lightcraft silent except for the faint crackling or popping sound associated with the plasma detonations. Operating at infrasonic pulse repetition frequencies has an advantage in that the time between pulses is long enough to allow full expansion and cooling of the heated plasma, which both increases the efficiency of each pulse and eases the load on the TMS cooling system. However, the total thrust of the PDE increases with the pulse frequency. The choice of an infrasonic pulse rate for purposes of stealthy operation will cause a significant loss of thrust, and greatly decreased agility of the Lightcraft.

6.4 Microwave Power Reception

The LTI-20 is powered by microwave radiation beamed from an orbital (or ground-based) power station with extremely accurate pointing and tracking capabilities to "paint" its beam on the cooperative Lightcraft target (see Section 20). In addition, a large-diameter transmitting antenna is required to produce a beam with the same diameter as the Lightcraft. The symmetry axis of the Lightcraft must be aligned within 1 to 2° with the incident microwave power beam whenever running on direct beamed power. The most recent rectenna arrays for the LTI-20 can support the use of two different microwave frequencies, specifically 35 and 140 GHz. Shorter wavelengths (i.e., higher frequencies) have a higher breakdown intensity threshold (see Figure 5.1.4) which decreases at all wavelengths as the Lightcraft gains altitude.

The shorter microwave wavelengths are absorbed and scattered more readily by moisture in the atmosphere, both by direct excitation of vibration of the water molecule (the microwave oven effect) and by Mie scattering from water droplets comparable in size to the wavelength of the radiation. Water and water vapor are both much more abundant in the lower atmosphere than at high altitudes.

As discussed in Section 6.6, impulse, thrust, and propulsive efficiency of air-breathing Lightcraft engines all decrease with increasing altitude. The LTI-20 is therefore designed to run on more than one microwave wavelength to allow it to exploit the great increase in performance that shorter wavelengths afford during high-altitude operation. Generally the LTI-20 employs a microwave wavelength of 8.57 mm (35 GHz) during PDE operation below 10 km altitude. After the Lightcraft passes the 10 km level while in PDE propulsion mode, the power beam wavelength may be switched to 2.14 mm (140 GHz) if a suitable power satellite (powersat) is available.

Note that the atmospheric transmission of 35 GHz power is limited by the breakdown intensity at 30 km altitude, which is 4 kW/cm^2. After allowing for the inefficiencies of conversion of microwave power into electricity, this beam intensity generates about 9 GW of electrical power onboard the 20 m LTI-20; the rectenna arrays are 18 m in diameter. During propulsion near sea level, the PDE generally absorbs only about 1.7 GW of microwave power. The PDE system on the LTI-20 can accommodate propulsive powers of 3.4 to 6.8 GW, depending upon the wavelength of the microwave energy.

The PDE system can occasionally draw power from the SMES, as discussed above. The twin perimeter magnets in the SMES unit can store over 900 MJ aboard the Lightcraft. During certain emergency procedures when the power beam is off or temporarily interrupted (see Section 6.7), the PDE can run independent of the beam for short periods of time (1 to 45 seconds, depending on the PDE thrust level and pulse repetition rate) on stored SMES energy.

6.5 PDE Operations and Safety

As with the ion propulsion mode, the PDE is most frequently used in non-orbital boost operations. The PDE mode can be manually operated from the bridge, or, when all crew members are inside the maglev landers or individual escape pods, it may be controlled directly by the FMS computer. From an engineering and safety standpoint, the principal concern is loss of beamed power from the orbital microwave station. Of nearly equal concern is thermal management of the internal waste heat load caused by excessive pulsing of the PDE thrusters without sufficient cool-down time. Both of these concerns are monitored continuously by the LTI-20's FMS computer, which relies on sensor data on the pressure levels at the Lightcraft rim, the temperature of the HeliOx coolant, and rim and lenticular hull temperature and heat loads. Brief "brownouts" or dropouts of the external power beam cause no threat to either Lightcraft performance or crew safety because of the redundancy provided by the SMES "battery." The FMS computer can instantly tap the energy stored in the SMES to power the rectenna array and drive the PDE propulsion system. With over 900 MJ of stored energy, of which 84% is available for PDE operation, the SMES can smooth over these brownouts with no loss of performance during subsonic, transonic, or supersonic fight maneuvers. When it appears that a power beam dropout will last long enough to consume the entire store of SMES energy, the FMS computer carries out a high-performance deceleration maneuver automatically. Even if the airspike fails during high-performance acceleration in the dense lower atmosphere, constant monitoring of flight systems status minimizes the probability of encountering excessive aerodynamic forces.

"Engine-on" maintenance of the PDE is limited to insuring that the SMES coils are full charged and functioning within normal parameters, that the rectenna arrays maintain their proper parabolic figure, and that the outer rim hull is free of scratches, dirt, and other physical defects that might compromise its strength and transparency. The PDE itself has few mechanical components, and requires a level of monitoring and care no different from those required by other LTI-20 systems.

Safety considerations directly related to PDE operation include (among others) monitoring of the pulse repetition frequency and the possibility of dangerous levels of microwave radiation surrounding the Lightcraft. Care must be taken that the PDE pulse repetition frequency not be in resonance with the motions of human internal organs or with nerve synapses. Although the PDE is generally operated outside the audible range for stealthiness, the computer also constantly monitors the pulse frequency to assure that no damage is done to the crew either inside or outside the Lightcraft. The computer also monitors the level of microwave radiation inside the Lightcraft hull to assure that if its Faraday cage is breached, human safety limits are not exceeded. Should microwave intensity rise to potentially dangerous levels, the computer can shut down the PDE and even trigger an evacuation of the crew in their individual escape pods.

6.6 Performance Specifications

The LTI-20 can achieve speeds of 0 (hover) to Mach 2.35 using the PDE propulsion system without an airspike. Normally, however, the MHD slipstream accelerator is employed for fight speeds above Mach 2.0. The top speed of the Lightcraft achievable by PDE thrust varies with altitude. As shown in Figure 6.6.1, the thrust drops off rapidly with increasing altitude, so that maximum thrust and acceleration are achieved at the lowest altitudes. This drop off of thrust with altitude, which is typical of all air-breathing propulsion systems, occurs largely because the density of ambient air drops off roughly exponentially with altitude. In addition, the exhaust thermal power developed by the PDE also decreases at high altitude because of the reduction in the breakdown intensity of the microwave beam (see Figure 5.1.4). As shown in Figure 6.6.2, the momentum coupling coefficient (the efficiency of conversion of beamed microwave power into engine thrust) increases

PULSED DETONATION ENGINE (PDE) MODE

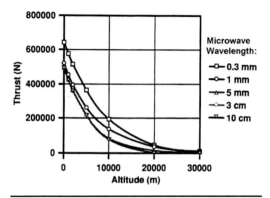

Figure 6.6.1: PDE maximum thrust vs. altitude for 10 m lightcraft (LTI-20 data is confidential).

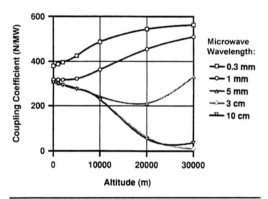

Figure 6.6.2: PDE momentum coupling coefficient vs. altitude for 10 m lightcraft (LTI-20 data is confidential).

with increasing altitude for the shorter microwave wavelengths (0.3 to 10 mm). The longer wavelengths (5 to 10 mm) have coupling coefficients that at first drop off with increasing altitude, then begin increasing again. Overall, even at low altitudes and longer wavelengths, the PDE is generally capable of providing the Lightcraft with enough thrust to achieve at least 160 g acceleration. The thrust level is well into the hyper-energetic regime and always above the minimum level required for a trans-atmospheric boost into orbit.

6.7 Emergency Procedures
Several emergency scenarios related to PDE thruster use are worthy of mention.

First, there is the possibility of structural failure and breach of the toroidal rim tube or lenticular double hull. The beamed energy needed to operate the PDE would then become hazardous to the crew if they were not in their protective gear or escape pods. If such a situation were to arise, the microwave power beam must be shut down instantly by the FMS computer or by a remote manual control link accessible from the bridge or from crew quarters. This manual shutdown command overrides any FMS computer command authority and automatically cuts off the beam from the orbital power station. The minimum time lag for beam cutoff is the round-trip time for a lightspeed signal between the Lightcraft and the LEO power station, approximately 10 microseconds (assuming the maximum range of 1100 km).

The most important safety-related procedure is the continuous monitoring of hull temperature and integrity and crew members' vital signs by the FMS computer system. If anything abnormal is detected the FMS computer automatically shuts down the PDE to locate the source of the problem while switching to the ion propulsion mode to achieve a survivable landing. The shutdown, depending on the severity of the damage, is accompanied by audio and visual signals to the crew to evacuate to the maglev landers or the personal escape pods, where the crew protected from both temperature rise and high microwave radiation levels.

Finally, the crew, when outside the protection of the Lightcraft hull, must be protected from potentially harmful exhaust from the PDE thrusters. The PDE system should not be engaged while crew members are working outside in close proximity to the grounded or hovering Lightcraft. Harmful microwave irradiation of the crew and possible exposure to high-temperature plasmas could result from improper use of PDE mode.

7.0 ION PROPULSION SYSTEM

The LTI-20 Lightcraft ion-propulsion system is used primarily for subsonic, low-performance maneuvers. These maneuvers include aerial taxiing, passenger retrieval, extended loiter, and covert operations. The maximum speed attainable in air-breathing ion-propulsion mode is approximately 160 km/hr (100 mph).

7.1 Theory and Application

The ion propulsion system aboard the LTI-20 is a device for converting beamed laser or microwave energy into electrical energy, thence into the kinetic energy of a relativistic electron stream, and finally the production and acceleration of ions created in atmospheric air.

The ion propulsion system is responsible for converting any initial onboard electrical energy into kinetic energy of the Lightcraft. The groundwork for this application of endoatmospheric ion propulsion was laid in the 1950s by de Seversky's successful flight of his ionocraft, which proved that efficient ion-powered flight was indeed possible. His experiments demonstrated momentum coupling coefficients in the range of 5 to 10 kN/MW for his "flying bed springs" apparatus, which used metal electrodes and accelerator grids.

The key to the success of the LTI-20's ion drive system is the large external surface area and partial buoyancy of the Lightcraft. The HeliOx gas mixture that pressurizes the Lightcraft also provides approximately 16.5 kN of buoyancy at 20° C, which means that the ion propulsion system needs to provide only 8 kN of lift to suspend an empty 2400 kg Lightcraft in the air. With a maximum vertical thrust of 11.4 kN (from the ion engine at sea level), the LTI-20 can suspend only 390 kg of payload in hover. However, if is first lofted airborne by a brief PDE burst (among other methods), the LTI-20 can indeed carry the mandated 1200 kg of payload in horizontal flight by exploiting dynamic lift in lateral ion-propelled, Frisbee mode flight.

As with conventionally powered airships, the flight physics are succinctly explained by the "Lift Triangle" in Figure 7.1.1. *Static Lift* is provided by the buoyancy of the HeliOx, and *Dynamic Lift* is that produced by motion of the Lightcraft through the air (i.e., aerodynamic lift). The third option is represented in Figure 7.1.1 as the

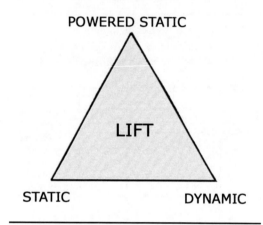

Figure 7.1.1: The "Lift Triangle."

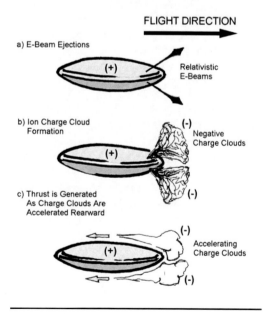

Figure 7.1.2: Thrust production process for ion propulsion mode. *(Courtesy of RPI.)*

ION PROPULSION SYSTEM

Powered Static corner of the lift triangle, which exploits both static lift and vertical thrust from the ion (or other) propulsion system.

The pulsed ion propulsion system that is integrated into the LTI-20 differs from its historic predecessors in that it lacks solid cathodes and accelerator grids. The craft emits pulsed 1-MeV electron beams from the rim into the atmosphere, lending itself a net positive charge as it forms negatively charged ion clouds in the desired direction of motion. Most of the electrons preferentially attach themselves to (previously neutral) molecular oxygen and water vapor molecules in the air to form negative ions, which are attracted toward the ship through a large electrical potential difference, collisionally entraining (largely neutral) air in the process. The newly created momentum of the air stream as it accelerates toward the positively charged Lightcraft is balanced by the acceleration of the Lightcraft toward the negative ion cloud (See Figure 7.1.2). This momentum exchange principal is the process by which any air-breathing engine develops thrust.

Usually the ion propulsion system must produce a vertical component of thrust to keep the Lightcraft airborne. This vertical force is a result of charged air rushing over or around the top of the craft as it accelerates downward through a large electrical potential. The wake left below the Lightcraft creates an "ion plasma cone" within which weakly ionized air is briefly entrained and recirculated. When the atmospheric humidity reaches high levels, condensation occurs at the vehicle rim (which is a region of reduced pressure) producing a fog-like "tail" hanging off the bottom of the

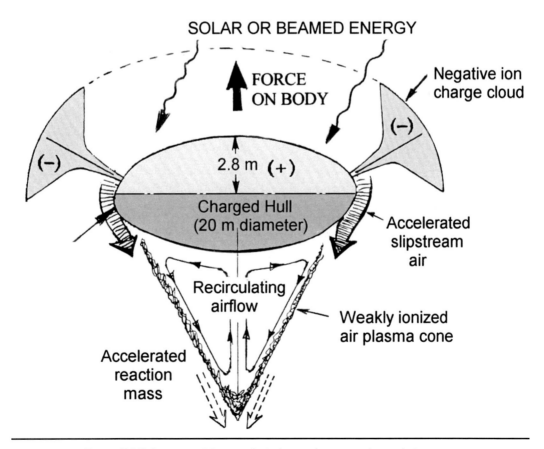

Figure 7.1.3: Ion propulsion mode in hover (note accelerated airstream and weakly ionized plasma cone).

Lightcraft (see Figure 7.1.3). Upon illumination by an orbital laser power beam (650 nm wavelength), the craft and its vapor tail become especially visible at dawn and dusk, as stray visible light (dull red in color) from the power beam is scattered into the fog. In lateral flight, the behavior of this weakly ionized plasma cone with flight speed and acceleration is indicated in Figure 7.1.4.

Because of the necessary vertical thrust component, the Lightcraft in ion-propulsion mode often maneuvers much like a conventional rotorcraft. A series of simple free-body diagrams in Figure 7.1.5 represents this similarity and the force balance that makes it possible. The thrust vector must have a large horizontal component during horizontal acceleration or deceleration to overcome inertia and aerodynamic drag forces.

The electrostatic pressure loads induced on the disc's hull by the ion propulsion system are important, but small in magnitude. Disc loadings generally run from 13 N/m² (electric field strength = 1.7 MV/m) to a maximum of 36 N/m² (field of 3 MW/m). These values, which are determined by the electrical breakdown strength of air, are 3X to 10X lower than the

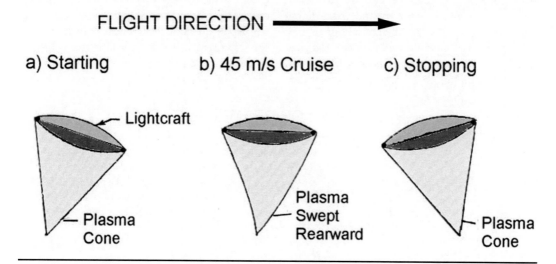

Figure 7.1.4: Behavior and shape of weakly ionized "plasma cone" in ion propulsion mode.

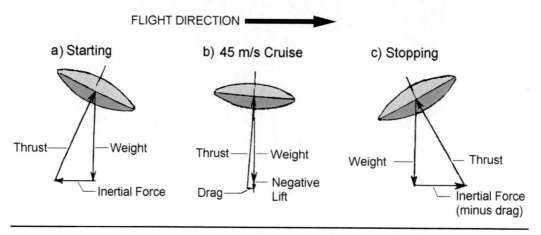

Figure 7.1.5: Free-body diagrams of starting, cruising, and stopping maneuvers in ion propulsion mode.

rotor loadings of manned ultralight helicopters, but fall exactly in the range of remote-piloted model helicopters.

7.2 Essential Physical Processes

The ion propulsion drive is made possible by three essential processes. First, the onboard electron accelerators ("E-guns") inject negative charge from the LTI-20 at relativistic velocities (1 - 5 MeV) into the surrounding atmosphere. Through collisional processes, the negative charges then slow down and attach to oxygen and water molecules, which then, as ions, are accelerated back toward the Lightcraft, entraining a large mass of air (the engine's working fluid). The electron beam is not ejected in a straight, collimated beam away from the ship. The beam whips wildly around within a funnel-shaped envelope and eventually disperses ("blooms") because of Coulomb repulsion, scattering, and Bremmstrahlung effects, in which the energetic electrons induce ultraviolet photon emission from atmospheric gases, photons which in turn cause further photoionization of the gas. In practice, the electrons form a trumpet or bell-shaped ion cloud that extends an average distance of 3- to 5 m from the 20 m diameter LTI-20 hull (see Figure 7.1.3). Many of the photons emitted by the Bremmstrahlung effect are cascade photons in the visible range, which explains the beautiful and often intense blue glow seen around the Lightcraft rim when the ion drive is used at night.

The penetration distance (stopping depth) of the electron beam in air is limited by molecular collisions, and hence by the local atmospheric pressure. The combined effects of virtual cathode instability, mode instability, and two-stream instability limit the performance of such electron accelerators. The balance between pressure sensitivity and decreasing air density at higher altitudes leads to an effective altitude limit for ion propulsion. The LTI-20 has a ceiling for ion propulsion of about 10- to 15 km altitude. In free air, the charge cloud has a decay time of about 1- to 4 msec. Therefore, in order for the cloud to produce useful "time average" thrust, it must be recharged frequently. Normally the LTI-20 electron accelerators pulse at a frequency between 50- and 1600 Hz in free air. It should be noted that, at close range, these high-current pulses can disrupt local non-optical communications, electrical household appliances, and unshielded motor vehicle ignition systems.

The electrons attach themselves to molecules in the air though one of two processes, each of which is represented below:

$$e^- + O_2 + X \rightarrow O_2^- + X^* \quad (7.1)$$
$$e^- + XY \rightarrow X + Y^* \quad (7.2)$$

where the asterisk denotes excited ("hot") atoms that carry off excess energy.

Once the ion cloud is formed it expands and simultaneously is attracted toward the ship, dragging much of the surrounding neutral air mass along. This process produces a low-pressure region across the front sloped portions of the lenticular hull which contributes additional forward thrust. The approximate magnitude of this thrust is given for three different ion-cloud configurations in Table 7.2.1 below, under standard sea level conditions. The corresponding maximum lateral velocities are also given. Sketches of these three cloud geometries are given in Figure 7.2.3.

The LTI-20 is also capable of hovering, with minimal maintenance energy expenditure, directly under large cumulus clouds (Figure 7.2.4). The volume-doubling (expansion dissipation) time of an ion cloud in dry air is about 4 msec, which governs how frequently the cloud must be recharged. In a cumulus cloud,

Table 7.2.1: Electrostatic thrust production for three forward charge ejection modes.

Mode	Q (mC)	F (kN)	F_v (kN)	M_c (kN-m^2)
Straight-head	2.55	1.440	0.000	0.000
Dual Symmetric Off-Angle	3.32	1.920	0.000	0.000
Off-Angle	2.85	1.359	0.531	3.781

Chapter 7

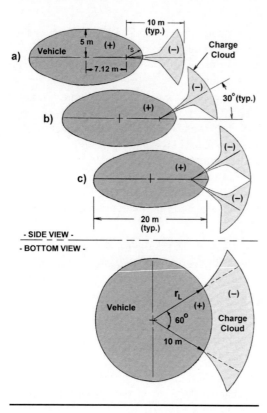

Figure 7.2.3: Three representative charge cloud geometries for ion propulsion mode.

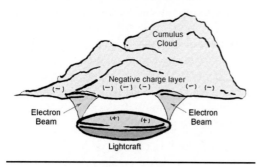

Figure 7.2.4: Lightcraft electrostatically supported in hover under large cumulus cloud. *(Courtesy of RPI.)*

however, this doubling time is nearly a week. In this environment the ion drive is so much more efficient (disc loading of 36 N/m2) that the Lightcraft can operate under solar power at low velocities, with its photovoltaic array scavenging light reflected from the ground below or the cloud above. Table 7.2.2 gives the LTI-20's ion propulsion performance when operating below

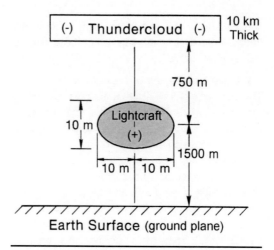

Figure 7.2.5: Lightcraft vehicle under 10 km thick thunderhead with negative charge density of -1.76 x 10-9 C/m³, producing mean electric field (E) of 1 MV/m between cloud and ground plane. *(Courtesy of RPI.)*

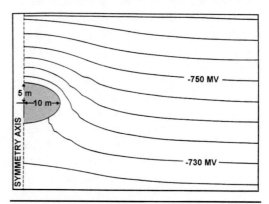

Figure 7.2.6: Equipotential lines for positively charged lightcraft carrying 1.14 x 10-2 C, under thunderhead in 1 MW/m atmospheric electric field. *(Courtesy of RPI.)*

a thundercloud (i.e., a high ambient potential gradient between the cloud base and the ground as in Figure 7.2.5); typical equipotential contours for this flight condition are given in Figure 7.2.6. The *under-thundercloud* propulsion mode enables a Lightcraft to "soar" laterally (i.e., x-direction in Figure 7.2.7) and "tack" (like a sailboat) by applying electrostatic thrust (F) and aerodynamic lift (L) simultaneously. For example, Figure 7.2.7 shows the disc oriented at an attitude of β = +20° (defined relative to

ION PROPULSION SYSTEM

Table 7.2.2: Performance parameters for three forward charge ejection modes, and for the "under thundercloud" mode (at optimum disc orientation).

Mode	V_c (m/s)	C_m (kN/MW)	η_D	F/A_{cl} (N/m²)	P_{beam} (kW)
Straight-ahead	20.4	48.3	0.985	14.4	29.8
Dual Symmetric Off-Angle*	31.0	32.1	0.995	11.1	59.9
Off-Angle	19.2	58.9	0.985	14.6	18.5
Under-Thundercloud (β=20°)	35.4	73.6	0.998	18.2	—

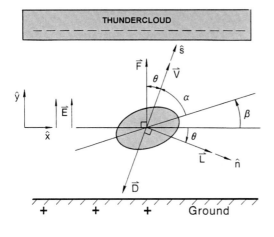

Figure 7.2.7: Orientation of charged lightcraft flying under thunderhead in ion propulsion mode. *(Courtesy of RPI.)*

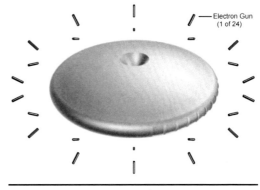

Figure 7.3.1: Exploded diagram of 12 pairs of electron guns, arranged symmetrically about disc circumference, ejecting at +/- 30°.

horizontal) as the positively charged hull is propelled upward to the right at V = 35.4 m/s (note opposing drag vector, D) at an optimum angle of attack (α) within the thundercloud's quasi-uniform electric field (E).

7.3 Ion Propulsion System Components

The ion drive system shares components with many of the other systems aboard the LTI-20. The superconductive magnets described in Section 5.2 can be used to store energy for operating the ion engine in stealth mode. Also, the rectenna array can be used to gather low-power microwave energy for the ion engine, and beam alignment is not critical; misalignments of ± 60% produce only a 40% degradation in output voltage. The electron accelerators that produce the ion cloud are also employed in the creation of the space plasma shield (Section 15.5), which requires only 30 keV electrons.

The electron accelerators were developed specifically for the LTI-20 and are 33 times more powerful than required by the plasma shield. They are the most efficient 1-Mev, low-mass electron beam guns ever produced. These accelerators fire beams of electrons away from the Lightcraft to create the ion cloud. There are 24 pairs of electron accelerators spaced evenly (130 cm) about the rim of the Lightcraft, pointing outward: 24 deflected 30° upward and the remaining 24 deflected 30° downward relative to the plane of the rim (see Figure 7.3.1).

The Indium-Gallium-Arsenide Nitride (InGaAsN) photovoltaic cells that cover the ventral surface of the craft can be used to collect either incident solar energy or beamed laser light (at 650 - 860 nm wavelength) to power the Lightcraft in the low power (inverted) flight and landing mode. The InGaAsN crystals that form these cells are grown on the inside SiC surface of the outermost hull layer to protect them from the hazardous space and atmospheric environments, and to allow them to be cooled by the Lightcraft's closed-cycle HeliOx TMS system (see Section 5.6). If the array were exposed on the outside of the hull, its performance would quickly degrade (see Figure 2.3.1).

The outer hull surface of the LTI-20 is covered with electric and magnetic field sensors that feed their critical data back to the FMS computer. The FMS uses data on the electrical potentials all over the external hull to help prevent unwanted electrical discharges and to permit rapid reconstitution of discharged areas of the ion cloud. When the electric potential between the ion cloud and the Lightcraft exceeds the breakdown strength of the local atmosphere, an arc discharge similar to atmospheric lightning occurs. Such a discharge destabilizes ion-cloud formation momentarily in regions near the arc because negative charge is "shorted" back to the positively charged hull. The FMS computer, upon sensing a potential drop, compensates as much as possible by rebuilding the desired charge cloud geometry. In the event of multiple discharges and subsequent attempts to repair the ion cloud, the flight may become erratic.

The outer surface of the Lightcraft hull is another major component of the system. The details of the SiC hull are given in Section 2.3. The polished hull surface must be kept very clean and smooth, since any imperfection greatly increases the possibility of a corona discharge and arc, with the ensuing complications spelled out above.

7.4 Laser and Solar Power Considerations

The previously described PhotoVoltaic (PV) array converts solar energy into electricity to directly run the ion drive at maximum performance under normal bright, noontime conditions. In the event that more power is required and especially for low-light or nighttime operations, the InGaAsN PV array can be illuminated by FEL laser light with a wavelength of 620 to 860 nanometers. (Note that the visible wavelength range is approximately 430- to 690 nm where eye sensitivity has dropped to 1% of its maximum value, with red perceived at 650- to 690 nm. The eye can detect radiation longer than 700 nm if the source is intense enough.)

This laser light is beamed directly from the same power station that delivers the microwave power beam (i.e., PDS-02 or PDS-03), or relayed through the array of Monocle™ or Binocle™ laser relay satellites deployed in optimum equatorial and polar orbits. The array of laser relay mirrors assures continuous global coverage, upon demand, by the entire LTI-20 fleet. Note that the "macro-pulse" from an RF-type FEL laser is comprised of 10 to 20 ps pulses at a gigahertz repetition rate. Within the macro-pulse, this is a high enough Pulse Repetition Frequency (PRF) that the InGaAsN array responds as if the laser is effectively in continuous wave operation.

Unlike the high power microwave beam required for ultra-energetic PDE and MHD maneuvers, however, the low-power laser beam need not be aligned with the axis of the Lightcraft. When insufficient laser (or solar) power is available to generate the desired ion thrust levels for flight, the SMES "battery" supplies the difference. Using the PV array, LTI-20's hybrid electric power system can charge up the SMES unit with beamed-laser or solar-derived energy, for later consumption by the ion drive. At normal incidence into the grounded Lightcraft, the 4 MW laser beam can completely recharge the 900 MJ SMES (from an 85% depleted state) in 5.6 minutes; with solar power (at high noon), it takes approximately 2.2 hours.

The InGaAsN array is approximately 29% efficient in the solar-powered mode, but exceeds 60% in the laser-powered mode. Incident power densities in the laser mode are, of course, far higher (10x) than in the solar mode, so the total waste heat flux in the laser mode is higher despite the greater conversion efficiency. Active HeliOx cooling of the double-hull is therefore extremely important under laser power. Table 7.4.1 gives the maximum electrical outputs of the InGaAsN array under both solar and laser power (normal beam incidence angle) and ideal environmental conditions. The maximum thrusts given are

Table 7.4.1 Maximum thrust of ion propulsion system on "real-time" beam power (assumes engine C_m of 5 kN/MW$_e$).

Power Source	Max Beam Intensity (kW/m²)	Beam Power (kW)	PV Array Efficiency (%)	Onboard Electric Power (KW)	Maximum Thrust (kN)
Laser	14.9	3800	60+	2280	11.4
Solar	< 1.34	342	29	99	~ 0.5

based on the available electric power, nominal engine performance of ~5 kN/MW (i.e., thrust / power conversion efficiency), and the standard 18 m diameter PV array with an active collection area of 255 m².

Actual thrusts under operational conditions are also dependent on atmospheric conditions, laser transmission losses (due to extinction, turbulence, thermal blooming), ion cloud geometry, beam incidence angle upon the array, and other variables.

Real-world flight conditions are seldom perfect. Half of the planet is in darkness at any time, and clouds, rain, snow and dust are common. Under unusual circumstances, frost and sheet ice can build up on the InGaAsN array and attenuate incident light levels. The emergency procedures for dealing with such problems are discussed in detail in Section 7.7.

Of course the laser beam intensity and spot size upon the photovoltaic array is continuously modified by atmospheric conditions, zenith angle, slant range, and beam / array incidence angle; at grazing incidence, much laser power may be reflected, even with the hull's anti-reflection coatings intact.

7.5 Performance Specifications

The above sections have covered the general theory, application, and essential components of the ion propulsion system. This section concentrates on specific usage of the ion engine. It is important to know not only the nominal flight performance, but also the limits to which ion propulsion technology can be pushed during maneuvers. The performance of the LTI-20 falls into the categories of vehicle orientation, thrust, and maximum velocity.

7.5.1 Vehicle Orientation

The orientation of the Lightcraft about the two axes orthogonal to the symmetry axis can be controlled by means of ion propulsion. Figure 7.5.1 presents the chosen reference axes and orientations for the purposes of this discussion. The Lightcraft is lenticular in shape, so that only roll (rotation about the flight direction axis) and pitch (nose up / nose down angle) apply. Yaw (angle between the nose of the vehicle and the direction of motion) is meaningless for a vehicle with axial symmetry, and the ion engine cannot affect yaw; nonetheless, the rate of rotation of the Lightcraft about its principal axis is important. (Spin torque is applied by the PDE mode in conjunction with azimuthal MHD forces.) The magnitude of the E-gun expelled charge and the direction in which it is expelled determine whether pitch or roll, or some combination of them, is produced.

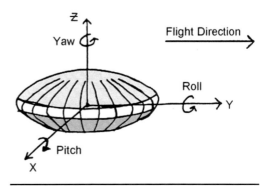

Figure 7.5.1: Standard reference axes and lightcraft orientation for lateral flight mode. *(Courtesy of RPI.)*

Pitch is produced by ejecting charge either in the forward or rearward flight direction (or both) at an angle to the plane of symmetry. Figure 7.5.2 shows a side view of a pitch-up

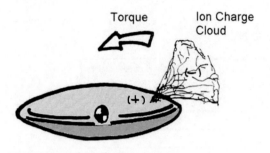

Figure 7.5.2: Lightcraft pitch or roll maneuver for ion propulsion mode – gives both torque and translation. *(Courtesy of RPI.)*

maneuver. Ejecting a charge in front of the Lightcraft at a positive angle to the symmetry plane produces a torque that pitches the vehicle nose-up, which is a positive pitch angle. Using an attraction force of 785 N at +30°, with a torque arm of 7.12 m, produces a torque of 5590 Nm and a pitch rate of 0.288 radians/second.

From Figure 7.5.2 one can also see that whenever charges are ejected forward at a negative angle to the plane of symmetry (downward), a clockwise rotation or pitch-down is produced. The positively charged edge of the Lightcraft is pulled toward the ion cloud wherever it is created. Besides torque, this maneuver also induces translation in the forward direction.

If charge is repeatedly ejected from both the leading and trailing edges, as in Figure 7.5.3, it is possible to pitch the Lightcraft 90° or even 180°, to accomplish a fore-to-aft "flip-over." If, for example, charges are ejected at an angle of

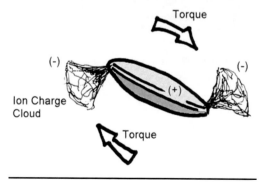

Figure 7.5.3: Pure pitch or roll maneuver in ion propulsion mode (gives torque only; no translation). *(Courtesy of RPI.)*

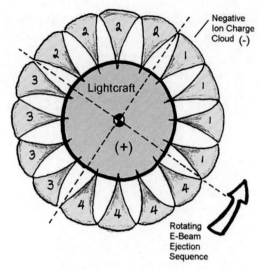

Figure 7.5.4: E-beam ejection sequence for performing "falling leaf" and "inverse falling leaf" maneuvers. *(Courtesy of RPI.)*

60° instead of 30° to the symmetry plane, a higher pitch rate will result because the vertical component of the force vector is larger, and hence more torque is exerted.

Rolling the Lightcraft requires charge ejection from both sides of the Lightcraft at right angles to the direction of flight. If the ejection angle is positive (as seen from the rear), a clockwise roll is produced. If the ejection angle is negative, the roll direction is counterclockwise. If a rolling maneuver is carried out with the same charge ejection from each thruster, and the same ejection angle of +30° used in the pitch example, a roll rate of 0.576 rad/s would result. To induce a 90° or full 180° "roll-flip," both the +30° and -30° ejections are repeated until the desired roll angle is attained. For even quicker roll rates, the ± 60° ejection angle should be selected.

Finally, if the asymmetric charge ejections in Figure 7.5.2 are fired upward sequentially around the Lightcraft rim (see Figure 7.5.4), a helicopter-like rotating lift vector is produced, resulting in a "falling-leaf" maneuver, as in Figure 7.5.5. In this flight mode, the partially buoyant Lightcraft has insufficient thrust to hover, so it settles into a controlled spiral descent to

ION PROPULSION SYSTEM

Figure 7.5.5: "Falling leaf" maneuver for descent in ion propulsion mode. *(Courtesy of RPI.)*

Figure 7.5.6: Inverse "falling leaf" maneuver for climbing (i.e., ascent) in ion propulsion mode. *(Courtesy of RPI.)*

landing. The sequence of frontal, then left side, then rear, then right side thruster firings, repeated smoothly and continuously, simultaneously lifts and rotates the Lightcraft rim. This spiral mode of flight requires timing of the firing of the ion clouds in coordination with the disc's inertial and aerodynamic behavior. With a light payload and ion thrusters at maximum, the LTI-20 can perform the inverse maneuver and climb spirally upward as in Figure 7.5.6, exploiting the disc's aerodynamic lift.

7.5.2 Ion Thrust Performance

The thrust produced by the ion drive is linked to rate of power consumption by a momentum coupling coefficient (C_m) which depends on local environmental conditions and on the mandated refresh rate for replenishing the ion clouds. Table 7.5.1 gives engineering values of these coefficients in newtons of thrust per megawatt of power expended.

Atmospheric conditions determine where the actual coefficient falls within the given performance range. The time-average thrust, of course, depends on the onboard electrical power available to the ion propulsion system (joules per ejection x PRF) and the configuration of the emitted ion cloud, and is limited by the electrical breakdown limit of the surrounding air. The maximum feasible thrust in clear air that can be produced by the LTI-20 ion drive in hover at sea level altitude is about 11.4 kN, assuming a breakdown strength of 3 MW/m.

In clear air, the C_m performance is generally 3 to 5 kN/MW at flight speeds of 15- to 45 m/s, but flying through low-altitude fog at reduced speeds, it can reach 10 to 30 kN/MW because the slowed doubling time of ion cloud volume in water vapor reduces the mandated refresh rate for a given thrust level. Hanging near motionless under large, moisture laden cumulus clouds,

Table 7.5.1 Ion drive performance vs. ion cloud refresh rate.

Refresh Rate (Hz)	Local Environment	Flight Speed (m/s)	Momentum Coupling Coefficient, C_m (kN/MW)
100 - 1500	Clear Air	15 - 45	3 - 5
10 - 100	Low Altitude Fog (light to moderate)	5 - 15	10 - 30
1 - 10	Under Clouds	0 - 5	30 - 100+

Lightcraft C_m performance can reach 30- to 100 kN/MW or more because the doubling time is nearly a week.

The power for this ion propulsion system can come from anywhere onboard the craft. It can come from the SMES system, beamed microwave power, laser power, or solar power. Laser and solar power were described in the preceding section. Each power source has its own advantages and disadvantages, and different methods of power collection and conversion yield different overall efficiencies. All these factors influence ion propulsion performance.

7.5.3 Maximum Velocity

Maximum velocities attainable using each of the previously described (see Section 2) ion cloud configurations are given in Table 7.5.2 below. These values are based on flight test data for maximum ion engine thrust in clear air at sea level with nominal LTI-20 mass and inertial properties, and rotational disc aerodynamics (Frisbee mode). Note that the *"dual off-angle"* ion cloud geometry with 90° azimuthal spread provides the maximum lateral flight speed of 44.6 m/s (~100 mph). The *single off-angle* and *straight-ahead* ion clouds give 15- to 20% lower cruise performance for the same cloud charge (i.e., measured in coulombs), separation, and azimuthal spread.

Table 7.5.2: Maximum flight velocity vs. ion cloud geometry.

Ion Cloud Geometry	Max. Flight Speed (m/s)	
	60° Span Ion Cloud	90° Span Cloud
Off-Angle	30.7	37.5
Dual Off-Angle	36.4	44.6
Straight Ahead	31.5	38.6

7.6 Evasive Maneuvers

The LTI-20 routinely executes a number of low-performance maneuvers that use the ion-propulsion system to evade detection by potential threats. These maneuvers include:

1) a low-noise low-visibility silent-running mode;
2) the microwave beam engagement maneuver; and
3) two methods for hiding among and within clouds.

The most frequently used of these maneuvers is the silent-running mode. Although the ion drive at maximum cruise speed is generally quiet when pulsing at 50 to 1500 Hz, using externally visible propulsion methods can give away the location of the Lightcraft. At night the 620- to 700-nm light from the tunable FEL laser power beam can easily be detected by the unaided eye as it reflects off the InGaAsN photovoltaic array. A *bright orange* color is perceived at 620-nm, changing next to *red-orange*, then *bright red* in the range of 650- to 650-nm, and finally dark *blood red* at 690-nm before "blinking out" at 700-nm. Depending upon laser intensity, wavelengths beyond 700-nm are invisible to the human eye, but are still easily detected by infrared sensors. The output performance of the InGaAsN array is maximized at the shortest wavelength (620 to 650-nm), so if forced to "blink out" by tuning into the 700-nm to 860-nm range, the LTI-20's flight speed must decrease.

As ion engine pulse rates fall from 1500- to 120 Hz, the E-gun's 1 cm diameter metal foils (mounted flush with the hull) decline in luminosity. Pulsing at the highest rates, the 48 E-guns' metal foils are each as "white" bright as a welding torch around the rim (130 cm spacing), but at 120 Hz they cool to a dull red, and finally blink out below 60 Hz.

Table 7.6.1 SMES-powered ion drive duration vs. lateral flight speed.

SMES Energy (MJ)	Flight Speed (m/s)	Lateral Thrust (kN)	Ion Drive Eff. (%)	Power Req'd (kW)	Duration (hrs)
750	3.0	5.40	50	32.4	6.43
750	4.0	9.78	50	58.7	3.55
750	5.0	15.0	50	150	1.39

ION PROPULSION SYSTEM

By using the 900 MJ of energy stored in the fields of the rim magnets, the SMES unit can power the ion drive for longer than 80 minutes at low thrust levels (see Table 7.6.1). This internal power source eliminates the possibility of detecting laser power beams and allows the LTI-20 to operate efficiently and autonomously over short distances.

When the high-power microwave beam must be engaged, such as for PDE use or for an MHD-boosted hyperjump, the ion drive can be used to orient the vehicle's rectenna to the beam immediately prior to the boost. The Lightcraft can perform this maneuver solely under internal power, or by a combination of external laser and internal SMES power. It is impossible to use laser power alone to orient the Lightcraft for microwave beam reception because the photovoltaic array and the rectenna array are on opposite faces of the Lightcraft disc.

One effective way to hide the LTI-20 is in a bank of clouds. This method is not as effective against radar detection as it against visible, line-of-sight detection. Nevertheless, "edge-on" the LTI-20 is so stealthy that its radar signature is no larger than a 1.7 mm BB (steel shot) at 80 km range. Since the Lightcraft toroidal rim is transparent to microwave radiation, only the outer edge of the annular (off-axis) rectenna gives a barely detectable glint. Section 7.2 discusses the Lightcraft's ability to support itself under or within a cloud for extended periods of time with very little energy expenditure.

If a trip into clouds is not practical, however, and the LTI-20 has enough water on board, steam can be ejected through the rim exhaust ports to create an artificial cloud "curtain." Alternatively, in regions of high humidity, the LTI-20 can actively cool its outer hull to below the dew point to condense water from the atmosphere and cloak itself with a cloud (see Section 16.16 for details). This may not be a wise choice if the Lightcraft is evading detection on an otherwise clear day, since a single little white cloud suddenly appearing in a completely clear sky would be a little too obvious. If background clouds are available, or if the Lightcraft is not being actively sought, this method of concealment is indeed feasible.

7.7 Emergency Procedures

There are a few possible emergencies that may arise while in ion propulsion mode, mostly related to a loss of the external power beam. An essential part of LTI-20 pilot certification is that they must be trained in the following ion-drive emergency situations: laser or microwave power loss caused by power station malfunction, degradation or loss of beam power caused by cloud cover, loss of solar power due to cloudiness or other environmental conditions, a failure of the high-voltage electron accelerators (E-guns), or power loss caused by hull contamination from dust or impacts with insect swarms.

In the case of a power satellite malfunction, the Lightcraft can be landed using the ion drive under either solar or SMES power. If the station is still delivering a low-power, megawatt-level microwave beam, stored SMES power or magnetic torquing (described in Section 17.5) can be used to flip the Lightcraft over to engage the rectenna arrays with the microwave beam. In the low-power mode the microwave beam need not be closely aligned with the axis of the Lightcraft.

If either the microwave or laser power beam is blocked by clouds, and under circumstances in which minimal hazard exists to life forms below, the pilot may request a high-power microwave pulse from the power satellite. High-intensity microwave power vaporizes water droplets and can "drill" a kilometer-sized hole through the clouds. The pulse can be continued until the line of sight from the power satellite to the Lightcraft is clear, at which point normal flight operations are resumed.

If solar power is suddenly lost or attenuated to a level insufficient to satisfy the Lightcraft's power demand, the captain may simply switch to an alternate external or internal power source. With 24 pairs of electron accelerators aboard, the rare failure of one or even a group of them can be compensated by reconfiguring the remaining accelerators to achieve balanced thrust. For example, the failure of one accelerator can be compensated by increasing the current in the two adjacent accelerators to permit smooth continuation with minimal performance loss. This procedure is especially effective when the LTI-20 is slowly rotating (Frisbee mode) as is normal for lateral flight.

Perhaps the most serious limitation on the LTI-20 is imposed by severe surface contamination. If, for any reason, the smoothness of the hull is compromised, corona discharges will jeopardize the stability of the ion drive mode. Constant charge loss through such discharges may make it difficult or even impossible to charge the hull to the potentials needed to achieve efficient ion thrust. If the contamination is light, the LTI-20 can use its in-flight ultrasonic cleaning system to eliminate dust or other contaminants from the hull surface. If hull contamination becomes excessive, the Lightcraft must make an emergency landing to clean its surface.

8.0 UTILITIES AND AUXILIARY SYSTEMS

The LTI-20 contains many systems dedicated to handling fluids and regulating environmental conditions that are essential to the survival and safety of the Lightcraft and its crew. These systems supply and regulate the cabin atmosphere and water for life support, waste removal, cryogens, food hydration, and artificial gravity.

8.1 Major Utilities Networks
The essential life-support systems of the LTI-20 and its crew escape pods are detailed below.

8.1.1 Cabin Atmosphere
The gaseous mixture breathed by the LTI-20 crew and used for inflation of the Lightcraft compartments is a mixture of Helium and Oxygen called HeliOx. At sea level, the internal pressure is held at a total pressure of 1.5 atmospheres (0.15 MPa) and circulated and regulated by a closed-cycle recirculation system; in the vacuum of space, cabin pressure is reduced to 0.5 atmospheres (0.05 MPa). The circulating HeliOx is monitored for excessive levels of water vapor, carbon dioxide, and trace gases, and is filtered to remove particulate contaminants. Mass spectrometers the size of a 1-cm cube monitor the gas composition down to parts-per-billion levels, keeping watch for toxic, flammable, or corrosive trace constituents. Since the Lightcraft may at times operate in zero gravity, where free convection cannot be relied on to circulate the internal atmosphere, cabin air is circulated by forced convection fans and positive displacement pumps.

8.1.2 Cryogenic Fluid Transfer
The cryogenic fluid transfer system supplies Helium and Oxygen for the HeliOx mixture breathed by the crew. The humidity is controlled for the health and comfort of the crew. The cryogenic helium that is used to cool the superconducting magnets in the SMES is also available to cool the cabin rapidly in emergencies.

8.1.3 Expendable Fluids
There are several types of expendable fluids on board the Lightcraft, although fluids are recycled whenever possible to minimize weight and replacement costs. Helium, Hydrogen, and Oxygen are required as propellants for the attitude control system and for trajectory modifications while the Lightcraft is traveling in space. Filtered, deionized water is required both for crew use and as an expendable coolant for the MHD propulsion system during high-power boost into space.

8.1.4 Food Rehydration
The microprocessor-controlled rehydration system maintains an inventory of the food options available. The crew member needs only to select the food. The computer feeds the dehydrated food item into the rehydrator, restores its moisture content, and delivers it at an appropriate temperature for consumption.

8.1.5 Gravity
Artificial gravity is created while in space by spinning the LTI-20 at 3 revolutions per minute (rpm). This rate of rotation produces an acceleration of about 0.2 g in the outermost corridor. While operating under artificial gravity, *real-time spin balancing* is essential to keep the Lightcraft's rotation and g-level steady. The crew, which makes up fully ¼ of the mass of the Lightcraft, can seriously affect the location of the center of mass, and hence the axis of rotation, by simply moving round in the interior. The LTI-20 is designed to detect such imbalance and correct it automatically by transferring water from one storage tank to another.

8.1.6 Waste Disposal
The weight restrictions on the Lightcraft mandate efficient removal of all waste. Liquid wastes are purified and recycled for use as drinking water or engine coolant. Solid wastes are reprocessed into recyclable and decomposable materials so that they can be reused or buried as necessary.

8.1.7 Hydrogen Electrolytic Production Unit
This auxiliary system produces gaseous

Hydrogen and Oxygen from on-board ultra-pure water using an electrolytic process. These gases are used to recharge the high-pressure bi-propellant tanks of the retro rockets in each escape pod and the maglev landers.

8.1.8 Water Purification System

The liquid-cooled rectenna arrays can only employ de-ionized water of the highest purity. Rectenna heat exchangers employ intricate microchannels that operate under elevated pressures, and fouling of these passageways cannot be tolerated. Hence water coolant scavenged from ground-water sources (freshwater lakes, rivers, streams, and ponds) or condensed from clouds, must first be deionized, then processed through reverse osmosis filters, and finally irradiated by intense UV light to eliminate microorganisms. This ultra-pure water (maximum load of ~2400 kg) is then stored in an annular tank attached to the high-pressure toroid near the rim.

8.2 Additional Utilities Systems

Resupply of consumables to crew quarters, survival pods, and maglev landers requires umbilical connectors.

The umbilical cords that connect the LTI-20 to its escape pods and landers are automatically detached when the pods eject from the Lightcraft or when the landers are deployed by the magnetic levitation system below the Lightcraft. Once the crew enters the escape pods, the FMS computer activates the air-lock and completely isolates each crew member from the outside environment. The interior of each escape pod is customized to the exact shape and size of its occupant, and each pod's internal environment is independently controlled. Section 14.9 provides more detail on the pod environment.

Each crew quarters is pressurized independently by the recirculating HeliOx system. Atmospheric sensors monitor the pressure inside the Lightcraft, which is regulated to a gage pressure of 0.5 atm (0.05 MPa). The environmental control system of the Lightcraft, discussed in Section 13, explains the reason for this choice of gage pressure. Section 14.3 describes the interior of the crew quarters in more detail.

8.3 Exterior Connect Hardpoints

At times the LTI-20 must be secured to the ground or to other objects. One dozen tie-down receptacles, specifically designed for this purpose, have been placed around the entire circumference of the vehicle (see Figure 8.3.1).

Figure 8.3.1: Reinforced hull areas for tie-down attachments. *(Courtesy of RPI.)*

In its partially buoyant *unloaded* state the LTI-20, like other airships, must be secured to the ground to keep it from drifting away in light winds and gusts. There is a reinforced ring-section on the rectenna side of the vehicle where the central (on-axis) rectenna array meets the outer (circumferential) parabolic antenna array. This ring-section is one area of the Lightcraft hull that can be stressed in either tension or compression, an ideal location for tie-down hardpoints. The skin around this area of the Lightcraft has also been thickened slightly to take the increased loads.

On the rectenna face of the Lightcraft, 12 additional tie-down fittings are distributed about the rim of the vehicle where the PV array meets the main toroidal tube. The 12 hatches of the escape pods have also been equipped with tow fittings. Three tie-down receptacles have also been placed around the inner rim of the Lightcraft's "donut hole."

The skin around the inner rim of the donut hole has been thickened considerably to allow a maglev lander to mate to a hovering Lightcraft; the lander's bell bottom exactly covers the external opening of the donut hole. In this mated position the superconducting magnets located on the maglev lander and the Lightcraft could attain a magnetic lock, allowing one

Lightcraft to lift another by using its maglev lander as a magnetic grapple to lift the downed Lightcraft (Figure 8.3.2).

8.4 In-Space Reaction Control System

The in-space reaction control system (RCS) provides the LTI-20 Lightcraft with 6 degrees of freedom (i.e., translation in any direction, plus precise angular rate adjustments in roll, yaw, and spin). All electric RCS rockets have a specific impulse exceeding 1000 seconds, and can be fired only when the space plasma shield is de-activated (see Section 15.5).

For emergency de-orbit from LEO, a maglev lander's gaseous H_2/O_2 retro rockets may be fired to provide the necessary ~100 m/s velocity change.

8.5 Airspike System Auxiliary Functions

The airspike system components include the central rectenna and the reflective outer surface of the docked maglev lander. During PDE and ion propulsion, when the airspike is not in use, the airspike system may be used for auxiliary purposes. Both central and off-axis annular rectennas can be used as phased array microwave radar systems, powered by the SMES "battery." This radar capability has both military and navigational (collision avoidance) uses. Known landmarks of navigational value and other vehicles will be identified from their radar and electromagnetic signatures by the FMS computer's database and Identification Friend or Foe (IFF) system.

For short-range radar, the microwave radar system is quite efficient and effective. However, because of diffraction effects and microwave attenuation by the atmosphere, laser radar (lidar) is recommended for long-range sensing. Although both systems are limited to sensing objects within line of sight of the Lightcraft, they are greatly enhanced by real-time databases provided by GPS and other orbital military assets.

8.6 Magnetic Tractor / Repulsor Fields

The powerful magnetic fields generated by the hull's superconducting magnets can be used to

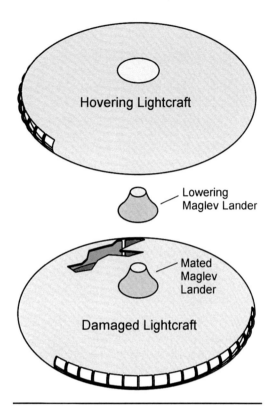

Figure 8.3.2: Rescue of downed / damaged lightcraft using Maglev lander as a magnetic grapple. *(Courtesy of RPI.)*

anchor the Lightcraft to ferromagnetic materials, such as the steel frame of a building or spaceport landing platform. Ferromagnetic objects (automobiles, etc.) can be picked up and transported (see Section 16.15). Using magnetic induction (i.e., ramping the current in Lightcraft magnets), forces can often be exerted upon non-ferrous metals by inducing electric currents in them. In an emergency, a Lightcraft may be used to lift another Lightcraft (Section 16.14), or a maglev-belt equipped crewmember. Of course, hovering Lightcraft are designed to lower their maglev landers to the ground on powerful magnetic fields. Reversing the polarity of the Lightcraft's magnetic field can also repel the vehicle away from a grounded maglev lander, providing a vertical launch-assist (e.g., for a slightly overweight Lightcraft).

9.0 COMMUNICATIONS

The LTI-20 is equipped with several types of communications systems, including some of the most sophisticated equipment available. The communications hardware is an integral part of the Lightcraft, the maglev landers, the escape pods, SAU (space activity unitard) crew uniforms, and in some cases implanted under the crew member's skin. The computer and communications technology used is not available to the general public.

9.1 Intra-Ship Communications

Communications between crew members aboard the Lightcraft can be achieved in two different ways.

The first way is through the use of the virtual reality system incorporated into the ultra-g personal protection suit, as described in Section 14.7. With this system running and linked to the central FMS computer, any crew member can easily access and relay information to other crew members or to the computer system itself. This is the only method of communication possible from inside the escape pod when the use of partial liquid ventilation prevents vocal communication. When using high-pressure HeliOx ventilation (good to 30 g), voice synthesizers automatically reduce the pitch into the normal range. Crew-to-crew communications by this method can only be effected when both parties are in the virtual-reality environment.

The second way is through the use of the FMS computer panels located throughout the ship. These panels can be activated either manually or by voice. Anyone on the Lightcraft can be contacted through the panels by either asking or manually entering the name, rank, or duty station of the intended recipient.

9.2 Ship-to-Ground Communications

Communication between crew aboard the LTI-20 and crew on the ground may be essential to the achievement of mission objectives. Ship-to-ground communication can be initiated from the computer panels (Figure 9.2.1) located

Figure 9.2.1: FMS computer access panel.
(Courtesy of RPI.)

throughout the ship by means of secure narrow-beam laser or microwave links.

Each crew member is issued an access code to contact anyone on the surface. This code can be entered vocally or manually. When available (and desired), a two-way video link can be established and displayed on the computer panel. All communications channels are secure.

During highly sensitive missions, ship-to-ground communications can be established only through the bridge. The computer panels throughout the Lightcraft are normally deactivated until the mission is complete.

9.3 Ship-to-Ship Communications

Many of the missions that the LTI-20 will undertake require ship-to-ship communication. Only the three highest-ranking officers assigned to the LTI-20 are allowed to contact other ships. Another ship may be contacted either from the FMS computer consoles on the bridge or from the escape pods of the three ranking officers. Contact can be initiated either by voice command or by manual input.

A standard procedure is carried out by all military personnel in command of an LTI-20 when contacting other ships:

1. To contact another ship, the captain inputs one of 3 codes. Each captain is issued a new emergency code, sensitive information code,

and normal communications code daily. This procedure minimizes the likelihood of hoaxes or spoofing, and alerts the officer receiving the communication to the type of communication which follows.
2. Captains identify themselves and the LTI-20 they command. Visual-audio communications (VideoCom) are required except during emergencies.
3. Information is exchanged on a highly secure encrypted channel. The FMS computer automatically secures the channel and encrypts the message to prevent interception by enemy forces.

Adherence to this procedure is vital in emergency operations and in highly classified mission operations. Any other personnel (such as the doctor or engineer) needing to contact other ships must receive clearance from the mission commander.

9.4 Personal Communicators

One of the newest and most innovative technologies created for the crew of the LTI-20 is the 2-way personal communicator. A communicator is issued to each LTI-20 crew member at the time of their assignment and is retained until discharge or death.

The communicator is a 2-mm diameter disk encased in a flexible plastic outer envelope that does not stimulate a human immune response. Depending upon the mission, the communicator may be surgically implanted under the skin near the vocal cords (Figure 9.4.1). Implantation of the transmitter allows vocal communication to be accomplished except during missions that require full liquid ventilation.

To activate the transmitter, the user says "communicate" and the name of the crew member or ship to be contacted. Any tone or level of voice, from whispering to shouting, is recognized and accepted by the transmitter.

The second essential part of the communicator is surgically implanted in the mastoid bone near one ear (see Figure 9.4.2). This device, a 1-mm disk encased in the same anti-rejection plastic as the transmitter, receives all communications and presents them as audio-frequency sound

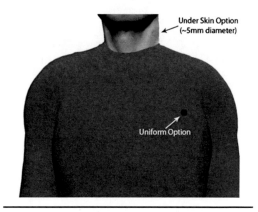

Figure 9.4.1: Transmitter options for personal communicator. *(Courtesy of RPI.)*

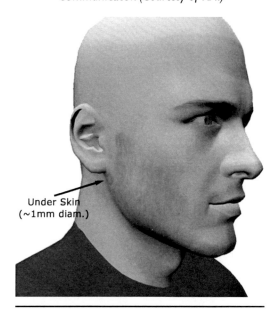

Figure 9.4.2: Receiver implant option for personal communicator. (Courtesy of RPI.)

vibrations in the ear. This state-of-the-art technology automatically adjusts its output level so that strong or weak received signals will not strain or damage the hearing.

Anyone receiving a message will first be alerted by a double tone. Once the tone is received, the transmission is activated by saying the word "receive." The communications system responds by activating the communications channel. To terminate the communication, either party says "end communication," which automatically terminates both send and receive channels.

Each receiver has a unique signature that is keyed to selected features in the DNA of the carrier. This signature feature expedites locating and identifying an individual, permitting the computers aboard the LTI-20 and at its home base to locate any individual, whether on the ground or aboard ship, at any time. The receivers thus function as electronic "dog tags."

The implanted transmitter also contains circuitry and sensors for electrocardiography, electroencephalography, respiration, and other medical parameters.

Each communicator has a code to activate distress signals. The distress mode has a range of 25,000 km through the use of zonal receptors, satellite-based SARSAT-type receivers, and long-range orbital relay facilities. This code is changed daily to prevent misuse in the event that a receiver is captured by an enemy. This distress signal can be activated only by voice control.

This form of communication is best suited for intra-ship, ship-to-ground, and intra-land links. This communicator cannot be used for ship-to-ship communication (see Section 9.3). The transmitter has a life expectancy of 3 years, whereas the receiver has a life expectancy of 5 years.

9.5 Long-Range Transceiver

The LTI-20 must be able to communicate over large distances while engaged in deep-space, interplanetary missions. For this kind of communication the Lightcraft uses the central parabolic rectenna in concert with the upper maglev lander at high frequencies (220 GHz). For cislunar space, low-frequency (35-140 GHz) microwave communications are entirely adequate without the lander.

9.5.1 High-Frequency Transceiver Mode

During transmission in high-frequency transceiver mode (220 GHz) the lander, using magnetic station-keeping control (MSC) methods, is held at a precise position relative to the central parabolic rectenna on the Lightcraft (Figure 9.5.1). The parabolic surface reflects and focuses the incoming signal into one face of the maglev lander, where the signal is processed and decoded. This geometry enables the Lightcraft to transmit and receive information simultaneously. These dual functions can coexist because the central parabolic rectenna is being used merely as a passive reflector. In the low-frequency transceiver mode (35 - 140 GHz), with the upper maglev lander not deployed, the Lightcraft's communications are limited to shorter cislunar ranges.

9.5.2 Low-Frequency Transceiver Mode

The Lightcraft transmits signals via the central parabolic rectenna at a frequency of 35 GHz (or 94 or 140GHz, depending on rectenna technology). The phased-array construction of the rectenna allows for precise directional control of the transmitted signal in any direction within 45° of the axis of the Lightcraft (Figure 9.5.2). To reach targets more than 45° from the zenith, it is necessary for the Lightcraft to pitch or roll to realign the target closer to the axis.

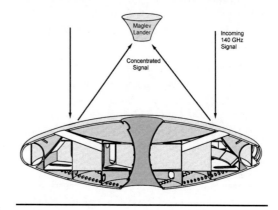

Figure 9.5.1: Long range microwave transceiver geometry employing maglev pod as secondary optic at 140 GHz. *(Courtesy of RPI.)*

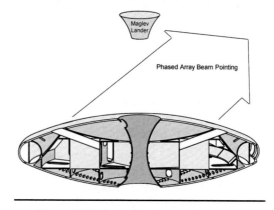

Figure 9.5.2: Low frequency microwave transceiver mode at 35 GHz. *(Courtesy of RPI.)*

10.0 MAGLEV SHIP BOARDING SYSTEM

Magnetic levitation is the use of magnetic attraction or repulsion to overcome gravity. The landers carried by the LTI-20 are deployed and recovered by means of magnetic levitation (maglev) forces.

10.1 Maglev System Introduction

The maglev system's primary role is provide easy and timely boarding in locations where landing a Lightcraft would be impractical or dangerous. Most often the maglev system comes into use when a Lightcraft must make several local stops to take on passengers prior to a transatmospheric boost. The system also provides an added factor of safety in that it allows for a greater distance between people and stray ferromagnetic objects on the ground and the most intense magnetic fields generated very close to the Lightcraft.

10.2 Maglev System Operation

The Lightcraft magnetic levitation and docking system consists of the maglev belt unit and the maglev lander system. Each is an independent means of boarding or descending from a hovering Lightcraft.

10.2.1 Maglev Belt System

Three main subsystems are responsible for the successful operation of the maglev belt system:

1) tracking sensors and control systems;
2) maglev belt;
3) superconducting magnets.

These are described below.

10.2.1.1 Tracking Sensors and Control System

These sensors on the Lightcraft's lower hull track the location and progress of the maglev-lifted individual and relay this information to the on-board computer. In addition, the maglev belt has an independent array of sensors that track its proximity to the superconducting magnets on the Lightcraft.

10.2.1.2 Maglev Belt

The maglev belt, which exploits room-temperature superconducting magnet technology, is usually integrated into each crew member's space suit. This standard-issue USSC equipment provides the magnetic field necessary to achieve levitation of its wearer. The field is generated by battery-powered coils arrayed circumferentially around the belt. The power source of the belt is a flexible ultra-capacitor and/or lithium battery that wraps around the front of the wearer's midsection. Under normal lift conditions this battery pack will provide enough energy for several activations. The belt is equipped with a load-bearing harness which wraps around the wearer's legs much like a rock-climber's rappelling harness (Figure 10.2.1). Lift capacity of the maglev belt scales with mass of the wearer, ranging from 40-60 kg (LCJs – see Section 3) up to 100 kg for regulation USSC crewmembers.

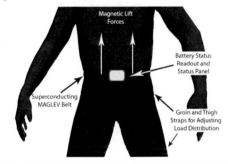

Figure 10.2.1: Details of Maglev belt system for personnel levitation and aerial transport. *(Courtesy of RPI.)*

10.2.1.3 Superconducting Magnets

The superconducting magnets in the Lightcraft's engine generate a powerful magnetic field that exerts a controlled lifting force on the maglev belt to pull its wearer up to the Lightcraft (Figure 10.2.2). At the greatest hover heights (~20 m) the outermost rim magnets (19.2 m diam., 4.4 MA each) perform the majority of the lifting. As the wearer of the belt approaches the bottom of the Lightcraft, the lifting is switched gradually to the inner magnet rings (e.g., 10 m diam., 1 MA each).

The most common scenario for passenger maglev boarding is:

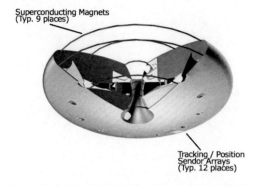

Figure 10.2.2: Nine superconducting magnets and tracking / position sensor array for Maglev docking system. *(Courtesy of RPI.)*

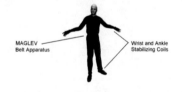

Figure 10.2.3: Personnel levitation and aerial transport via Maglev belt to hovering lightcraft. *(Courtesy of RPI.)*

1. The prospective passenger stands in a large open area, preferably more than 10 meters across and necessarily devoid of any ferromagnetic material. The Lightcraft, already informed of the passenger's location, hovers directly above the passenger.
2. The passenger activates the maglev belt device. Sensors on the dorsal side of the Lightcraft track the exact position of the belt relative to the Lightcraft. Sensor data are sent to the FMS computer, which then "powers up" the superconducting magnets.
3. With the FMS computer relaying corrections and field intensity adjustments to both the coils and the belt, the passenger is gently pulled from the ground and drawn upward at a uniform velocity by the main engine's magnets.

This process continues until the passenger is within 4 meters of the undersurface of the Lightcraft (see Figure 10.2.3), whereupon levitation is switched to the inner 10 m diameter magnet. Thereafter, the velocity of the passenger is reduced to permit safe entry through the dorsal hatch and air-lock. At this point the FMS computer relays an instruction to the maglev belt to deactivate itself.

Figure 10.2.4 shows the magnetic field generated by a 10-m diameter Lightcraft magnet carrying 1 MA given as a function of range to the belt wearer. Note that the perpendicular field component peaks at 16 Webers/m^2 for a range of roughly 3 meters. (At 15 m range, the perpendicular component of the magnetic field falls to 0.57 gauss, roughly equal to Earth's geomagnetic field.) The levitation force vs. belt current and range (H) is displayed in Figure 10.2.5 for a 30-cm maglev belt carrying 1 MA of current. The handoff range for a 1 MA belt current lifting a 100 kg crewmember is 4-5 meters; 75-kg at 7 m; 40 kg LCJ at nearly 10 m.

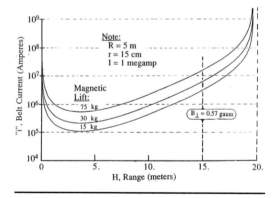

Figure 10.2.5: Levitation force vs. range upon 15 cm diameter Maglev belt carrying 1 MA. (Assumes standard applied magnetic field from 10 m diameter lightcraft coil conducting 1 MA.)

MAGLEV SHIP BOARDING SYSTEM

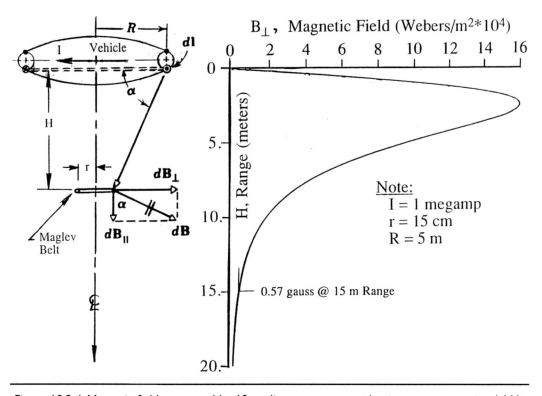

Figure 10.2.4: Magnetic field generated by 10-m diameter superconducting magnet carrying 1 MA.

To transfer a passenger to the ground, the maglev belt is activated, the air-lock door is opened, and the wearer of the belt steps out. The FMS computer locates the passenger and runs the aerial de-boarding program which executes the opposite sequence described earlier. (Note: LTI-20 superconducting magnets carry up to 4.4 MA each, which greatly extends their levitation range.)

10.2.2 Maglev Lander

The second principal maglev subsystem consists of two small plug-shaped landers that fit snugly into the circular channel ("donut hole") though the center of the Lightcraft (Figure 10.2.6). These landers are equipped with two circular superconducting magnets that circle the top and bottom ends of the lander (Figure 10.2.7). These magnets are used to make the lander hover, as well as to lock it into its berth on the axis of the ship. Maglev landers can be used to shuttle passengers or cargo not equipped with maglev belts between the ground and an airborne Lightcraft. The space within the lander is also

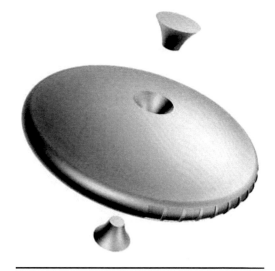

Figure 10.2.6: Lightcraft with twin Maglev landers deployed. *(Courtesy of RPI.)*

used as an air-lock for passengers entering through the dorsal hatch.

Cargo such as water coolant can be loaded into the landers at ground level, then magnetically

Chapter 10

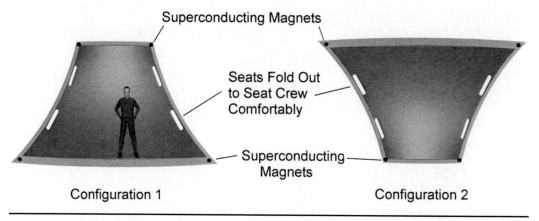

Figure 10.2.7: Reconfigurable Maglev landers can transport personnel in both orientations. *(Courtesy of RPI.)*

lifted up for transfer into the Lightcraft storage tanks (Figure 10.2.8). The levitation and lifting operation of these landers is exactly the same as that of the maglev belt system, except that the landers are magnetically locked in place by the inner rings of superconducting magnets in the Lightcraft hull.

Figure 10.2.8: Scavenging and transport of water coolant using Maglev lander. *(Courtesy of RPI.)*

10.3 Limitations of Maglev System

There are certain limitations on the maglev system that must be recognized to assure its safe and effective use. The primary limitations of the LTI-20 maglev system are as follows:

1. The maximum height from which a 100-kg person equipped with a maglev belt can be plucked off the ground is 20 meters. Above this height the maglev system would require raising the magnetic field of the belt to levels that are dangerous to humans.
2. The maximum service weight of cargo that can be lifted by a maglev lander is 2400 kg at a maximum range of 5 m. This is the limit on the mass of expendable water coolant that can be taken aboard in a single maglev lander sortie. Landers have an internal tank volume of 2.4 m^3 which can be loaded in 9.5 seconds.
3. The maximum height from which a fully loaded lander (e.g., 6 LCJs or payload of ~300 kg) can be lifted by the Lightcraft magnets and successfully locked into its berth is 20 meters.
4. Under normal conditions, both the Lightcraft and the object being levitated should be momentarily stationary relative to each other. In emergencies, this condition can be modified to expedite extraction from danger. This exception is discussed further in Section 10.5.

10.4 Maglev Landers in Mated Mode

As describe previously, the landers are locked magnetically into their berths in the center of the ship by the fields generated by the superconducting magnets in both the lander and the Lightcraft. Any cargo or passengers to be taken aboard the Lightcraft must first be pressurized to 1.5 atmospheres with the HeliOx mixture. Actuators in the Lightcraft deploy and engage an O-ring seal to the lander hatch as soon as a magnetic lock has been achieved, thus forming a stable, air-tight boarding passage between the Lightcraft and the lander. The hatches between the two craft can then be opened safely. While docked in this manner, the landers may serve as air-locks for use either within the atmosphere or in space.

In Earth's atmosphere, during lateral flight operations, the Lightcraft is spun around its vertical axis to achieve gyroscopic stabilization (i.e., Frisbee mode). The ship may also be spun while in orbit to provide $1/6$ G artificial gravity for the crew. In the former case, the rotation may be uncomfortable or disorienting to the crew, who may choose to occupy the maglev landers. The landers can be despun by the magnetic bearing system (the "motor drive" function) in the docking bay, alleviating the discomfort of the crew. For extended lateral flight maneuvers, bridge functions are normally transferred to one of the landers.

10.5 Emergency Maglev Pick-Up Operations

During military missions the Lightcraft may occasionally be needed to extract personnel on the ground from hazardous circumstances, such as medical evacuation (MEDEVAC) for patients in critical condition, using a maglev lander. In cases in which the usual pickup time of several minutes might put the patient or crewmember at serious risk a rapid extraction procedure can be executed. In such emergencies the computer tracks the grounded lander to a range of 300 meters, then takes over direct control of Lightcraft piloting. At a distance of 25 m from the lander the Lightcraft slows to 15 m/sec and the lander is snatched from the ground at 2 - 3 g acceleration. As soon as the lander is safely docked within the ship, the Lightcraft accelerates at the maximum (medically) possible rate, upward and away from the pickup point (see Figure 10.5.1).

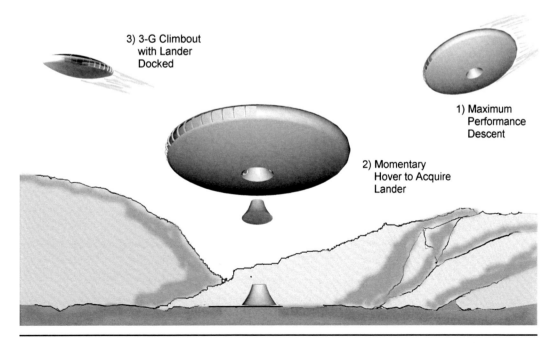

Figure 10.5.1: Emergency crew extraction via Maglev landers. *(Courtesy of RPI.)*

Chapter 10

This sequence is planned to minimize the time spent in proximity to the lander pickup point, and hence minimize exposure of the Lightcraft to hostile action. It is important to understand that this sequence can be used only when the person requiring rescue is in severe and immediate danger, since this extraction maneuver may expose its subject to hazardously high acceleration and magnetic field levels. Obviously the condition of the person requiring evacuation must be known well enough to assure that the extraction process itself does not inflict additional injuries.

11.0 REMOTE SENSING SYSTEMS (CLASSIFIED — Section Removed)

11.1 Sensor Systems
11.2 Long-Range Sensing
11.3 Navigational Systems
11.4 Lateral Sensor Arrays
11.5 God's-Eye View: Imaging Holograph

12.0 TACTICAL SYSTEMS

The LTI-20's greatest tactical advantages are speed, covertness, and range. The all-weather, multiple propulsion system Lightcraft is capable of accelerations in excess of 200 g in the PDE mode and 30+ g in the MHD mode. It is designed to carry a crew of 6 - 12 on both atmospheric and space operations, flying to any destination on Earth in less than 45 minutes. In tactical use, the LTI-20 takes advantage of its high speed and low radar cross-section by traveling rapidly to its target, entering its landing zone undetected, performing covert operations, and returning to its base without leaving evidence of its visit. Tactical weaponry on the Lightcraft includes high-power microwave beams, pulsed lasers, weapons use of the propulsion systems, and personal stun weapons.

12.1 Tactical Doctrine

The typical LTI-20 mission profile is designed to avoid armed combat and weapons use. The Lightcraft has the ability to outrun conventional kinetic energy weapons and evade radar search very effectively; nonetheless, the LTI-20 is able to defend itself at need. Typical Lightcraft missions include:

1) USSC commando or SEAL insertion and recovery;
2) search and rescue operations;
3) reconnaissance;
4) interdiction of terrorists;
5) automobile abduction;
6) satellite attack or defense;
7) astronaut rescue;
8) recovery of covert military personnel.

All tactical activities are overseen by the Communications / Tactical Officer (CTO) as directed by the Mission Leading Officer (MLO), or, under special circumstances, the Second in Command officer (SCO). The MLO briefs the entire crew prior to every mission, and delivers the mission debrief to Space Command headquarters.

12.2 High-Power Laser Relay Mirror

On command of the MLO, a 10 MW remote laser power beam (selectable 620- to 860-nm wavelength), sent directly from the orbital power station or through a LEO relay satellite (Figure 12.2.1), may be reflected at a target by using the adaptive-mirrored lower surface of either maglev lander as a "fighting mirror." In providing this weapons function, the lander can either be docked or free-flying (Figure 12.2.2). This system can deliver lethal beam intensities to targets as far as 1000 km from the Lightcraft, assuming suitable atmospheric and line-of-sight conditions. Such a relay of laser energy prevents utilization of the airspike, but allows simultaneous reception of beamed microwave power for conversion to onboard electric power.

12.3 Pulsed Laser Weapons System

The LTI-20 is equipped with three pulsed lasers that can be used as a last-resort defensive weapon. The triply redundant pulsed laser weapons system (PLWS) is fully automated and aimed by a closed-loop control system which tracks incoming targets using a completely self-contained laser radar (LIDAR) system. Fire control solutions are updated every few milliseconds as the PLWS monitors the position and velocity vectors of the target and projects its future motion. The PLWS fires one of its three lasers at the projected target location at maximum cycling rate until the threat is destroyed or disabled. Each laser is capable of 300 kW average power, with a peak power of 100- to 150 MW. If necessary, the PLWS can be turned on hostile aircraft, missiles, satellites, and even personnel (as a last resort). Although the PLWS run time is limited by the 900 MJ SMES "battery," it is still capable of neutralizing even armored targets. Nonetheless, its primary role is defensive.

12.4 Microwave "Active Denial" System

LTI-20 Lightcraft are equipped with an 95 GHz active denial system that can project 50 to 100 kW beams to ranges exceeding 1 km. Figure

TACTICAL SYSTEMS

Figure 12.2.1: "Monocle" laser fighting mirror deployed in low earth orbit. *(Courtesy of SDIO.)*

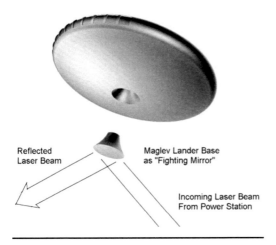

Figure 12.2.2: Use of Maglev lander base as laser relay mirror.

12.4.1 reveals the essential features of this nonlethal antipersonnel weapon, which is intimately integrated with the Lightcraft's rectenna arrays (operating in transmit mode). A 1 - 2 second burst of millimeter-wave flux penetrates the skin to $1/64$ of an inch (0.4 mm) causing a burning sensation as the skin is rapidly heated to 130° F. The heating effect, which is

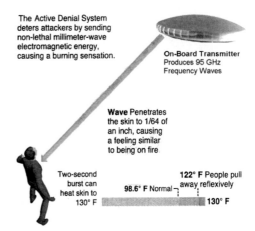

Figure 12.4.1: Microwave "active denial" system (95 GHz beamed from lightcraft).

caused by excitation of water molecules in the skin, is intense enough to make combatants feel that their clothes are about to ignite. Combatants hit by the beam immediately jump out of its path. Normal skin temperature is 98.6° F, and when it reaches 122° F, they pull away reflexively from the terrifying sensation of a sudden blast of heat through the body.

12.5 Space Power Station Engagement

If necessary, a target can be engaged directly by the space power station by bombarding the target with laser power levels up to 10 MW, and/or microwave powers up to 20 GW. Direct engagement by the power station or LEO relay satellite, however, momentarily deprives the Lightcraft of its source of beamed power and forces it to operate temporarily on the SMES unit. This use of "battery" power in turn limits both the propulsion mode of the Lightcraft and the availability of power for other onboard functions, such as PLWS firing. Since direct power station engagement of targets may compromise the covertness of the mission, it is likely that it would be not attempted except under extreme circumstances.

12.6 Personal Stun Weapons

Lightcraft crew members participating in extravehicular activities are equipped with hand-held and/or belt-mounted stun weapons that derive power from their maglev belt SMES unit, and/or lithium battery pack. When directed at a hostile person, the personal stun weapon (PSW) discharges a small portion of the energy stored in the maglev belt "battery," sending a laser-initiated electrical current through the target's body at one of three power settings selectable on the weapon itself:

1. low current, causing extreme discomfort but no lasting effects,
2. medium current, causing temporary impairment of the nervous system and loss of consciousness for several minutes, and
3. high current levels, causing potentially lethal paralysis of the cardiopulmonary system.

The PSW uses two laser precursor "micro-pulses" to trigger electrical breakdown of the atmosphere along two separated, high-conductivity plasma filiaments that extend ("line of sight") to the target plane. Acting much like fine copper wires, these channels are opened for only a few nanoseconds to complete the electrical circuit, just long enough to deliver a powerful, high-voltage discharge at the designated range into the adversary's body.

12.7 Lightcraft Auto-Destruct System

As a matter of tactical last resort, the LTI-20 is equipped with a fully automatic self-destruct system (ADS). The 900+ MJ of energy stored in the SMES unit is sufficient to shred the entire Lightcraft, thus rendering it useless to the enemy. The destruct sequence must be initiated by the MLO and confirmed by the SCO. After selection of an ADS countdown time sufficient for crew pod ejection, an audible alarm and countdown is initiated. The sequence can be terminated at any time by the MLO or SCO.

13.0 ENVIRONMENTAL SYSTEMS

The LTI-20's environmental systems include those responsible for managing and regulating the HeliOx inflatant and coolant, generating artificial gravity in space via rotation, atmospheric humidity and purity regulation, emergency environmental control systems, and waste management control systems.

13.1 Environmental Control System

The environmental control system (ECS) is the most critical of the LTI-20's major systems. This system is designed with triple redundancy to maximize crew safety and protect them even in the unlikely event of multiple life support and propulsion system failures. A HeliOx plenum management system (PMS), integrated into the LTI-20's pneumatic (inflatable) structure, stores and circulates HeliOx throughout the interior of the Lightcraft. This gas is breathed by the crew and also used for active cooling of rectennas (low power only), photovoltaic arrays, and the external hull. In addition to attendant ductwork, heat exchanger and compressor components, the HeliOx PMS is largely comprised of two interrelated plenum chambers: a high-pressure perimeter toroid, and the low-pressure lenticular hull envelope. Of course, the perimeter toroidal pressure vessel effectively serves as the structural "backbone" of the craft.

At sea level, the LTI-20 lenticular envelope is inflated with HeliOx at 1.5 atm pressure, whereas the perimeter toroid is maintained at 7.5 atm. Whether climbing or descending, the lenticular envelope's inflation pressure is constantly adjusted to maintain a gauge pressure of 0.5 atm. relative to the local external static pressure. During a transatmospheric flight to space, HeliOx is scavenged from the envelope, cooled, and pumped into the toroid, finally raising its pressure to 10 atm as the ship reaches the vacuum space environment; in the process the lenticular envelope pressure falls to 0.5 atm.

A schematic of the combined environmental control and cooling system (ECCS), given in Figure 13.1.1, shows how the PMS and ECS subsystems are interconnected to provide the designated functions with closed-cycle HeliOx (nominally regulated at 21° C). Note that HeliOx is circulated through the lenticular double-hull which normally serves as a heat exchanger to reject waste heat into the external environment by means of radiation and/or convection. Recirculated HeliOx may also be used to cool specific sections of the hull (e.g., leading edges in supersonic flight) and reject excess heat out through other hull sections (i.e., aft-facing sections).

For a grounded or stationary Lightcraft ECS functions dominate, so excess heat is eliminated by circulating heated HeliOx through the entire double-hull, or just the sun-shaded lower hull – again rejecting to the ambient atmosphere.

In like manner, closed-cycle HeliOx cooling of the dorsal laser photovoltaic array (e.g., in lateral Frisbee mode flight) is possible because the upper and lower lenticular hulls can be isolated from each other by valves in the PMS ductwork. Hence, waste heat from the upper hull photovoltaics is collected by the HeliOx and routed through the lower double hull to be carried off by the cooler external slipstream.

Similarly, at low microwave power beam levels (10 MW maximum) recirculated HeliOx can provide adequate rectenna cooling for extended periods by engaging the hull heat exchanger system (Figure 13.1.1). However, for very energetic hyperjump maneuvers that power up PDE and MHD engines, rectenna waste heat levels far exceed the limited ECCS capacity and can only be extracted by open-cycle water coolant, ejected as steam into the atmosphere (see Figure 13.1.2).

Before the cooled HeliOx can be used for human respiration it must pass though the life support control system, where it is cleansed of impurities such as dust, carbon dioxide, and potentially toxic trace gases. A contamination sensor is attached to the life support system to monitor the quality

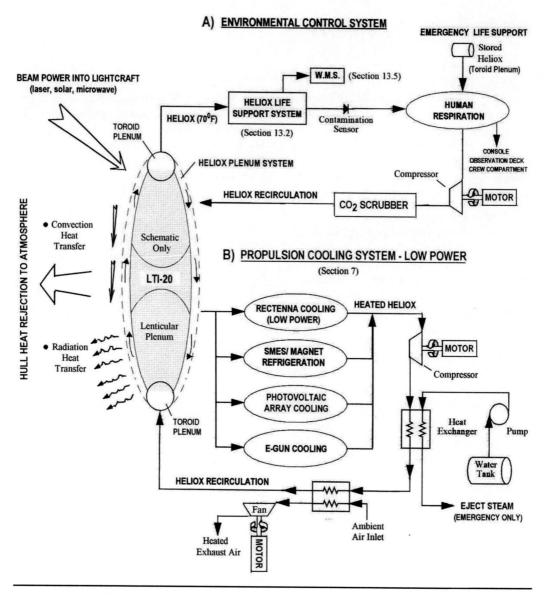

Figure 13.1.1: Closed-cycle environmental control and low-power propulsion cooling system.

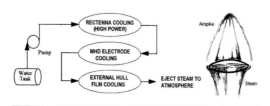

Figure 13.1.2: Open-cycle MHD slipstream accelerator cooling system of LTI-20.

and flow rate of HeliOx to the observation deck, bridge, and crew compartments. When the contamination sensor reports a failure in the life support system, the FMS computer automatically shuts down the defective component and alerts the crew. There are three parallel HeliOx life support systems in the environmental control system, each of which operates at 33% of its rated capacity under normal service. This practice affords 3:1 redundancy to protect the crew from even the highly improbable event of multiple life support and propulsion system failures. The HeliOx plenum management system and environmental control system are operated by the FMS computer.

There is a pair of sensors for approximately every 1 m^3 of internal volume of the Lightcraft. The life support system receives updated data from each sensor every 0.1 second. The FMS computer analyzes these data and accordingly controls the temperature and humidity of the HeliOx stream throughout the observation, bridge, and crew compartment. These sensors are also used for fire detection inside the Lightcraft (see Sec. 18.7).

An emergency backup life support system provides high-pressure reserve tanks of HeliOx located at most corridor junctions so as to maintain a breathable atmosphere long enough for the crew to evacuate to their escape pods.

13.2 HeliOx Life Support System

The HeliOx life support system is responsible for maintaining the oxygen, water vapor, and carbon dioxide concentrations in the cabin at safe levels. It also filters out particulates and removes chemical contaminants while condensing perfluorocarbon vapors for reuse.

Water is vital to the survival and health of the Lightcraft crew. Some of the water is required in the form of water vapor to keep the sensitive mucous membranes in the eyes, nose, and mouth moist. Water vapor levels in the cabin are therefore closely monitored and controlled.

The HeliOx mixture is normally breathed in the main cabin and maglev landers at a pressure of 1.5 atm at sea level. Upon landing, the lenticular envelope pressure may be reduced to 1.1 atm, but any lower level would compromise the structural integrity of the Lightcraft. The 1.5 atm helium-oxygen mixture in the Lightcraft has a partial pressure of 160 mm Hg of oxygen, a mixture commonly used by deep-sea divers. Helium is used in place of the nitrogen in air to provide fire retardance while avoiding the problem of nitrogen narcosis; nitrogen is sufficiently soluble in the bloodstream so that exsolution of nitrogen bubbles can occur during rapid decompression, causing a painful and dangerous condition called "the bends."

The HeliOx life support system is regulated by FMS computers linked throughout the Lightcraft. These computers run the life support system, which recirculates cabin HeliOx and filters out its impurities, maintaining the composition and temperature of the gas within safe limits. HeliOx circulates through a network of ducts and cleaning units located throughout the plenum system. Excess HeliOx is pumped into the toroidal tank at the rim of the Lightcraft or into other auxiliary tanks. Because HeliOx is essential for buoyancy and pressurization, it is never vented to the atmosphere except under extreme emergency conditions.

13.3 Artificial Gravity Generation

Because humans have evolved under Earth's gravitational acceleration, the body's systems require gravity to maintain the skeletal strength and circulation of the blood necessary for re-adaptation to Earth's surface conditions. In space the LTI-20 generates artificial gravity by spinning around its symmetry axis at 3 to 4 RPM. This rotation produces an acceleration of about 1/6 g in the outer corridors of the Lightcraft, similar to the surface gravity of the Moon.

Figure 13.3.1 shows the orientation of LCJ personnel (statures of 132 to 162 cm) inside the LTI-20's rotating crew compartment. The rotation allows the crew to walk on the outer curved wall of the Lightcraft. Exercising in this partial-gravity environment helps maintain the crew's physical fitness while they are in space. Under these circumstances the normal walking speed is limited to 0.44 m/s, or about 1.6 km/hour.

A study of the architecture of artificial gravity environments by Hall (see Figure 13.3.2) reveals the relationships between the human "comfort zone" and the gyration radius and rotational speed. Based on a 10-m radius for the LTI-20, with the crew members walking on a surface 8 m from the axis, the nominal rotation rate is 4 RPM, at which the acceleration of the crewmember is 0.15 g at the bridge on the observation deck. For USSC commandos (statures of 163 to 196 cm), this is clearly outside their comfort zone, deep in the vertigo region; for such personnel there IS no comfort

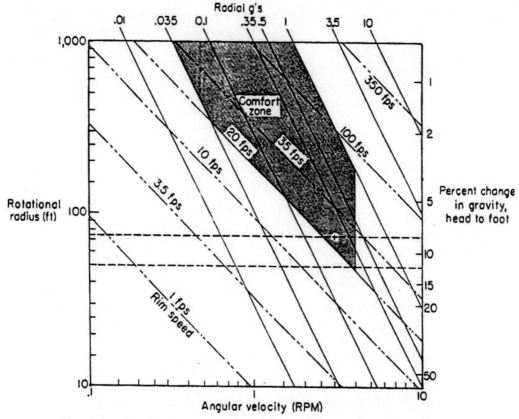

The graph defines the rotational characteristics that would be used in conjunction with interpretation of physiological response (comfort zone) to size a manned space station.

Figure 13.3.2: "Comfort chart" of artificial gravity.
(Courtesy of Hill and Schnitzer, 1962.)

zone with gyration radii less than 15 m. The difference between the accelerations at the average commando's head and feet (e.g., 6.2 and 8 m from the axis for a 180 cm stature, respectively) is 25%, which further contributes to vertigo and disorientation.

In contrast LCJ crewmembers find the LTI-20's 1/6 g artificial gravity quite tolerable, even comfortable. Accommodations at the bridge include reconfigurable seating from upright to semi-supine, menu-selectable for crew preference.

13.4 Emergency Environmental System

Critical environmental systems aboard the LTI-20 are built for high levels of reliability and are designed with a high level of redundancy to alleviate the impact of point failures. Even with these precautions, however, the potential for system failure still exists.

The emergency environmental support systems were designed to protect the crew and prevent the total loss of a system or of the Lightcraft itself. Several emergency alert levels can arise, each level requiring an appropriate and timely response.

The outer corridor of the Lightcraft is divided into sections by pressurized bulkheads. Access to these sections is through walkways that pass through the bulkheads. All such passages can be sealed off and made air-tight by doors. As in a submarine, breach of the hull in any section triggers automatic closing of the bulkhead doors, isolating the affected section from the rest of the Lightcraft. This precaution

ENVIRONMENTAL SYSTEMS

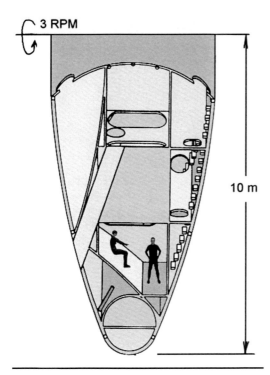

Figure 13.3.1: LTI-20 artificial gravity environment for space. *(Courtesy of RPI.)*

protects the integrity of the Lightcraft's inflated tensile structure.

The flight engineers and FMS artificial intelligence systems built into the ship must evaluate the emergency and attempt to rectify the situation before a disaster occurs. If the situation cannot be corrected quickly, the crew evacuates the outer portion of the Lightcraft and enters their escape pods. The MLO, pilot, and flight engineer, who remain to repair the damage, are the last to leave the ship. Emergency HeliOx respirators are located at various stations around the Lightcraft to provide short-term life support in the event that the main atmospheric life support system fails. Emergency extra-vehicular activity (EVA) equipment, including helmets and gloves, is located in these kits in case the atmosphere in the Lightcraft is not one in which the crew can safely work.

If complete failure of the Lightcraft seems imminent the escape pods are ejected to remove the crew from harm as quickly and safely as possible. Release of the escape pods is the last resort. More detailed information about escape procedures can be found in Section 18.

13.5 Waste Management System

The LTI-20 sustains a partially-closed environmental system to support its crew members. The main purpose of the waste management system is to make optimal reuse of waste products in order to minimize the volume and mass of expendables. The Lightcraft, a 20-m diameter super-pressure balloon, must meet a limit of 3600 kg of launch mass (neglecting the expendable H_2O coolant load), of which 1200 is considered payload – i.e., crew and escape pods. Optimization of the use of interior space and minimization of the mass of expendables are therefore essential.

Liquid and solid wastes are held in separate holding tanks. There is also a designated hazardous-waste containment tank to store toxic, radioactive, or biologically hazardous substances. These hazardous wastes are removed after the Lightcraft lands or docks with a space station. Gaseous and liquid hazardous waste can be vented overboard in space, where photoionization by ultraviolet sunlight and magnetic sweeping by the solar wind quickly removes these wastes from the solar system at a speed of several hundred kilometers per second.

The liquid waste recycling unit uses a series of electrical and mechanical filtration processes to separate the solid component of liquid waste. Also, the liquid waste recycling unit isolates and purifies evaporated perfluorocarbons used in partial-liquid ventilation systems (the escape pod's ultra high-g breathing aid). In the solid waste recycling unit the solid waste is compressed, and its liquid component is extracted and transferred to the liquid-waste recycling unit. A containment unit holds the remaining solid waste until it can be discarded.

After completing the recycling process, the recycled liquid goes through a microbial treatment, and the quality of the liquid is monitored. Liquid that cannot be adequately purified and reused is dumped in the direction opposite to the direction of motion of the Lightcraft. The crew. Figure 13.5.1 is a schematic of the waste recycled liquid is now ready to be used by the management system cycle of the LTI-20.

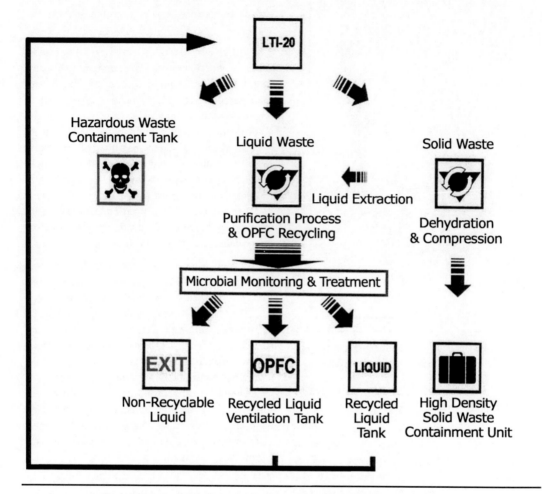

Figure 13.5.1: Waste management system schematic. *(Courtesy of RPI.)*

14.0 CREW SUPPORT SYSTEMS

The health and safety of the 6 to 12 members of the crew is essential for any Lightcraft mission. Each crew is carefully selected to provide the knowledge and experience necessary to execute a specific mission profile. The crew requires food, air, sleep, medical care, and efficient interaction with the Lightcraft systems and each other so that they may be at their best in carrying out their duties.

14.1 Crew Support

The purpose of the crew support systems is to enable the crew to perform at 100% of capacity, whether engaged in routine operations or dealing with emergencies during highly stressful military missions. Since the ultimate success or failure of each mission rests solely on the performance of the crew, human factors and ergonomics have high priority. The crew must be able to live comfortably and work effectively within a small volume without interfering with each other. Although mission times may not usually be long, occasional operations may require extended periods of time aboard the Lightcraft.

Each crew billet requires an individual with special training and skills appropriate to the mission. The personnel in the crew include the MLO and SCO (or two pilots), navigation officer, communications / tactical officer (CTO), load master / scanner (LMS), flight engineer, and mission specialist, all of whom are focused on the execution of the mission (see Section 3.2 for a detailed description). Some missions may include a flight medical officer who is a general practitioner and skilled surgeon.

The Lightcraft also offers a lounge area for crew relaxation and socialization, where individuals can eat and talk together. There are also many forms of diversion, learning, and entertainment available via the virtual reality system. The LTI-20 is designed to accommodate essential human needs so as to safeguard the morale and health of the officers whose welfare is vital to mission success.

14.2 Medical Systems

The medical system on board the Lightcraft uses technologically advanced, highly miniaturized equipment, computerized expert systems, and a highly trained and skilled physician to diagnose and treat injured or ill crew members. Medical functions are assigned to two small rooms near the axis of the Lightcraft. The inner room of the medical bay is used for examinations and surgery, and the outer serves as the doctor's office. The examination room has collapsible beds that can be used as either examining tables or sleeping quarters for crew members with injuries or contagious diseases. All beds are equipped with restraints for the safety of the patient during maneuvering and low-gravity operations.

Although there is only one fully trained medical officer on board, every Lightcraft crew member is trained in basic emergency medical functions (first aid) such as CPR, the Heimlich maneuver, splinting, and treating burns. Especially on military missions, it is essential that the crew be prepared for medical emergencies that arise while they are away from the Lightcraft.

The doctor is assisted not only by the expert systems resident on the FMS computer, but also by a mechanical arm that assists during emergency surgery. This arm is computer actuated and carries sensors to enable it to assist with minimal voice control by the surgeon. The medical bay is also equipped with body-scan equipment that examines the injured area and displays its findings for use by the medical officer. This Magnetic Resonance Imaging (MRI) equipment takes full advantage of the 1- to 2-Tesla field produced by the Lightcraft's superconducting magnets.

The medical bay also is stocked with a variety of drugs and vaccines that may be required in the course of a mission. Special equipment also includes molding splints, which use sensors, actuators, and the main computer to conform to and support a broken limb in the manner of a splint. The medical bay is designed to handle several simultaneous emergencies, but time and resources are limited on a military mission. A crew member who needs major surgery or reconstructive work, but is stabilized and not in danger of getting worse, can be placed on life support systems within an escape pod. The computer continually monitors the individual and keeps the medical officer informed of the patient's status. Upon arrival at a base with full medical facilities, the patient is transferred to a hospital.

14.3 Crew Quarters System

The LTI-20 Lightcraft is equipped with individual quarters for all 6 to 12 crew members (Figure 14.3.1). Each room is set up to accommodate the crew member's sleeping needs under conditions of both atmospheric flight and space flight.

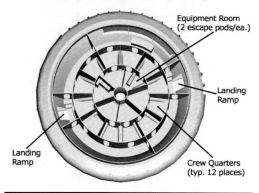

Figure 14.3.1: LTI-20 crew quarters.
(Courtesy of RPI.)

Each room is equipped with a lightweight "air mattress," inflated by HeliOx, that retracts into a wall when not in use. The bed can be deployed in two different orientations to compensate for the different directions of effective gravity felt during atmospheric flight and space flight.

During space flight the bed functions in a vertical or upright position, covered by a cushioned "sleeping bag" that restrains and warms the occupant. The crew member simply slips into the bag and zips it closed. The sleeping bag is attached to the wall to prevent unintended "sleep-floating" of the occupant during micro-gravity (or zero g) space missions.

When traveling in the atmosphere, the bed is deployed horizontally, but restraint is still useful to protect against sudden trajectory changes or turbulent buffeting.

14.4 Food Rehydration Unit

As with all life forms, humans must maintain top physical shape to perform at their best. To do this, they must receive three nutritionally complete meals per day. Life is no different aboard a Lightcraft. The crew can select from a variety of dehydrated foods which are packaged to minimize their launch mass. Nutrients lacking due to the absence of fresh fruits and vegetables in this diet are made up by appropriate supplements. The rehydration of food is accomplished using the purified and deionized water that is stored on the ship for multiple purposes.

Food is available in the main cabin area. The diner selects a meal from the menu, and the packages of the individual courses are loaded into the rehydration unit. The computer automatically injects the appropriate amount of water and bakes or cools the item as appropriate. The diner removes the food once it has been transformed into a sumptuous repast.

All food items are fortified with vitamins and minerals essential to human health. Food preferences are solicited from the crew before each mission to assure that the food selection is palatable to each. Subject to space limitations, the crew's favorites can generally be accommodated. Selected snacks are also available to the crew, subject to the same considerations of nutrition and compactness as the other foods.

14.5 High-Quality VR Environment

One of the most advanced features of the LTI-20 is the way it incorporates a high-quality virtual reality (VR) environment into everyday operations. Any crew member can access the VR system through the headgear of the ultra-g suit (officially called the space activity unitard, or SAU). Authorized personnel can control the ship in VR mode via the VR system's linkage with the FMS computer.

Virtual reality also assists in the performance of training exercises. All military personnel are

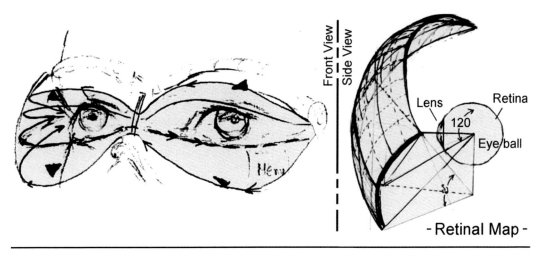

Figure 14.5.1: Virtual reality goggles with direct retinal projection. *(Courtesy of RPI.)*

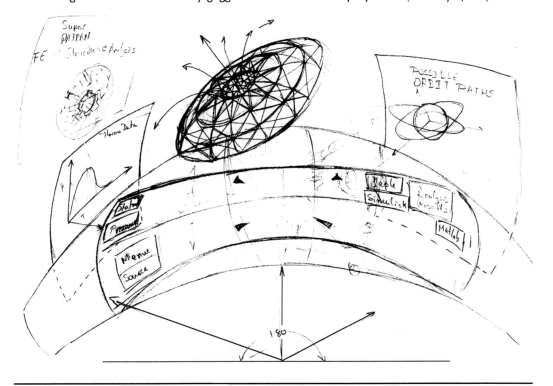

Figure 14.5.2: Sample virtual reality display link to FMS computer. *(Courtesy of RPI.)*

required to perform training exercises during long missions.

The VR system is also used for relaxation during extended missions. A large number of programs, selected to accommodate the crew members' preferences, including interactive games, educational subjects, travelogues, and feature-length movies, are stored in the computer's database.

To enter the virtual reality system, a crew member must put on the ultra-g suit and the VR goggles through which the VR world can be viewed (Figures 14.5.1 and 14.5.2). The suit is covered with tiny sensors and actuators to enable the wearer to feel and manipulate the virtual world. With the suit on, the suit control panel responds to a spoken command to activate the VR system. Once inside the system, all features of the system are accessible from a series of menus that can be selected either by suit touchpad or by voice command.

To operate the ship from the VR world in emergencies, proper authorization must be granted. Unauthorized personnel are locked out of this program. The recreational and training features are available to everyone, subject only to classification clearance.

14.6 Life Support Options

Of all the major systems on board the LTI-20, the life-support systems are among the most important. The loss of these systems could result in the loss of the entire crew and their mission. For this reason, several options are available for life support.

In the main cabin of the Lightcraft and in the maglev landers the crew breathes the HeliOx mixture, in which the nitrogen component of natural air is replaced by helium. As described in Section 13.2, this substitution is effected to avoid the problem of nitrogen bubble formation in the blood stream (the bends) as a consequence of overly rapid decompression.

Within the escape pods the crew members have a choice of breathing apparatus. If very high accelerations are not required (beyond 30 g), the high-pressure HeliOx system can be used. However, if ultra-high accelerations (e.g., 50 to 200 g) are planned, liquid ventilation is mandatory for the protection of the crew. Breathing liquids under extreme acceleration places excessive strain on the chest muscles and requires the assistance of a respiratory unit that senses when the body is trying to inhale and assists by forcing perfluorocarbon fluid in and out of the lungs.

The crew members have two choices of military standard liquid ventilation systems: partial liquid ventilation (PLV), or total liquid ventilation (TLV) systems. Partial liquid ventilation fills only the deepest portions of the lungs with a perfluorocarbon liquid in which oxygen is highly soluble. HeliOx gas fills the remainder of the lung volume. The more commonly chosen method requires two tubes to be inserted through the nasal passages, down the back of the throat and between the vocal cords into the lungs. The location of these tubes, which prevent normal vocal communication, is shown in Figure 14.6.1. These tubes fill the lower lungs with the oxygenated perfluorocarbon liquid and remove it when the high accelerations are over. A photograph of the nasal tubes can be seen in Figure 14.6.2. In this option, all communication is via the VR link.

The second option involves surgical implantation of removable connections for the liquid breathing tubes. The tubes are placed in

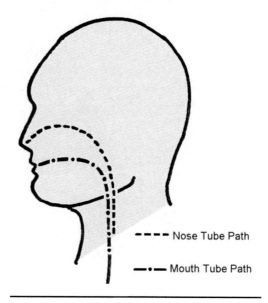

Figure 14.6.1: Liquid ventilation tube path. *(Courtesy of RPI.)*

the trachea, just below the vocal cords, and run down into the lungs. The positioning of the neck connection can be seen in Figure 14.6.3. This option keeps the face free of tubes, offers greater head mobility, and does not interfere with verbal communication.

14.7 Ultra-g Personal Protection System

The body protection required for all military personnel aboard the LTI-20 is the revolutionary liquid-filled ultra-g suit: the Space

CREW SUPPORT SYSTEMS

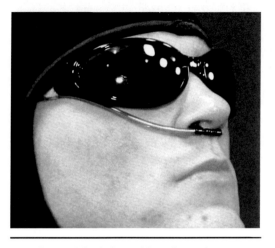

Figure 14.6.2: Partial liquid ventilation nose tubes. *(Courtesy of RPI.)*

Figure 14.6.3: Tracheal implant option for partial liquid ventilation. *(Courtesy of RPI.)*

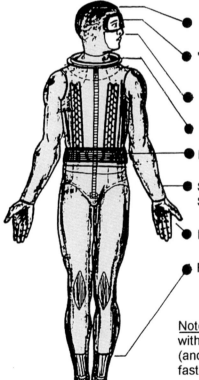

- Skin-tight flexible "scuba diver" pull-over head gear (microwave reflective)
- "Heads-up" virtual reality "wet" goggles with direct retinal projection
- Microwave reflecting mesh pull-over for nasal, mouth, and ear cavities
- Optional collar for "fish bow" space helmet
- Maglev belt with flexible, rechargeable battery
- Skin tight "divers" suit with microwave-reflecting, Spandex material; allows EVA with proper head gear.
- Microwave reflecting gloves
- Form-fitting integral boots (microwave reflecting)

Note: All clothing is devoid of exposed zipers, pockets with flaps, etc. that could trigger microwave breakdown (and subsequent high temperature plasmas). Velcro fasteners are used exclusively throughout the SAU.

Figure 14.7.1: Features of ultra-g space suit (space activity unitard).

Chapter 14

Activity Unitard, or SAU. This SAU suit is to be used when traveling at high transatmospheric speeds or in space.

As shown in Figure 14.7.1, the suit has a skintight flexible scuba-diver's pullover headgear that is microwave-reflective. The gear also contains virtual reality goggles with direct retinal projection by LEDs or with HUD visor display (Figures 14.5.1 and 14.5.2). The goggles are made of material that becomes opaque at very high light intensities, thus protecting the eyes from damage by laser beams. High-reflectivity coatings are used as well. The suit also is equipped with microwave-reflecting, form-fitting integral boots and removable gloves. Velcro fasteners are used instead of buttons or exposed zippers which can initiate microwave breakdown and high temperature air plasmas.

A microwave-reflecting grid (an almost invisible, flexible metalized, mesh with high electrical conductivity) is provided to protect the face, as well as mouth, nose, and ear cavities from incident microwave radiation (Figure 14.7.2). Optional equipment includes a collar designed to accommodate a fish-bowl helmet. An optional maglev belt with a flexible battery pack is also available for special missions.

The suit itself is made of a skin-tight, bullet-proof, spandex-type microwave-reflecting material. The military-issue suit has a liquid-filled liner that inflates against the armor of the escape pod so as to leave no voids and no room for motion. This procedure, along with partial or full liquid ventilation, enables the occupant to withstand high accelerations.

All military personnel assigned to the LTI-20 must wear this SAU suit at all times during missions unless specifically authorized to remove it.

14.8 Lightcraft Boarding Options

Boarding of the Lightcraft can be accomplished in several ways. The most common and convenient method of boarding a hovering Lightcraft is by means of the maglev belt (Figure 10.2.3). Passengers can also be shuttled to and from the Lightcraft in a 6-person maglev lander (Figure 10.5.1). Section 10.2 contains further details on these methods of boarding.

When a Lightcraft is resting on the ground using its escape pods as landing gear, passengers may board by entering the hatch of their own escape pod (Figure 14.8.1). After the craft is airborne the pods are retracted, allowing passengers to exit their pods into the interior of the Lightcraft.

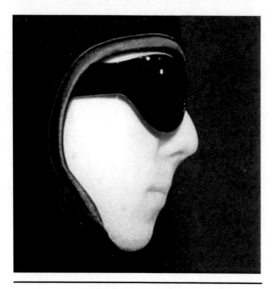

Figure 14.7.2: Microwave reflective metallic mesh for face protection. *(Courtesy of RPI.)*

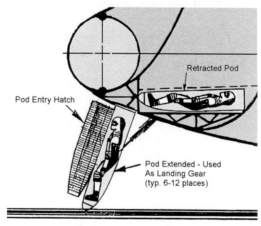

Figure 14.8.1: Pod utility for immediate entry / egress, when used as landing gear (LTI-15 configuration).

CREW SUPPORT SYSTEMS

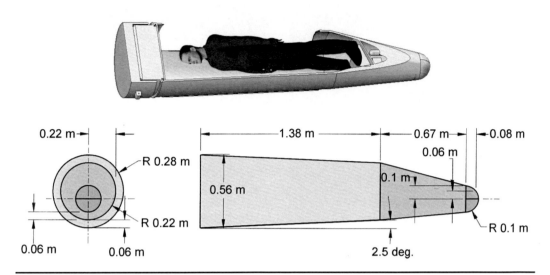

Figure 14.9.1: Individualized, custom-fit escape pod. *(Courtesy of RPI.)*

The simplest way to board a grounded Lightcraft is by the ventral (i.e., with rectennas facing downward) boarding ramp (Figure 3.5.1). This ramp deploys when the craft is less than 4 m above the ground. Once a section of the rectenna array is retracted within the craft, a stiffened section of the Lightcraft hull simply swings downward to provide a boarding ramp.

14.9 Escape Pod Architecture

Each LTI-20 crewmember is provided a custom-fit escape pod (Figure 14.9.1) with full reentry capability. In the event of imminent catastrophic structural failure of the mothercraft, all 6 to 12 pods are ejected simultaneously in a radially outward direction as shown in Figure 14.9.2. The following subsections describe the escape pod architecture inclusive of human factors (ultra-g protection, VR system, etc.), normal pod boarding and de-boarding procedures, pod propulsion systems, emergency ejection events, and autonomous pod flight operations. More details on LTI-20 escape pod technology are covered in Sections 3.4, 4.2, 14.7, 16.11.2, and 18.3.

14.9.1 Custom Crew-Tailored Pod Environment

The interior of an escape pod is custom-fit to each crew member. The pod is lined with a form-fitted rigid, lightweight shell with a liquid-inflated cushion that accommodates a specific person's body shape (Figure 14.9.3). The liquid cushioning layer, roughly 1 cm thick, assists in minimizing injury to the human body during ultra-high-g maneuvers. The lining shapes itself to conform to the body, thereby making the escape pods universal in use. The lining provides shock cushioning as well as optimum support for the body. The 1-cm liquid layer also provides adequate protection against the soft x-ray radiation that is released when the Lightcraft disables its space plasma shield.

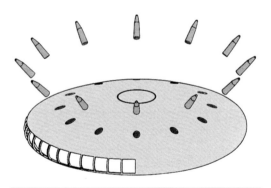

Figure 14.9.2: Emergency simultaneous ejection of all 12 escape pods. *(Courtesy of RPI.)*

If the need arises, military personnel are able to remain within the escape pods for periods of an hour or more. The VR system in combination with the inner lining provides for comfort during extended missions.

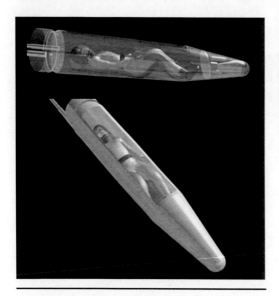

Figure 14.9.3: Custom fit escape pod.
(Courtesy of HR Pictures.)

Each pod has a temperature control to accommodate the preferences of the crew member within. The VR system allows complete control of the environment inside each pod, and links directly into the FMS computer.

A latch is provided on the inside for manual opening of the pod's hatch. The LTI-20's escape pod is designed to accommodate the full range of USSC flight personnel (see Section 3.2).

14.9.2 Pod Propulsion System

The pod propulsion system consists of a nose-mounted retro thruster and four attitude

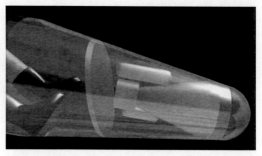

Figure 14.9.4: Pod transparent view of nose-mounted retro thruster and propellant tanks.
(Courtesy of HR Pictures)

control (RCS) rockets attached to the rear bulkhead next to a retractable aerodynamic flap. The retro thruster is sized for a 3 g de-orbit maneuver, as well as terminal descent braking upon landing. When 6 to 12 pods are deployed as landing gear, their retro engines may be fired simultaneously to provide emergency liftoff thrust for the mothercraft. All rocket thrusters are fueled with gaseous H_2/O_2 propellants, burning at a fuel-to-oxidizer ratio of 1:6 by mass. Figure 14.9.4 gives a cutaway view of the retro thruster and propellant tanks at the nose of each pod. Retro rockets are continuously throtteable from 50% through 110% rated power level.

Table 14.9.1 gives the performance data for the pod's H_2/O_2 propulsion system, which delivers an impressive specific impulse of 420 to 427 seconds. The gaseous, pressure-fed propellants are stored in graphite-epoxy filiament-wound,

Table 14.9.1: Performance data for pod H_2/O_2 rockets.

Parameter	Metric
Mixture ratio of hydrogen to oxygen (by mass)	1:6
Ambient combustion chamber pressure	3.45 MPa
Pressure and temperature of incoming gasses:	
Hydrogen =>	4.31 MPa at 25 K
Oxygen =>	4.31 MPa at 83 K
Ignition system	Electric spark torch
Temperature of thruster ignition	2200 K - 3330 K
Ignition Time	20 ms
Spark energy level	10 mJ
Duration from electrical signal to 90% thrust	75 ms
Specific impulse (Isp)	420 - 427 sec

Table 14.9.2: De-orbit specifications for pod retro rockets.

Parameter		Metric	
		LTI-20A Pod	LTI-20B Pod
LEO Orbital Velocity		~ 8,000 m/s	~ 8,000 m/s
Delta-V for De-Orbit		100 m/s	100 m/s
Nominal Thrust		5500 N	2940 N
Maximum Burn Time		3.3 s	3.3 s
Propellant Mass:	Hydrogen (gas) =>	0.6 kg	0.32 kg
	Oxygen (gas) =>	3.6 kg	1.92 kg
Propellant Tankage Mass		4.2 kg	2.24 kg
Total Retro Propulsion System Mass		12.6 kg	6.7 kg
Gross Mass (pod + occupant + propellant)		~ 187 kg	~ 100 kg
Nominal De-Orbit Deceleration		3 g	3 g

aluminum-lined tanks at 43.1 atmospheres; the combustion chamber operates at 3.45 MPa. An electric-spark torch ignites the bipropellant in 20 ms, requiring only 10 mJ of energy from the pod's battery. The 90% peak thrust level is attained 75 ms after transmission of the electrical signal.

Table 14.9.2 presents the specifications for retro-rockets installed in LTI-20A and LTI-20B pods. Both deliver the nominal 3-g deceleration for 3.3 seconds, and 100 m/s delta-V necessary for de-orbit. Each of the four attitude control thrusters produce ~200 Newtons of thrust and share the same propellant, tankage, and electric spark ignition system as the main retro rocket. This RCS system enables accurate adjustments in roll, pitch, and yaw in the vacuum of space, as well as during reentry. The full reentry flight control system, which includes the rear aerodynamic flap, enables precision trajectories to be flown with ease.

14.9.2.1 Propellant and Tankage

Both the LTI-20 rectenna array cooling system and pod propulsion system rely on water that is frequently scavenged from the environment. Fresh water is preferred, but salt water can also suffice since all impurities are first extracted (then discarded) using an elaborate filtration process. The pod's hydrogen and oxygen propellants are often manufactured on-board the LTI-20 using its own electric power systems and any available purified water source. The electrolysis process is both simple and highly reliable, requiring little maintenance. Once generated, the oxygen and hydrogen gases are individually collected, compressed, and stored in dedicated high-strength composite tanks aboard the LTI-20 mothership, or pumped directly into depleted pod propellant tanks.

14.9.2.2 Pod Nose Fairings

The pod's hemispherical nose incorporates a small circular fairing (or hatch) that rotates open expose the retro rocket's exit nozzle. The pneumatically activated hatch is opened immediately prior to firing the retro rocket, and closed / locked when the burn is complete. The hatch system is doubly redundant with manual backup valves. Retro hatches are normally closed when pods reside in their launch tubes, and remain closed after emergency pod ejections to preserve the sleek biconic aerodynamics. Retro-hatch seals are sufficiently tight to survive reentry with negligible leakage.

14.9.3 Pod Retro Thruster Operations

The retro thruster in each biconic pod serve one primary function (independent escape pod de-orbit, and retro-braking before touchdown), and two auxiliary / backup functions for the LTI-20 (vertical take-off thrust augmentation, and mothership de-orbit from LEO). Pod retro thruster operations for each function are described below.

The principal requirement for pod retro thrusters is to de-orbit the manned biconic vessels following emergency ejection in low

Chapter 14

Earth orbit. Hence, the thruster and on-board propellant load are sized to deliver the necessary ~100 m/s delta-V to leave LEO.

Auxiliary / backup function #1: By firing all 6 to 12 pod retro thrusters, the LTI-20 can ascend to an altitude of ~20 meters to engage the ion propulsion mode. Once the mothercraft has reached the desired altitude, all pods are retracted before subsequent flight maneuvers are initiated.

Auxiliary / backup function #2: Aided by retro rockets in the Maglev landers, the 6 - 12 pod retro thrusters can de-orbit the mothership from LEO. This requirement places significant demands on Maglev pod's chemical propulsion system, and effectively sizes its H_2/O_2 propellant load. Normally, the LTI-20's electric RCS system provides this de-orbit function with a specific impulse of 2x to 4x greater than pod H_2/O_2 chemical rockets.

14.9.4 Pod Boarding and De-Boarding

Each regulation USSC pod has a hinged rectangular hatch in its biconic aeroshell "skin," sized to accommodate statures of flight personnel assigned to LTI-20A and LTI-20B Lightcraft. The hatchway is sufficiently large that a crewmember can quickly and easily enter or exit the pod. The hatch O-ring seal encircling this opening remains airtight in the vacuum of space. Figures 14.9.5 and 14.3.1 show the central LTI-20 compartments where crewmembers enter the pods, prior to insertion into the launch tubes.

a) Pod retracted

b) Pod extended as landing gear

Figure 14.9.6: Pod deployment for use as landing gear on LTI-20. *(Courtesy of RPI.)*

Figure 14.9.7: Pod detail with actuation and guideway mechanism. *(Courtesy of RPI.)*

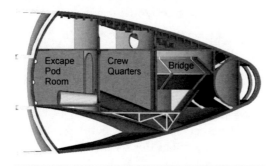

Figure 14.9.5: LTI-20 cross-section showing escape pod room, crew quarters, and bridge. *(Courtesy of RPI.)*

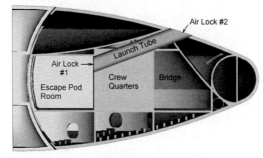

Figure 14.9.8: Air-lock mechanism at breach and muzzle ends of launch tubes. *(Courtesy of RPI.)*

CREW SUPPORT SYSTEMS

One of the standard LTI-20 vertical-decent-to-landing procedures employs several pods deployed as shock-absorbing, landing gear "legs." Each launch tube has a linear EM motor that smoothly extends the pod out through the Lightcraft hull, then rotates the pod downward to contact the ground (see Figures 14.9.6 and 14.9.7). In the process, all crewmembers previously loaded into their sealed escape pods are effortlessly lowered out of the craft. With the Lightcraft firmly resting on the ground, crewmembers can exit their pods at will. Of course, used in reverse, ground personnel can be transported into the Lightcraft interior with ease.

14.9.5 Launch Tube Seals

An air-lock mechanism is provided at both ends of the launch tube (see Figure 14.9.8) to prevent depressurization of the LTI-20 interior, which is maintained at 0.5 bar above local ambient. The process of loading a biconic pod is akin to loading a ballistic shell into a cannon: Basically, the breech end of the launch tube is opened, the pod is inserted, and the breach is closed and sealed. Each escape pod is also equipped with a pneumatically inflated seal that fills any gaps between the launch tube and the pod's aft bulkhead.

When the biconic pods are deployed as landing gear, elliptical air-locks (with O-ring seals) at the Lightcraft hull are opened just prior to "gear" extension. However, in extreme emergencies (e.g., catastrophic Lightcraft structural failure), these same air-locks are explosively separated just prior to pod ejection.

14.9.5 Emergency Biconic Pod Ejection

USSC regulations for ultra-energetic LTI-20 flight maneuvers require that crewmembers be confined to the safety of their escape pods and ultra-g suits. The threat of catastrophic structural failure is imminent if the 0.5 bar inflation pressure for the LTI-20 hull is severely compromised. The FMS computer will perceive this threat and begin ejection countdown procedures. Within the sensible atmosphere, directed-energy airspike support will be sustained by the FMS long enough for escape pods to separate, even if emergency electrical power must be pulled from onboard superconducting magnets.

Pod ejection is performed by a rocket gas generator that pressurizes the launch tube section between the pod and the breech. The rocket gas generator, which is integrated with the breech end, is sized to eject the pod to a safe distance from the mothership (see Figure 14.9.9) with a nominal acceleration of 15 to 18 g. Once the pod has cleared the mothercraft the crew member must negotiate a controlled de-orbit, hypersonic reentry and aerobraking, staged deployment of the ballute and main parachute (Figure 14.9.10), and gliding descent to pre-selected landing spot. The pod's ultra-g suit can protect the occupant against a maximum 200-g impact.

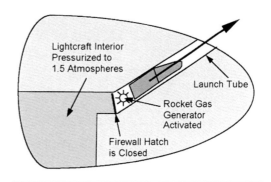

Figure 14.9.9: Pod ejection by rocket gas generator option. *(Courtesy of RPI.)*

Chapter 14

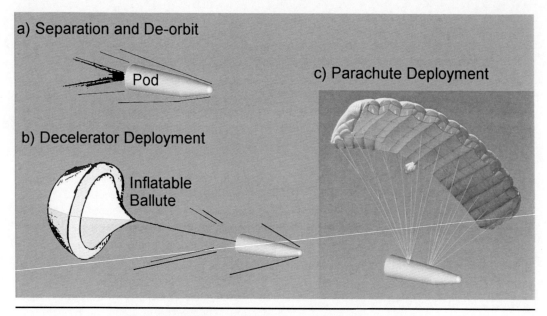

Figure 14.9.10: Escape pod ejection and recovery sequence.

15.0 AUXILIARY LIGHTCRAFT SYSTEMS

The auxiliary systems on the LTI-20 Lightcraft comprise a variety of specialized equipment and functions that exploit the ship's superconducting magnets and high external magnetic fields. These functions and equipment include the maglev landers, magnetic docking and "magnetic crane" functions, space radiation shielding, and EVA recovery by means of the magnetic belt.

The lander has several purposes and uses. It can retrieve coolant water from remote sources, function as a shuttle craft for transporting personnel and cargo between the Lightcraft and the surface, and serve as an alternate emergency escape system. The maglev lander's unique shape is designed to provide specific directed-energy airspike functions. It is also capable of assisting in shielding against dangerous space radiation and storing additional energy to supplement the capacity of the superconducting magnetic energy storage (SMES) system in the Lightcraft. Finally, since the Lightcraft is designed to lift an external mass equal to the maglev lander's mass of 2400 kg (PDE mode only), the lander can serve as an aerial crane to rescue another downed Lightcraft or abduct an automobile from a highway.

The magnet system on the LTI-20 can also magnetically deploy and retrieve the crew by means of the maglev belts in their SAU suits. However, extravehicular activity in this manner will be used only if it becomes absolutely necessary to deal with an emergency, such as micrometeoroid damage to the hull.

Figure 15.0.1 shows a maglev lander descending from the Lightcraft within the atmosphere. This scenario could take place if the maglev lander were being lowered to take on a load of water (Figure 10.2.8), to pick up passengers (Figure 10.5.1), to rescue a disabled Lightcraft (Figure 15.0.2), or even to abduct an automobile (Figure 15.0.3) – all using PDE propulsion.

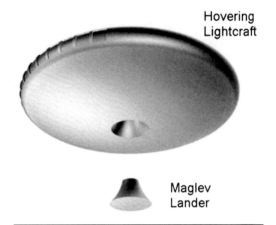

Figure 15.0.1: Hovering lightcraft lowering Maglev lander with SMES magnets. *(Courtesy of RPI.)*

Figure 15.0.2: Maglev lander rescuing a disabled lightcraft. *(Courtesy of RPI.)*

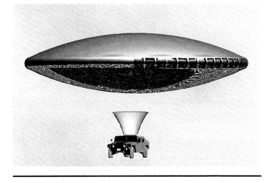

Figure 15.0.3: Automobile abduction using Maglev lander as magnetic grapple.

15.1 Maglev Lander Operations

The maglev lander is a vital part of the Lightcraft. It is designed to accomplish four main functions while in atmospheric flight.

First, the maglev lander can extract 2400 kg of water (2.4 m^3) from a nearby lake, river, or reservoir and lift it to the Lightcraft for use as expendable coolant in the MHD propulsion mode. This aspect of the lander is covered in detail in Section 16.10.

Probably the most important function of the maglev lander is to serve as a shuttle craft for personnel. The lander can descend magnetically to a person's doorstep. A person can step into the lander (see Figure 15.1.1) and in a few seconds step out into the Lightcraft. For this "elevator" function, the lander is used over vertical lift distances that are typically less than one vehicle diameter.

Figure 15.1.1: Maglev lander as shuttle craft for personnel transport. *(Courtesy of RPI.)*

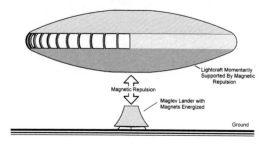

Figure 15.1.2: "Launch assist" by magnetic repulsion against Maglev lander. *(Courtesy of RPI.)*

In situations in which a quick and silent takeoff is vital, the maglev lander can be used almost like a catapult to launch the Lightcraft while the ion propulsion system is charging up (see Figure 15.1.2). To launch the Lightcraft in this way, the superconducting magnets in the Lightcraft and lander are aligned to repel each other. This procedure exerts a substantial force on the lander as the Lightcraft pushes the lander into the ground, thereby propelling the Lightcraft away from an immediate threat and enabling it to engage another propulsion method. As soon as the ion propulsion system is engaged, the lander is magnetically recovered and docked before the Lightcraft flies away.

Finally, in military operations, the maglev lander can fly separately from the Lightcraft, although it must be close enough to remain magnetically attached to it, and use its powerful superconducting magnets to pick up ferromagnetic objects. One such scenario might involve the Lightcraft hovering over a vehicle driving down a road. The maglev lander is deployed from the Lightcraft, magnetically picks up the steel frame automobile, and abducts the vehicle and occupants (see Figure 15.0.3).

15.2 Magnetic Docking Bays

One unique aspect of the maglev lander is that it can dock and undock with the Lightcraft magnetically to load and unload cargo. Magnetic docking requires far fewer mechanical parts, which entails higher reliability, longer service life, lower mass, and ease of operation.

Both the Lightcraft and the maglev lander contain superconducting magnets (nine on the Lightcraft and two on each lander) that encircle the craft. These magnets are extremely powerful, with a stored energy exceeding 900 MJ on the Lightcraft alone, all of which can be turned on or off at will. Normally the magnets in the lander and Lightcraft are set to attract each other for use as an elevator. This magnetic attraction lifts the maglev lander into the central "donut hole" in the Lightcraft, filling the entire opening. The exterior surfaces of the lander and Lightcraft are completely flush with each other while the lander is docked. The exterior mating surfaces of the Lightcraft and lander are coated with a material with a very low coefficient of friction to aid the docking process. The primary centering forces during docking are provided by the magnetic bearing system (MBS). The MBS can also be used to exert torques between the two craft in the manner of an electric stepping motor, making it possible for the Lightcraft and lander to spin independently with minimal friction and mechanical stress on either structure. Besides aligning the doors on the two craft, this electric motor can despin the maglev landers so that the Lightcraft can rotate independently for gyroscopic stabilization in lateral flight mode (see Sec. 10.4).

As detailed in Section 16.10, the upper portion of the maglev lander is equipped with an in-flight refueling system to offload 2400 kg of water coolant quickly into the Lightcraft storage tanks. These plumbing connections in the Lightcraft are located just above the rectenna dish (see Figure 15.2.1), with pipes running to the annular water storage tank in the Lightcraft rim.

The maglev lander can also be used as a transport vehicle for up to six people. After the lander has attached itself to the Lightcraft, it is magnetically rotated to align the doors of the two craft. At this point, an accordion-type, double O-ring sealed corridor is extended to join the doors and sealed against the lander. The lander is then pressurized with 1.5 atm of HeliOx (at sea level) to match the pressure inside the Lightcraft. After sealing and pressurization are complete both doors can be opened. Passengers then exit from the lander into the upper corridor of the Lightcraft, which is also designed to act as an air-lock. In this usage, an unpressurized lander could be docked to an air-lock at ambient exterior pressure. The passengers would exit the lander and enter the air-lock, the air-lock and lander doors could be closed, and the air-lock then pressurized to 1.5 atm while the lander departs for another sortie. Egress from the air-lock into the Lightcraft is by means of a hatch and ladder in the air-lock ceiling.

15.3 Maglev Lander Configurations
The maglev lander is designed to serve multiple functions so as to minimize the overall launch mass of the Lightcraft. Both landers are identical

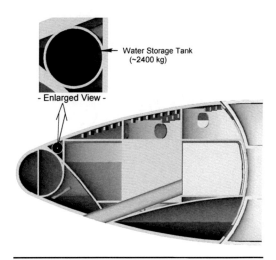

Figure 15.2.1: Location of 2400 kg water storage tank in LTI-20. *(Courtesy of RPI.)*

to provide valuable redundancy.

As mentioned in Section 15.1, the lander can pick up either 2400 kg of water or six passengers. The interior design of the lander accommodates both functions. Seating for six passengers can be folded away, and passengers can be carried safely no matter which end of the lander is up (see Figure 15.3.1).

The lander's shape is specifically designed to conform to the contours of the Lightcraft. The smaller top end of the lander also serves as part of the long-range communications system (see Sec. 8.5). It fits snugly in the "donut hole" in the Lightcraft's central rectenna. The exterior surface of the bottom hull of both landers is microwave-reflective to focus light forward efficiently for use in air spike formation with directed energy, but it is also laser-reflective. This concave parabolic lower surface of the lander is coated with high-reflectivity films tuned to 650 nm laser light (see Sections 5.2 and 5.4). The surface is optically adaptive, controlled by computer-driven actuators that can be adjusted to direct the remote 5 to 10 MW laser beam onto different targets.

Since the figure contour and surface quality of the bottom mirror of the maglev lander is so sensitive, it must be kept free of contamination and therefore cannot be allowed to contact the

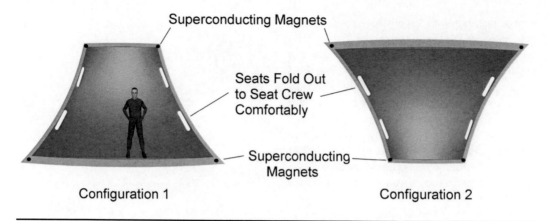

Figure 15.3.1: Reconfigurable Maglev landers can transport personnel in both orientations. *(Courtesy of RPI.)*

Figure 15.3.2: Maglev landers are equipped with retractable tri-pod gear. *(Courtesy of RPI.)*

ground. A small tripod landing gear can be extended from the bottom of the lander to keep it 30 cm off the ground (see Figure 15.3.2). Each lander is also equipped with a parachute and small retro rocket for emergency use.

15.4 Extravehicular Activity in Space

The Lightcraft is designed to enable extravehicular activity (EVA) in space for repair of the hull or other systems. EVAs are normally reserved for emergencies.

Leaving and reentering the Lightcraft in space can be dangerous. As mentioned in Section 2.4, there is an array of sensors throughout the hull to monitor radiation levels. During normal space operations the space plasma shield is on to protect the passengers from energetic solar flare protons (see Sec. 15.5). However, it would be fatal for a crew member to exit the Lightcraft while the shield is activated. The shield must be turned off to allow an EVA.

Generally, when a hull problem arises such as a sensor report of a small leak, the flight engineer is assigned the task of repairing it. First, the shields must be slowly shut off. During this process the collapsing radiation shield bombards the hull with keV electrons, which produce a cascade of soft X-rays. For protection against this hazard, and to enable prompt ejection in the event of an emergency, the crew must be in their protected escape pods during shield shutdown. If, as the shields are shutting down, an unsafe level of radiation is detected inside the Lightcraft hull, the EVA is temporarily aborted and the shield is restabilized. If the shield cannot be safely lowered and the integrity of the hull is compromised, the crew may be forced to eject in their escape pods.

The engineer assigned to an EVA puts on a special EVA suit equipped with micrometeoroid armor, a maglev belt, and the proper tools. The space suit is physically or magnetically tethered to the Lightcraft to prevent the engineer from drifting off. The engineer then enters the maglev lander air-lock or one of the 12 pod launch tubes. Doors are shut, and the section is slowly depressurized. The hatch is then opened, and the engineer may exit to fix the problem. Upon completion of the repair the engineer reenters the lock, the procedure is reversed, and operations return to normal.

While the engineer is outside the ship during an EVA, the engineer's position is controlled by magnetic forces through the interaction of the engineer's maglev belt with the Lightcraft's magnetic fields. The EVA suit also contains a reaction control system similar in function to NASA's manned maneuvering unit (MMU), but much smaller. If both positioning systems on the EVA suit fail, the Lightcraft still has a backup retrieval system. The maglev landers can be magnetically deployed from the Lightcraft and sent out a small distance to rescue the drifting crew member.

15.5 Space Plasma Shield Operations

On Earth we are protected from solar flare proton irradiation by the 1 kg/cm² of shielding afforded by our atmosphere. In space the only protection against energetic protons is afforded by the space vehicle itself. The entire configuration of the LTI-20 is designed around an ultra-light Space Plasma Radiation Shield (SPRS). The SPRS is able to protect human bodies from protons with energies up to 200 MeV as well as micrometeoroids. Meteoroids of masses up to 1 tonne, which would encounter a Lightcraft on the average of once every 200 million years, could be avoided in the extremely unlikely event of a threatened collision. It was not until the development of SPRS protection against radiation and micrometeoroids that space travel truly became routine.

The basic principle behind the LTI-20 plasma radiation shield is that the spacecraft envelops itself with an external positive charge that repels protons (see Figure 15.5.1). The Lightcraft's superconducting magnets are used to set up a strong magnetic field whose polar axis circles the Lightcraft rim. Electrons from the solar wind plasma are attracted to the positively charged Lightcraft hull, but cannot cross the magnetic field lines to reach the hull and neutralize the charge. Instead, the electrons drift zonally around the Lightcraft rim. The negative charge on the trapped electron cloud is kept equal to the positive charge on the Lightcraft hull by the conductivity of the solar wind, making a sort of cylindrical capacitor. The charge separation sets up a strong radial electric field around the Lightcraft.

When bringing the field strength up, special linear accelerators (LINACs) eject beams of electrons

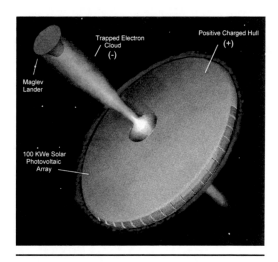

Figure 15.5.1: SPRS shield activated with lightcraft hull charged to 200 MV With high SMES magnetic field.

with energies of 300 keV to 1 MeV to leave the Lightcraft hull positively charged as the strengthening magnetic field captures the electrons. The solar cells that cover half the surface of the Lightcraft supply enough electricity to power the shields. A total of 30 kW is needed to energize the magnetic field and to run the cryogenic magnet cooling system. A photovoltaic array with an efficiency of 29% on the non-rectenna side of the Lightcraft is capable of supplying 100 kW from sunlight alone. The extra power is stored in SMES unit for later use when solar power might not be available. The SPRS has a radial leakage-current density of about 1.25×10^{-11} amp/cm^2, which is small enough to have little effect on the performance of the shield. Note that the "donut hole" in the center of the Lightcraft is absolutely necessary for the electron cloud to pass through. When the shield magnetic field is turned off the electron cloud collapses unrestrained onto the Lightcraft hull, striking with enough energy to generate soft X-rays and force the crew into their escape pods. The 1-cm liquid layer in the pods provide ample protection against the electron and X-ray emission.

To open the axial hole through the Lightcraft and allow the free circulation of the electrons, the maglev landers must be separated from the Lightcraft and kept at a safe distance from the electron cloud. The landers, which continue to magnetically track the Lightcraft, can be used for long-distance communications (see Section 8.5).

16.0 FLIGHT OPERATIONS

The LTI-20 Lightcraft is capable of performing a great range of low-energy and hyper-energetic flight modes and maneuvers. These modes and maneuvers enable the Lightcraft to accelerate, cruise, hover, and operate at low power to avoid detection. These features make it possible for the Lightcraft to function in many different environments and to execute a wide variety of military missions.

16.1 Introduction to Flight Operations

Flight operations may be categorized as either performance-related or mission-related.

Performance-related operations include all the *flight modes* and *maneuvers* required for successful functioning of the Lightcraft hardware, including all the flight systems necessary for the Lightcraft to take off, accelerate, maneuver, land, and so on, in the course of executing military missions in the atmosphere and in space.

The Lightcraft has three basic quasi-continuous, transatmospheric flight modes:

a) Subsonic ion propulsion (IP) flight
b) Supersonic pulse detonation engine (PDE) flight
c) Hypersonic magnetohydrodynamic (MHD) flight

Cruise propulsion modes are sustainable, efficient methods of flight through a planetary atmosphere. Ion propulsion is used for the covert hover and low-observables cruise modes. For example, reconnaissance missions conducted beneath cumulus clouds in the atmosphere require the "covert hover" mode. The "reduced power" mode is used for silent running on internal SMES power, when the nature of the mission requires that the Lightcraft be difficult to detect or recognize. While in the PDE flight mode, the Lightcraft may either hover or accelerate rapidly to supersonic cruise speeds, according to the requirement of the mission. Hypersonic cruise demands MHD slipstream propulsion, but the PDE mode must first accelerate the craft through Mach 1 before the MHD engine will start.

In addition to these three basic flight modes, the LTI-20 can perform 12 basic types of maneuvers:

a) Vertical takeoff, hover, and landing
b) Conventional pitch, roll, and yaw
c) Gyroscopic stabilization
d) Pendulum motion or "falling leaf"
e) Circular flight (loiter)
f) Low observables modes
g) Soaring / thermaling mode (riding thermals)
h) High-g acceleration in almost any direction (lateral or axial)
i) High performance "hyperjump" (to evade detection)
j) Hypersonic atmospheric skipping
k) Orbital boost and reentry

By combining the different cruise modes with the various flight maneuvers, the Lightcraft can complete a wide variety of mission profiles, as covered in Section 16.2 below.

16.2 Mission Types

The LTI-20 has been designed for hyper-energetic Space Command missions. The extreme speed, ultra-light-weight design, and maneuverability of the Lightcraft make it suitable for a wide range of covert military operations, including surveillance, reconnaissance, rescue, and limited strike and defensive roles. Typical missions for the LTI-20 fall into one of the following categories:

16.2.1 *Retrieval of a downed Lightcraft.* The LTI20 has the capability of lifting its own mass (2400 kg) with its maglev coils using the PDE mode. This ability enables the Lightcraft to pick up another intact but disabled Lightcraft and transport it to a safe location for repairs (see Figure 15.0.2).

16.2.2 *Retrieval of sensitive "black boxes."* In circumstances in which a Lightcraft is damaged beyond repair, it may have to be destroyed. Military Lightcraft are designed to self-destruct if necessary, principally by shorting the SMES units while fully charged. However, the "black boxes" are designed to remain intact in

Lightcraft accidents. The LTI-20 is equipped to locate and retrieve critical black boxes from crash debris, along with flight data recorder, computers, and other sensitive components.

16.2.3 *Rescue of downed personnel.* In the event of a catastrophic emergency, a Lightcraft will eject the entire 12-member crew in their escape pods or maglev landers. The LTI-20 can be deployed to remote territories with less than a full crew to rescue the downed aviators from hostile territory.

16.2.4 *Transport of emergency supplies.* The LTI-20 and its crew are capable of sustaining accelerations up to 30 g with pressurized HeliOx ventilation, and 200 g or more with liquid ventilation systems. In the event of an emergency, the Lightcraft can immediately bring supplies to anywhere on Earth in less than 45 minutes. This feature enables the Lightcraft to travel to the Moon in about 5.5 hours, leaving Earth at 22 km/s.

16.2.5 *Abduction of small military targets.* The Lightcraft is capable of lifting its own empty mass of cargo (2400 kg). The lift of the Lightcraft in PDE thruster mode, combined with the maglev coils, gives the Lightcraft the ability to abduct ferromagnetic objects of a wide variety of weights, shapes, and sizes. Such objects might include a steel-frame automobile on a road (Figure 15.0.3), or any hostile item of equipment of similar mass.

16.3 Flight Modes and Maneuvers

There are three basic cruise modes and 12 basic types of maneuvers of the LTI-20, as described in the following sections.

16.3.1 Cruise Propulsive Modes Under normal conditions the LTI-20 Lightcraft is capable of sustaining quasi-continuous cruise over a wide range of speeds, ranging from subsonic to hypersonic. The crew can select, in conformity to the mission profile, any of three propulsive systems to achieve the desired velocity and performance.

16.3.1.1 Subsonic Cruise

The Lightcraft generally cruises at subsonic speeds when operating for long periods within the atmosphere. In subsonic flight the Lightcraft uses its ion propulsion system to achieve a maximum speed of about 150 km/hr. As shown in Figure 16.3.1, the principle of ion propelled forward flight for the Lightcraft is vaguely similar to that of a 20th century helicopter, except for the absence of rotor noise (see Section 7.0 for further details).

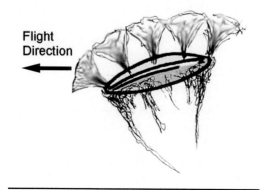

Figure 16.3.1: Lateral flight in ion propulsion mode. *(Courtesy of RPI.)*

16.3.1.2 Supersonic "Cruise"

The LTI-20, with the help of its pulsed detonation engine (PDE), can easily accelerate through the speed of sound in 1 second (~30 g); hence PDE "cruises" of just a few seconds, applied intermittently, are the norm. The Lightcraft can achieve supersonic speed either laterally, using a linear airspike in front of the leading edge of the rim (Figure 16.3.2), or with the axi-symmetric airspike axis parallel to the direction of flight (Figure 16.3.3). However, unless operating on SMES power (very short PDE periods only), the incident microwave beam must be exactly aligned with the vehicle axis lest the Lightcraft receive insufficient power (see Sec. 6.0 for details).

16.3.1.3 Hypersonic "Cruise"

The MHD propulsion system is principally used to achieve hypersonic velocities sufficient to leave Earth's atmosphere; transatmospheric "cruises" to orbital speed last only a few minutes. In addition to providing the thrust necessary for escaping Earth's gravity, the MHD engine is also capable of accelerating the Lightcraft in hypersonic dashes through the

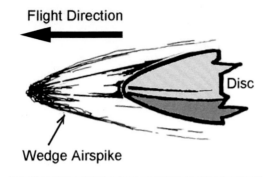

Figure 16.3.2: PDE acceleration with lateral airspike engaged. *(Courtesy of RPI.)*

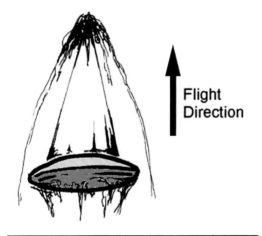

Figure 16.3.3: Axial MHD flight with airspike engaged. *(Courtesy of RPI.)*

atmosphere, assisted by the airspike (Figure 16.3.3). In this propulsive mode the airspike is activated to reduce drag, and the MHD slipstream accelerator is used to thrust the Lightcraft to hypersonic speeds. Both lateral (cross-beam) and axial (along-beam) flight modes are accommodated by this system, which uses the airspike to alleviate the bow shock, annihilate the sonic boom, and reduce drag and heat transfer dramatically. Section 5.0 treats the MHD engine in detail.

16.3.2 Basic Flight Maneuvers

The LTI-20 Lightcraft is designed to fulfill a broad range of Space Command missions, including reconnaissance, surveillance, and rescue. With its exceptionally agile thrust-vectoring ability, the Lightcraft is capable of a number of exotic hyper-energetic maneuvers.

16.3.2.1 Pitch and Roll Maneuvers

Pitch and roll maneuvers are performed by thrust vectoring the three primary propulsion systems. The LTI-20, while using its ion propulsion engine, experiences low pitch and roll rates typical of conventional fixed-wing aircraft. In contrast, the PDE and MHD thrusters can give lightning responsiveness in pitch and roll.

16.3.2.2 Flip Maneuver

The flip is especially useful immediately after takeoff to acquire the high-power microwave beam from a power satellite (see Figure 16.3.4). The ion propulsion system can perform the 180° rotation (pitch or roll) in approximately three seconds, whereas the PDE system is 10 times (or more) quicker.

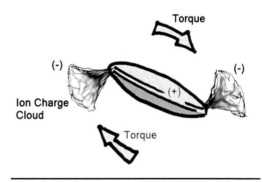

Figure 16.3.4: Pure pitch / roll maneuver to "flip" with ion propulsion (torque only, no translation). *(Courtesy of RPI.)*

16.3.2.3 Spin Stabilization

Using elements of both PDE and MHD propulsion systems, the LTI-20 Lightcraft can initiate and maintain a desired rate of axial spin (yaw rate) to provide gyroscopic stabilization (see Section 17.3.3). The Lightcraft may also be "spun up" prior to liftoff when supported on a de-spun maglev lander (containing the crew), by switching its magnetic bearings into the "drive motor" mode.

16.3.2.4 Pendulum Maneuver

The "falling leaf" maneuver (Figure 16.3.5) is used when the Lightcraft is directly above the landing site and the normal glide-slope approach

Figure 16.3.5: "Falling leaf" maneuver for controlled descent. *(Courtesy of RPI.)*

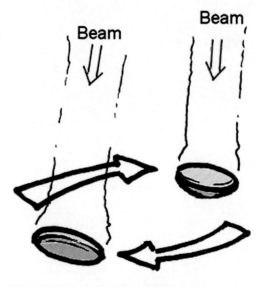

Figure 16.3.6: Circular-flight loiter maneuver. *(Courtesy of RPI.)*

path is obstructed by trees, buildings, or rough terrain; hence, a controlled "pendulum oscillation" descent must be performed with the ion propulsion system. Applied in reverse for ascending flight, the vertical "cork-screw" maneuver allows a Lightcraft to climb efficiently out of a restricted space.

16.3.2.5 Circular-Flight Loiter

Station-keeping is often important during high-altitude surveillance and reconnaissance missions to provide onboard sensors with a stable platform (see Figure 16.3.6). The Lightcraft performs this circular-flight loiter mode while tracking the microwave or laser beam, using its low subsonic motion to generate aerodynamic lift, while cutting engine thrust to the minimum.

16.3.2.6 Ion-Propulsion in Low-Observables Mode

The ion propulsion system is used for subsonic travel in hostile territory and for cruising at low altitudes, below the vision of radar. This is known as low-observables mode (LOM). The LTI-20 has a minimum radar signature equal to a 1.7 mm diameter BB at a range of 1.6 km. The perimeter toroid and half the lenticular hull (high-power side) are microwave transparent. In stealth mode, the craft is oriented edge-on to hostile radar, so as to minimize the radar signature.

There are several options for propulsive energy sources, including solar power, beamed microwave or laser energy from the remote USSC power stations, and the on-board SMES "battery." Each power source has specific advantages and disadvantages at each point in a mission profile. Avoidance of detection is the central principle of LOM flight.

Solar energy, because of its availability, is the propulsive method of choice in daylight operation. Solar power, however, has a number of tactical limitations. The LTI-20 must flip upside down to direct its photovoltaic array at the Sun. Although operated in such a way as to be invisible to radar, the LTI-20 is still a large and bright object hovering in the sky, readily visible to ground observers. Under solar power alone the LTI-20 is limited to a speed of ~5 m/s (about 11 miles per hour). A Lightcraft spotted by a ground observer may be unable to reach its destination before it is intercepted by the defenders. Terrain-following flight over a forest canopy or in valleys is a logical tactic to reduce the chances of discovery. Further, solar energy at an intensity of 1 kW/m^2 is required to sustain LOM flight at 5 m/s. Cloud cover, haze, dust, and Rayleigh scattering of light when the Sun is low in the sky will frequently limit the Lightcraft to even lower speeds.

Beamed laser or microwave energy, by comparison, can move the Lightcraft much faster than solar power because of its tenfold higher power density. Although it is possible to detect the scattered microwave power that is not absorbed by the rectenna, in practical reality the enemy must be actively scanning for that frequency to have any hope of detecting the Lightcraft before it has stealthily passed over the target area. Finally, beamed microwave power is an all-weather, round-the-clock option, whenever the power satellite is in direct line of sight to the Lightcraft.

Laser power is highly effective under certain conditions. Red laser light of about 650 nm wavelength is used because of its high photovoltaic efficiency (~60%). The laser beam itself is scattered very little by air, and hence is not highly visible. However, the photovoltaic array glows a bright red when it is illuminated by the laser, unless the threat of detection demands a frequency shift into the near IR (e.g., 860 nm) where the beam is invisible to the naked eye. Low observables operation in the near IR reduces the electric power output from the photovoltaic array, and hence cruise speed. As with solar power, the Lightcraft must fly upside down with its photovoltaic array facing the laser (i.e., the normal orientation for landings). The red glow of the top side is bright enough to be readily visible at night from the ground, but simply tilting the Lightcraft away from a potential observer may hide the light from detection. The performance advantage of laser energy over solar energy is a 5-fold speed increase, a direct result of the 23-fold increase in available propulsive power (see Sec. 7.4). The choice between solar and laser power is simply a tradeoff of risks and benefits.

A typical mission flight plan might call for the use of microwave power outside a 15-km radius from a site where ground observers and radar equipment are expected, followed by a flip over and a switch to solar or laser power for use closer to the site. Depending on local conditions and flight geometry, use of visible laser power (i.e., 620 - 700 nm) may or may not be practical at night.

16.3.2.7 Reduced Power Mode for Silent Running

The on-board SMES units can also be used during low-observables flight, subject to the limited amount of energy that can be stored by the magnets, normally exceeding 900 MJ. At a flight speed of 5 m/sec, SMES power can propel the Lightcraft a distance of about 25 km. At a speed of 80 m/sec, it can propel the Lightcraft less than 2 km. The SMES unit can be used at any time and in any weather without fear of detection.

16.3.2.8 Covert Hover Mode beneath Cumulus Clouds

An alternative means of achieving low-observables flight without flying below radar coverage is to hover inside clouds. The ion-propulsion engine is especially effective in clouds because water vapor molecules ionize easily and remain stable for hours. Since minimal power is expended by the Lightcraft while hiding in a cloud, it is common to target a hyperjump on a cloud formation (see Sec. 16.3.2.11). Hovering inside a cloud is an effective means of cloaking the presence of a Lightcraft until the coast is clear.

16.3.2.9 Riding Thermal Updrafts

The LTI-20 has a very low disc loading ("wing" loading) of only 11 kg/m^2, which makes riding thermal updrafts (soaring) an attractive loiter mode if favorable atmospheric conditions are prevalent. This maneuver exploits the circular-flight mode, initially with PDE engines running on several megawatts of remote microwave power; under rare conditions, ion propulsion with SMES, laser or microwave power can be used. The heated engine exhaust can initiate and sustain large thermals with powerful updrafts in the atmosphere, often rising to 3 km altitude. The LTI-20 can then "soar" within these buoyant and rising air masses, at greatly reduced power levels, by seeking the regions of highest vertical velocity. After the thermal dissipates a new thermal can be created, and the maneuver is repeated.

16.3.2.10 Rapid Acceleration in PDE Mode

At sea level, the ion-propulsion system produces only enough vertical lift (11.4 kN) to support half the Lightcraft's empty mass of 2400 kg. The rest of the force needed for hover flight, approximately 12 kN (without a payload), comes from the natural buoyancy of the helium-rich inflation gas. To also lift the mandated 1200 kg of payload, the LTI-20 must resort to the "dynamic" corner of the Lift Triangle (see Section 7.1): it must exploit aerodynamic lift generated by the LTI-20's lenticular "wing" in lateral, horizontal flight.

Since the Lightcraft is an inflatable disc with fixed volume, it displaces a fixed volume of air. At higher altitudes, the density of the air is less than at sea level, and the buoyancy force is correspondingly reduced. At altitudes as low as a few kilometers, the buoyancy is reduced to the point that the thrust of the ion engines can no longer keep an empty Lightcraft hovering. To compensate for the decreased buoyancy, it is necessary to switch to PDE propulsion.

To maintain a hover, the PDE exhaust gases must be vectored straight down as in Figure 16.3.7. The PDE thrust, like that of the of the MHD engine, can be vectored in almost any direction, allowing great agility and lightning-fast maneuvering ability. For this reason, PDE propulsion is the best choice in combat and self-defense situations: speed is one of the LTI-20's greatest assets. From a stationary hover, the PDE thrusters can accelerate the Lightcraft at 200 g or more, reaching Mach 2 in a heartbeat and demonstrating the ability to evade any opponent successfully. The Lightcraft's rapid thrust-vectoring capabilities have forever eliminated the classical aerodynamic problems of stall and loss of control that would otherwise arise in attempting such energetic and violent maneuvers.

The PDE thrusters can be engaged from ion-propulsion mode or cruise flight, using either the SMES units or beamed power. It is easy to switch from one propulsion mode to another.

It is not necessary for the crew to be in their escape pods during PDE-mode flight at any acceleration less than about 3 g. Control of the

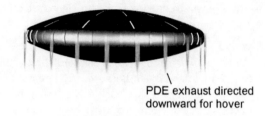

Figure 16.3.7: Lightcraft maintained in hover by directing PDE exhaust downward. *(Courtesy of RPI.)*

Lightcraft can be maintained through the personal access displays. However, in tactical situations, the crew is required to remain in their escape pods to allow a wider choice of propulsion and maneuver options. In friendly territory, where rapid evasive maneuvers are not expected, the escape pods normally remain unoccupied.

16.3.2.11 Hyperjump to Evade Detection

While operating in hostile airspace, the LTI-20 can create the illusion of disappearing instantly in flight. This hypersonic "jump" maneuver, often called a hyperjump, is normally accomplished under external beamed power. For distances of less than 11 km, however, magnetically stored energy from the SMES unit can be used. There is insufficient stored energy to move the Lightcraft very far, or to use the airspike for more than 1 to 2 seconds. The hyperjump can be oriented in any direction, but it is most often performed upward.

The primary requirement of the hyperjump is that it is faster than the eye can follow, which requires an acceleration of at least 20 g. As always, the Lightcraft begins its motion with the PDE thrusters, which are vectored so that the Lightcraft is in position to use the airspike and MHD accelerator. Typical jump distances are 2 to 10 km vertically, which is sufficient to reach the nearest clouds about 90% of the time.

The pilot can choose an acceleration rate for the hyperjump. At 20 g acceleration, a 2-km jump requires 4.5 seconds. At 100 g, the same jump takes 2 s, and at 200 g, it would take only 1.4 sec. The choice of acceleration, which comes down to a tradeoff of energy vs. time, is at the discretion of the MLO or pilot.

Occasionally on extended ventures in hostile airspace a high-power microwave beam may not be available when needed, leaving the MHD accelerator inoperable (except for very brief bursts on SMES power). A 1-km lateral hyperjump can still be made using the PDE exclusively, although the blunt vehicle cannot be pushed much past Mach 2 (about 680 m/s) without the airspike. If the acceleration rate is limited to 23.5 g, the jump duration is 3 seconds.

To make the jump undetectable and to avoid a visible contrail, no expendable water coolant can be ejected as steam. Fortunately, the hyperjump is very short in duration. Cooling can be accomplished by injection of a small amount of liquid helium coolant into the pressurized hull after absorbing the waste heat from the engines, but without venting gases from the Lightcraft.

While the LTI-20 accelerates at up to 30 g during a hyperjump, it is essential for the crew to be installed within their escape pods and breathing pressurized HeliOx; beyond 30 g, partial or total liquid ventilation is mandatory. It is impractical for the crew to be constantly climbing in and out of their escape pods while traversing hostile territory and such activity may consume precious time needed for escape. Since the purpose of a hyperjump is to evade immediate detection, the crew would normally remain in their pods while in hostile airspace so as to make the hyperjump available at any time without delay. On an emergency mission to rescue a downed pilot, a hyperjump cannot be initiated the moment the pilot is taken aboard. The best option to accelerate the Lightcraft at only 1 to 3 g to a speed of Mach 2, and to travel at this velocity using PDE propulsion until the new passenger is able to get into an escape pod, at which point the hyperjump can be executed.

16.3.2.12 Hypersonic Atmosphere-Skipping

Hypersonic atmospheric skipping is similar to the "skip trajectory" described in Section 17.5.2, except that the MHD engine is briefly fired up during the skip to maintain the LTI-20's kinetic energy. The objective of this maneuver is to "cruise" at the top of the atmosphere at high suborbital speeds (~ Mach 12 or more) under the minimum time-average beam power. The trajectory resembles a stone skipping on water – i.e., multiple passes through the thin upper atmosphere. After the initial loss of velocity upon re-entering the sensible atmosphere, the Lightcraft uses both aerodynamic lift and MHD thrust to skip out of the atmosphere. The maneuver is repeated until the range objective is satisfied.

16.3.2.13 Orbital Boost and Reentry

Making a high-performance flight into space is one of the primary design criteria for the LTI-20 Lightcraft. This function is performed by using the PDE to break Mach 1, then switching to MHD mode and accelerating toward the microwave power-station or relay satellite. Figure 16.3.3 shows the LTI-20 in the axial MHD boost mode. Boost trajectories and orbital paths are controlled so that the Lightcraft never collides with the power-beaming station in orbit, but instead travels ballistically through space to its intended destination (see Section 19.9.5).

As discussed in Section 7, the Lightcraft's ion-propulsion engine can receive a low-power microwave beam from any angle of incidence on the rectenna. However, both the PDE and MHD thrusters require that the microwave beam be exactly aligned with the axis of symmetry in order to operate in a high-power mode. The PDE is used to accelerate the Lightcraft axially or laterally to the beam at speeds exceeding Mach 2, with accelerations reaching 200 g (maximum performance level). Such high acceleration levels (0 to Mach 2 in 0.35 s) require partial or total liquid ventilation of the crew in their ultra-g suit equipped escape pods.

If two power satellites are in view, the LTI-20 may switch seamlessly from one powerlink to another, in rapid succession. For example, the Lightcraft may climb and accelerate laterally in PDE mode with the rectenna kept aligned to the first power beam until Mach 2 is reached. Then PDE thrust is vectored to pitch the Lightcraft over rapidly until its axis is aligned with the next power station (see Figure 16.3.8). At this point the airspike is activated and the Lightcraft switches to MHD slipstream accelerator mode resuming an acceleration of up to 30 g (maximum) into space. However, if only one

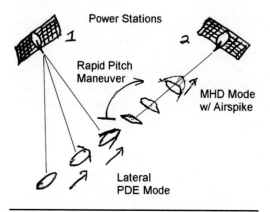

Figure 16.3.8: Seamless transition between orbital powerlinks using lateral PDE to axial MHD propulsion mode. *(Courtesy of RPI.)*

power or relay satellite is available, the axial PDE acceleration mode boosts the Lightcraft beyond Mach 1, whereupon the MHD slipstream accelerator takes over, as described in Section 19.9.5. If acceleration along the transatmospheric trajectory never exceeds 30 g then the crew may breathe pressurized HeliOx in their escape pods, simplifying orbital flight logistics.

It is necessary to take on 2400 kg of water coolant before initiating high-g acceleration to orbital velocities. The water is expended as steam during acceleration. Because the rectenna requires considerable cooling during high-performance flight, escape from hostile territory cannot be accomplished without taking on a full load of cooling water. If a fresh water river, lake, or reservoir is available in the operation area, replenishment of the cooling water can be accomplished readily (Figure 16.3.9). If there is no water nearby, or if the proximity of hostile forces makes it temporarily unsafe to collect water, then the Lightcraft must be moved to a better location or into hiding without any use of coolant.

When there is no immediate danger from the enemy, the ion-propulsion system can be employed in a low-observables mode (see Section 16.3.2.6 - 16.3.2.7) to reach water undetected. If an enemy is nearby and water cannot be collected safely in the present location, the best option is to make a hyperjump (see Section 16.3.2.11) to another location with water. If the enemy is aware of the necessity of water for LTI-20 operations they may guard the nearest water source. A hyperjump to a more distant water source would then be advisable. After such a hyperjump the ion engine is re-engaged to complete the retrieval of water using the PDE mode.

The orbital position (i.e., altitude and inclination) of the LEO power satellite is obviously very important in governing Lightcraft operations (see Section 19.9.8). Since the powersat is orbiting at ~8 km/s, it is impossible for it to remain stationary above a hostile country, and yet the Lightcraft must receive beamed power at

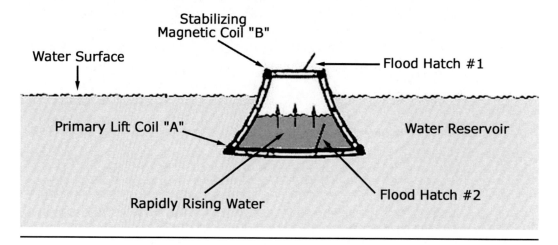

Figure 16.3.9: Maglev lander picks up water. *(Courtesy of RPI.)*

FLIGHT OPERATIONS

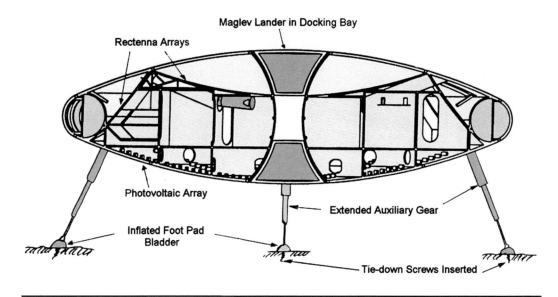

Figure 16.4.1: Deployment of auxiliary tripod landing gear. *(Courtesy of RPI.)*

even the most remote locations. This logistical problem is solved by engaging multiple transmission links through power relay satellites. With a number of power satellites (PDS-03) and GBM stations (PDS-02) in operation, a power transmission path from any powersat (within range) can be linked via mirrored relays that bounce the beam to the Lightcraft by direct line-of-sight. Section 19.9 covers powersat beaming logistics in much greater detail.

16.4 Takeoff and Landing

The LTI-20 can land by means of any of three "landing gear" options; the auxiliary tripod landing gear, the use of three or more extended escape pods, or a maglev lander deployed as a "foot."

16.4.1 Auxiliary Tripod Landing Gear

The LTI-20 is equipped with auxiliary tripod landing gear to be used when the Lightcraft is partially buoyant and carrying no water ballast, with a flight mass up to 3600 kg. The tripod gear is designed for emergency use when the photovoltaic array is non-functional or disabled and the Lightcraft must still land and lift off under remote microwave power. Hence, the photovoltaic array faces the ground, and the high-power rectenna array faces space. The tripod landing gear can also be used to "anchor" the Lightcraft close to the ground under wind gusts as high as 100 km/hr (aided by its ion engines). Each landing gear leg is a lightweight telescoping assembly that extends from the Lightcraft far enough to keep the Lightcraft hull level at up to 5 m from rugged and uneven terrain. Each leg has an inflatable foot pad that prevents the landing gear from sinking into soft ground.

The landing gear is deployed through portholes in the photovoltaic array, with the gear extension actuators supported by the rectenna truss structure (Figure 16.4.1). The landing gear is extended outward by a telescoping pneumatic piston assembly that extends the landing gear until the desired length is reached and the foot pads are inflated and locked into place.

Inflatable bladders on the footpads are constructed of a thin, lightweight, puncture-proof material that can support contact pressures of up to 0.3 atm (Figure 16.4.2). Each bladder is divided into six independently inflated subsections that conform to the ground underneath. The inflatable bladders, once fully deployed, have a hemispherical shape and are bonded to a central ankle-boot and hinged foot pad. Two different footpad diameters, 1.2 m and 1.5 m, are in current use on the LTI-20. The small footpad assembly is for use when the expected gear load is 2400 kg (3600 kg less

Chapter 16

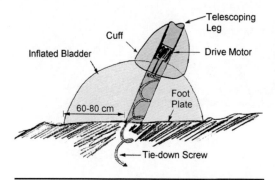

Figure 16.4.2: Inflatable foot-pad with tie-down screw. *(Courtesy of RPI.)*

1200 kg buoyancy). The larger footpad bladder assembly is for use when there is no buoyancy and the gear load is 3600 kg. These large footprints allow the LTI-20 to be set down on very soft ground without settling. The hemispherical shape prevents water or mud from collecting on top of the footpads. At takeoff, each footpad bladder is retracted into a protective cowling before the gear leg is retracted into the Lightcraft.

Once the landing gear have touched down, an internal screw equipped with a burrowing ultrasonic tip works its way into the ground to secure the vehicle. The screw can be extended up to 0.6 m into the ground to act as tie-downs (Figure 16.4.2). The landing gear can be rotated at their point of attachment to the Lightcraft, allowing the entire Lightcraft hull to be tilted toward an incoming power beam to secure optimal power reception.

The pneumatic deployment systems of the telescoping landing gear can be activated in such a manner as to spring the vehicle into the air. The gear is first shortened to bring the Lightcraft close to the ground, then extended rapidly to propel the vehicle upward. Once fully extended, the landing gear is immediately retracted into the vehicle so that the hull surface can be fully charged for ion propulsion mode. It is possible, but not recommended, to land the LTI-20 on water or wetlands with the tripod landing gear assembly fully deployed. The footpads have enough displacement to float the Lightcraft, but care must be taken lest waves and wind overturn the vehicle. This wet-landing procedure can be used in emergencies, but runs the risk of losing the charge on the Lightcraft's hull into the water. The only ways to lift off from water unassisted and begin ion propulsion are to either extend and submerge a maglev lander and use it as a trampoline to magnetically bounce the Lightcraft into the air, or to extend and fire the escape pod reentry rockets.

16.4.2 Escape Pods as Landing Gear

All 12 individual escape pods are designed to function as landing gear (see Figure 14.8.1). The pod deployment actuators provide a shock-absorbing action upon contacting the ground. This system expedites deployment or extraction of personnel in time-critical missions that prohibit depressurization of landing ramp compartments. Once the "gear" is fully extended, crew members simply exit or enter their designated pods (see Section A.9). Of course, a minimum complement of three pods deployed in an equilateral triangle can provide stable gear geometry.

16.4.3 Maglev Lander Deployed as a Pedestal

As discussed in Section 15.1, a maglev lander may also serve as "landing gear" (see Figure 15.1.2). Magnetic repulsion forces between the lander and LTI-20 mothership can be varied to provide shock damping upon ground contact. Each lander is itself equipped with three retractable legs that can support the LTI-20 under light and variable wind conditions (see Figure 15.3.2). When a quick take-off is mandated, the maglev lander can catapult the LTI-20 vertically into the air by activating its superconducting magnets in repel mode.

16.5 PDE Pickup of Water-Filled Maglev Lander

Water is needed aboard the LTI-20 as a source of consumable liquid for crew life support and for the open-cycle cooling system. About 9 GW of electrical power is generated by the rectenna during transatmospheric boost to orbit, requiring the use of 2400 kg of water coolant to remove the 1 GW of rectenna waste heat extracted by the microchannel heat exchanger system. The mass of water that must be carried for crew life support varies with the mission. The Lightcraft meets its water requirements by means of onboard storage, filtration, and purification and by retrieval of water during the mission.

One of the maglev lander's basic roles is to serve as the primary means of water retrieval. The Lightcraft, while in PDE propulsion mode, is positioned less than two vehicle diameters (40 m) above a natural water source. By reducing the magnetic force on the lander exerted by the Lightcraft, the lander is lowered gently into the water. Just before contact with the water surface, two flood hatches are opened in the top and bottom of the lander to allow water to surge into the lander at a flow rate of 0.25 m^3/s (see Figure 16.3.9). A coarse screen in the lower hatch prevents ingestion of debris. After closing both hatches to contain the water payload, the lander is magnetically retracted into the Lightcraft (see Figure 15.0.1). During lift, the payload weight is calculated from the magnetic force needed to lift the lander. Any excess water over the Lightcraft tank capacity is vented through drains in the lower surface of the lander.

Once the lander is securely docked to the Lightcraft, the water is passed through the Lightcraft's filtration, desalination, and microbial purification system and thence is pumped into the main perimeter tank.

The pumping and purification systems are located on the inner wall of the Lightcraft's "donut-hole" (see Figure 16.5.1). Using two pumps for redundancy, water from the lander is filtered and purified by Reverse Osmosis (RO) through a semi-permeable membrane. The RO process removes some 98% of dissolved minerals and virtually 100% of suspended particulates. The membrane is a light, strong thin-film material that can withstand pressure differentials as high as 2500 N/m^2, which is sufficient to desalinate sea water. When the process is complete, salts, minerals, sediments, organic particulates, and other debris are flushed off the membrane and ejected overboard. (To leave no visible trace, such debris is widely distributed over a large area from the air.) Freshwater sources are far easier and quicker to purify than sea water, and also minimize the load on the filtration system while maximizing the

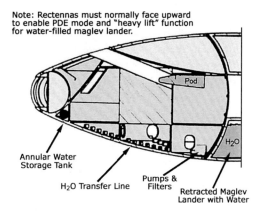

Figure 16.5.1: On-board pump, filtration, and water storage tank system. *(Courtesy of RPI.)*

lifetime of the osmotic membrane. Salt water is used only in emergencies.

The purified water from the RO unit is pumped through tubes along the hull structure into a toroidal tank which is directly below and attached to the perimeter superconducting magnet ring (see Figure 15.2.1), an ideal location for the tankage because flight propulsive forces in MHD mode are applied directly to the perimeter magnets. The toroidal water tank is elaborately baffled to prevent sudden and dangerous shifts of its center of mass under high and rapidly changing accelerations.

16.6 Water Collection from Cumulus Clouds

Sections 16.3.2.6 - 16.3.2.8 discussed the logistics of evading detection by hovering in clouds. One useful feature of the LTI-20 is its capability of filling its water storage tank using the very pure water contained in cumulus clouds. Because making a hyperjump into a cloud (see Section 16.3.2.11) is often necessary in escaping hostile territory, and because it is necessary to take on 2400 kg of water before making a high-g MHD boost into space, mining water from clouds effectively kills two birds with one stone.

Clouds are composed of water droplets, typically micrometers to millimeters in size, in equilibrium with water vapor. Cloud droplets are formed by the condensation of water vapor on tiny hygroscopic particles called Cloud Condensation Nuclei (CCN). The most common CCN are crystals of the salts sodium chloride and ammonium sulfate. In clouds in continental airspace there are usually about 500 to 1000 tiny water droplets per cubic meter of air, aggregating about 0.5 g/m³. To obtain 2400 kg of water, it is necessary to sweep all the cloud droplets out of a cubic volume 150 to 350 m on a side. *However, in a thunderstorm cloud the volume to be swept could be as small as a 90-m cube.* Ideally, the cloud chosen for mining should be large enough so that it is not substantially depleted by the water-collection process.

This cloud mining procedure, which is performed while using the PDE thrusters, begins by actively regulating the hull temperature to aid in the condensation process. The droplets typically freeze when supercooled to about -5° C. At such temperatures, the hull must be heated to prevent external icing. In warmer weather, larger droplets can be formed by cooling the hull below the dew point of the ambient air. The hull temperature is controlled by taking in a small quantity of water and heating or cooling it as required. A small fraction of the water condensate taken in during cloud mining is used to maintain the desired hull temperature.

The next step in cloud mining is to tilt the Lightcraft to a negative angle of attack. The ion guns are used to charge the forward water droplets, and the maglev lander is positively charged in the ship's central cavity, and the negatively charged water droplets are pulled into the ship (see Figure 16.6.1). The water is then filtered, deionized by the RO unit to remove the CCNs, and collected in the rim water tanks. Section 5.6 provides additional information on the use of water in the Lightcraft's open-loop cooling system.

The FMS computer estimates the time required to fill the water tanks by monitoring the cloud density, vehicle speed, water capture rate and tank content. The water content of the cloud is conveniently measured by a microwave probe beam (i.e., monitoring attenuation). Since clouds are notoriously inhomogeneous, estimates of fill-up times require substantial averaging over

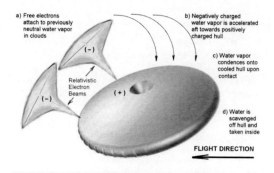

Figure 16.6.1: Cloud mining procedure (water collection). *(Courtesy of RPI.)*

large and noisy data sets, but these computations are routine for the FMS computer with appropriate sensor input.

At an average flight velocity of 20 m/s, which is the most efficient speed for cloud mining, it could take 9 to 90 minutes to scavenge 2400 kg of water. With ideal cloud conditions and sophisticated cloud seeding techniques (e.g., silver iodide dispersal) to trigger heavy downpours, mining times can be cut to 1 or 2 minutes. In most operational contexts, nine minutes is a prohibitively long time, but less than two minutes is judged sufficiently compatible with orbital transit (i.e., flyover) times of LEO microwave power satellites and relays.

Nevertheless, cloud mining is therefore available as a technique for use at need, but is rarely the optimum means of water retrieval. In a mission in which there is immediate danger of detection, cloud mining should not be attempted. A better approach would often be to take on enough cloud water to make a hyperjump to a suitably remote and secure lake or pond, using the MHD thrusters for lateral travel.

16.7 Procedure for Rescue of Downed Lightcraft

The LTI-20 Lightcraft is designed to lift a load equal to its own empty weight of 2400 kg. This ability is valuable in the event that a Lightcraft should crash or be unable to take off under its own power, and thus require rescue. Any LTI-20 has the ability to carry another Lightcraft to safety within the atmosphere, but not to orbit or on a transatmospheric boost.

When a Lightcraft is in imminent danger, its FMS computer activates all applicable safety procedures to preserve the vehicle and its crew. In the unlikely event that the computer determines that the Lightcraft cannot be saved, it ejects the crew in their escape pods and allows the ship to crash. The FMS computer determines the least destructive crash configuration for the falling Lightcraft and attempts to minimize the impact damage. If the computer remains operational after the crash it will assist in crew rescue and ship recovery operations.

The survival of the FMS computer and subsequent recovery of any downed or damaged Lightcraft has very high priority. The craft and its components have therefore been designed to facilitate retrieval. The loss of a Lightcraft to a hostile power could prove extremely dangerous because of the highly advanced technology contained in it, including the hardware responsible for its extreme agility and speed. To preserve national security and technological superiority, another Lightcraft must be deployed immediately to retrieve the downed craft.

The retrieval process can be done as a flyby, as shown in Figure 8.3.2. The rescue Lightcraft first positions itself over the downed Lightcraft so that its lander contacts the upper hull. Next, the maglev coils in the downed craft's upper hull (and/or its lander) are activated. Acting in concert with the coils in the rescue craft, enough magnetic force can be exerted to lift an empty craft and carry it to safety using PDE engines.

If recovery of the intact downed Lightcraft is impossible, due for example to failure of its superconducting magnets, then an attempt must be made to recover its black boxes and flight computer, after which the vehicle would be destroyed.

16.8 Abduction of Ground Transport Vehicles

The magnetic field generated by the superconducting magnets can be used to pick up any magnetized or ferromagnetic object up to approximately the empty mass of the Lightcraft itself (~2400 kg). Under field conditions, abduction of a steel-frame automobile or small truck may be possible (see Figure 15.0.3). Care must be taken lest the abducted object, or armed personnel in it, may threaten the hull integrity of the Lightcraft. An armed abductee could fire into the Lightcraft at close range, possibly causing catastrophic hull failure and loss of both vehicles. Hence, USSC procedures require that abductees be quickly immobilized or rendered unconscious using non-lethal weapons (e.g., stun gun, ultrasound, low-power density microwave burst, etc.).

16.9 Fleet Maneuvers / Operations

Constellations of Lightcraft must occasionally be deployed to "hot spots" in global rapid-response, highly classified missions. Such missions require specialized flight formations and precise maneuvers that efficiently timeshare orbital and terrestrial power-beaming assets in order to "fuel" real-time fleet operations. Figures 16.9.1 through 16.9.5 reveal "representative" Lightcraft squadron maneuvers and operations, selected to illustrate fundamental concepts and features of time-shared power transmission.

In Figure 16.9.1, a Lightcraft constellation is shown descending from high altitude or LEO, in single file formation, with each craft sequentially taping intermittent power from the orbital beam for active braking before ejecting laterally into multiple 500 m spaced, autonomous hover altitudes.

Figure 16.9.1: Lightcraft constellation descending from high altitude or LEO, in single file formation. *(Courtesy of RPI.)*

Figure 16.9.2 shows a squadron of Lightcraft time-sharing a single orbital power beam, wherein the *mission-leading* craft is reinforced and supported by its squadron which is waiting "in the wings," should complications arise. Figures 16.9.3 and 16.9.4 portray the developing sequence for an automobile abduction wherein the *mission-leading* craft and supporting squadron timeshare the orbital beam, all flying in "low observable" modes. In Figure 16.9.4, the *mission-leading* craft links with the orbital beam to descend and carry out the abduction, since high thrust and power are needed to lift and transport the automobile off the highway.

Figure 16.9.5 illustrates the "loiter mode" suspension of a Lightcraft constellation in natural or artificially created and augmented

Figure 16.9.4: Mission-leading lightcraft descends to carry out abduction. *(Courtesy of RPI.)*

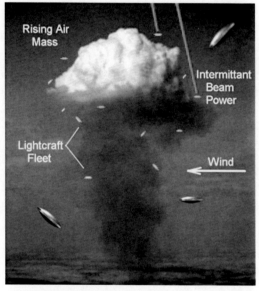

Figure 16.9.5: Lightcraft fleet suspended in thermal, drifting with local winds.

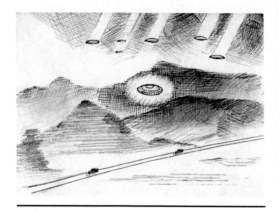

Figure 16.9.2: Lightcraft squadron time-sharing single orbital power beam. *(Courtesy of RPI.)*

Figure 16.9.3: Lightcraft flying in "low observable" modes, stalking automobile. *(Courtesy of RPI.)*

thermals in the troposphere. Such formations enable the fleet to drift effortlessly with prevailing winds, under minimal sustaining beam power, both at low altitudes within the convective boundary layer (0 - 1 km), and up to 16 km in extreme cases. The objective is to create rising air masses with super-adiabatic lapse rates that significantly exceed the local dry adiabatic lapse rate (e.g., ~1° C temperature change per 100 m altitude increase). Plumes of waste heat from individual Lightcraft (i.e., formation flying in predetermined trajectories) coalesce into strong, organized thermals with diameters exceeding 100 m. When the thermal plume rises through the convective boundary layer, it becomes entrained in whatever wind currents exist at the time. As the thermal floats upwards and drifts downwind, Lightcraft soar up the thermal's core (i.e., corridor of maximum vertical velocity), and lodge into the thermal's head or "bubble." Although natural mid-day thermals generally dissipate in 3 - 10 minutes depending on the lapse rate and wind velocity, beam-energized thermals can last as long as the beam is *on*. Four factors determine the size, shape, and strength of thermals:

a) atmospheric density,
b) lapse rate,
c) temperature difference,
d) moisture content, and of course,
e) input beam power.

As long as a supply of warmed air is available, a thermal will continue to exist. Natural and artificial cumulous clouds present additional opportunities for concealing the Lightcraft fleet, while subtly enhancing the natural thermal activity.

16.10 Stealth Procedures

Although LTI-20 Lightcraft stealth features and procedures remain highly classified, two representative examples are provided here. Figure 16.10.1 shows the LTI-20 hovering in radar-stealth mode, with the high power rectennas facing the earth. Note that both the perimeter high pressure toroid and lower hull's lenticular envelope (see exploded diagram in Figure 16.10.2) are microwave transparent. The dual nested rectenna array (shown facing the ground) can be electronically configured to absorb common radar frequencies and hence make the LTI-20 effectively invisible to long-range microwave sensor sweeps. The minuscule residual radar signature is undetectable by radar systems in common use throughout the planet.

Arguably, perhaps the most highly effective stealth procedure in the LTI-20's arsenal

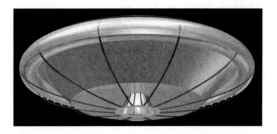

Figure 16.10.1: LTI-20 hovering in radar-stealth mode, with the high power rectennas facing the earth. *(Courtesy of Media Fusion.)*

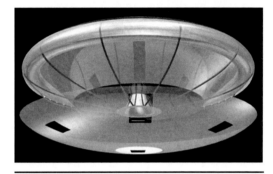

Figure 16.10.2: Exploded diagram of LTI-20, with lower microwave-transparent hull removed. *(Courtesy of Media Fusion.)*

involves the creation of *air plasma* decoys to confuse, mislead, threaten, and/or evade the aggressor force. The procedure invokes the projection of high power laser or microwave energy (from on-board or remote sources) to initiate and sustain stationary or moving "plasma fireballs" or "ball-lightning" that are deliberately detectable by radar and/or direct visual sightings. The procedure has been effectively applied in a wide variety of missions to:

a) project the illusion of an overwhelming force (e.g., the "presence" of a large squadron when in fact, none exists),
b) fake Lightcraft departures after transitioning into "low observable" modes,
c) threaten aggressors with simulated close fly-bys or head-on collision threats; and the like.

17.0 FLIGHT DYNAMICS

The LTI-20 Lightcraft must achieve both stability and control during atmospheric flight, including both lateral and axial flight modes (see Figure 17.0.1). Stability and control must be maintained in hovering, subsonic, supersonic, and hypersonic flight regimes.

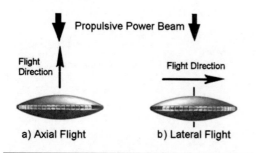

Figure 17.0.1: Lightcraft in lateral and axial flight modes (referenced to power beam axis).

17.1 Introduction to Flight Dynamics

The LTI-20 aeroshell is a simple lenticular disc with no traditional control surfaces to stabilize or direct its flight path. The absence of wings, fins, tail, elevators, ailerons, elevons, flaps, canards, winglets, trim tabs, etc. jutting out beyond the smooth lenticular hull surface is a direct consequence of design requirements for the space plasma shield (Section 15.5). Also, in the subsonic ion-propulsion mode such surface irregularities would trigger massive corona discharges off these edges and prevent the hull envelope from charging up for flight (see Section 7). In space, the hull would not be able to reach the 200 MV potential needed to protect against solar proton storms.

17.1.1 Disc Aerodynamics

Attachment of any physical control surfaces to the thin, pressurized hull of the Lightcraft would be extremely difficult, if not impossible. Such fins and stabilizing surfaces would compromise the structural integrity of the hull during the high-Mach-number maneuvers for which the Lightcraft is designed.

In the absence of external control surfaces, the Lightcraft must stabilize and control its flight by means of active thrust vectoring from its ion, PDE, and MHD propulsion systems. For example, the pitching moment produced by aerodynamic lift in lateral Frisbee-mode flight must be nulled out by active thrust vectoring (see Figure 17.1.1).

17.1.2 Disc Flight Dynamics

The dynamical flight behavior of a lenticular disc is determined by a real-time interaction of weight, lift, drag, thrust, acceleration, and vehicle moments of inertia. The response of the LTI-20 Lightcraft to flight control inputs is modeled by a set of equations stored within its FMS computer. The LTI-20's flight characteristics set it apart from all conventional aircraft types in several fundamental ways.

a) The craft can hover and perform vertical takeoffs and landings.
b) The craft is able to fly in both axial and lateral directions.
c) The craft can accelerate at 200 g or more in PDE flight, and 30 g in MHD flight.
d) Supersonic drag on the vehicle can be reduced at will by using Directed Energy Airspike (DEAS) technology.
e) Inclement weather and other adverse conditions have little effect on the Lightcraft, but may necessitate changes in either the choice of power beam frequency (laser or microwave), or propulsion mode, or both.
f) In subsonic and supersonic regimes, the crew is despun in the maglev landers; for the hypersonic regime, the crew is housed in their custom-fit escape pods.

17.1.3 Stability

The LTI-20 Lightcraft is dynamically unstable in several flight modes. Nonetheless, it is designed for outstanding controllability and maneuverability in hovering, subsonic, supersonic, and hypersonic flight regimes. With the aid of its FMS computer, active flow control, and thrust

FLIGHT DYNAMICS

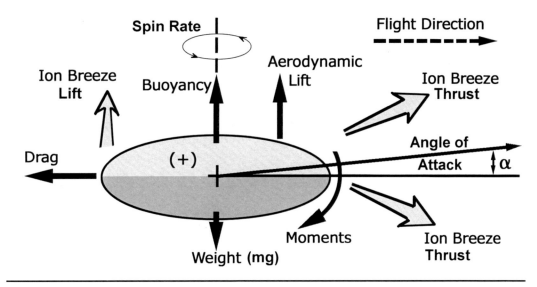

Figure 17.1.1: Free-body diagram of disc in lateral subsonic flight, using ion propulsion mode.

vectoring engines, the LTI-20 enjoys stable and controlled flight throughout its trans-atmospheric design envelope. Some important aspects of Lightcraft flight stability are:

a) Hovering at low altitudes is aided by the craft's natural buoyancy.
b) The unpowered lenticular hull is intrinsically unstable in lateral flight.
c) Rotating the lenticular Lightcraft or its maglev landers about the principal axis provides additional gyroscopic stability as required during flight.
d) Active flow control dramatically alters the vehicle stability.
e) Aggressive thrust vectoring in all three propulsion modes is closely linked with vehicle flight stability.

17.1.4 Control

The Lightcraft's motions must be under positive control at all times to permit safe transportation of the crew during missions. Several features of the control system are important:

a) The FMS computer receives information from sensors both inside and outside the Lightcraft and makes the adjustments necessary to maintain fight stability and control.
b) Control and maneuvering of the Lightcraft is accomplished through manipulation of the surrounding atmosphere using thrust vectoring and geomagnetic torqueing.
c) The ion, PDE, and MHD propulsion systems provide active thrust vectoring for control in subsonic, supersonic, and hypersonic flight, respectively.
d) Active flow control (on a massive scale) significantly affects vehicle aerothermo-dynamics, control, and maneuvering authority.

17.2 Subsonic Flight Regime

The LTI-20's ion propulsion system enables flight speeds up to 44 m/s, whereas the pulse detonation engine (PDE) can quickly push the Lightcraft to supersonic speeds. Under calm atmospheric conditions at low subsonic velocities (hover to 5 m/s), the ion propulsion mode does not require hull gyroscopic stabilization, but the maglev landers (docked in their magnetic bearings) must be rapidly spun. The crew may therefore remain on the bridge on the observation deck unless cruise speeds must exceed 5 m/s. At flight speeds of 18 to 160 km/hr (maximum), hull gyrostabilization becomes necessary. At this time the crew must move into the two maglev landers, which are despun in flight. The FMS computer automatically compensates for the rotational inertia of the rotating disc and the despun maglev landers.

The Lightcraft is thus capable of performing the low-altitude flight maneuvers necessary for rescue or reconnaissance missions.

17.2.1 Subsonic Stability and Control

The ion thrusters on the Lightcraft use electromagnetic fields to vector the ejected relativistic electron beams that produce ion charge clouds and engine exhaust. Under ion propulsion, the Lightcraft can be programmed to fly in any direction without the necessity of pitching or rolling. Figure 17.2.1 shows the Lightcraft using its ion engines to accelerate, cruise, and decelerate to a stop, much like a helicopter.

High winds are the most common form of adverse weather conditions facing the Lightcraft while operating under ion propulsion. The Lightcraft, with partial buoyancy at sea level, can be blown around by wind gusts. Without compensation, a strong gust could tilt or flip the craft, or throw it off course. The ion engines are capable of compensating for gusts of up to 160 km/hr. Once the craft's sensors detect even a minute unintended change in either the craft's attitude or lateral position, the ion-propulsion unit compensates by projecting the ion clouds at the necessary angles to maintain the craft's planned position and orientation (see Section 17.4). The Lightcraft can also compensate for gusts by geomagnetic torqueing in Earth's magnetic field when the alignments are correct.

At flight speeds greater than 160 km/hr, the propulsion system must transition from the ion engine to the PDE thrusters, and gyro-stabilization is mandatory. For accelerations exceeding 3 g, the crew must relocate to their escape pods. The FMS computer automatically compensates for changes in the rotational inertia of the Lightcraft and unoccupied maglev landers.

17.2.2 Adverse Flight Conditions

Slow, low-flying vehicles, such as helicopters, airships, and ultralight aircraft, can be seriously threatened by both natural and artificial adverse flight conditions. The Lightcraft is no exception. The most common adverse natural conditions include hail, rain, snow, icing, dust storms, high winds, lightning storms, and poor visibility. Man-made threats range from artificial weather disturbances such as smoke and smog to hostile military action.

Light rain and mists cause few problems for Lightcraft operations in the ion-propulsion mode when the propulsive power is drawn from a low-power microwave beam. Beamed laser power, however, is not an option in dense fog or rain because of severe attenuation and scattering of the light beam by rain and clouds. Heavy rains must be strictly avoided. The cumulonimbus clouds that produce heavy rains may be impenetrable even to the 35 GHz microwave power beam, forcing the use of onboard SMES power for short periods of time. Lightning storms pose a serious threat to the Lightcraft because the high positive charge on the hull during ion propulsion mode causes it to

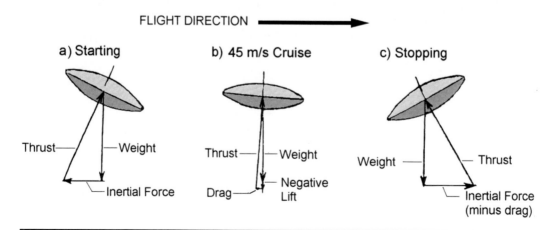

Figure 17.2.1: Free-body diagrams of starting, subsonic cruising, and stopping maneuvers.

attract lightning strikes. Aside from the obvious possibility of structural damage from a lightning strike, such a discharge could also kill the thrust (momentarily) over a large part of the Lightcraft's surface, causing it to sideslip and fall abruptly. The best protection against lightning is to avoid the vicinity of electrical storms.

Fair-weather cumulus clouds, in contrast, present an opportunity instead of a threat. A Lightcraft in ion-propulsion mode can, by depositing negative charge in the base of the cloud, actually reduce the power needed to sustain flight or to hover. This effect is described more fully in Section 16.3.2.8.

Dense cumulus clouds do, however, hinder hyperjump departures. The surrounding cloud must be evaporated, possibly at a substantial energy cost, before a remote-powered hyperjump can be performed with either the PDE or MHD engines. This cloud evaporation procedure may be avoided by running the hyperjump engines on internal SMES power.

17.2.3 Geomagnetic Torqueing
There is an alternative way to torque the Lightcraft to maintain level flight or tilt the Lightcraft without using ion or PDE propulsion. The Lightcraft can pitch or roll inside Earth's geomagnetic field by using the powerful magnetic dipole field produced by the pair of superconducting magnets in the Lightcraft rim. The interaction of the Lightcraft's field with Earth's field produces a torque comparable to that achieved by using the ion thrusters.

The amount of torque available depends on the local strength of the Earth's magnetic field (B_e ~ 0.5 gauss), the magnetic moment of the Lightcraft's dipole field (see Figure 17.2.2), and the included angle between the two fields:

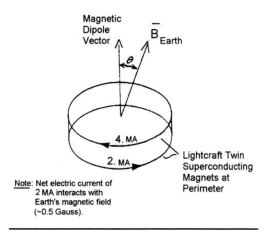

Figure 17.2.2: Lightcraft rim magnets for pitch / roll control using geomagnetic torque. *(Courtesy of RPI.)*

$$T = (I_N A\, B_e) \sin \Theta$$

where T is the magnetic torque in Newton-meters (Nm), I_N is the "net" electric current in millions of amperes, $A = \pi\, r^2$ is the area enclosed by the loop (for an LTI-20, A = 289.5 m²), and Θ is the "alignment" angle between the local direction of Earth's field and the magnetic moment of the Lightcraft's magnets. Figure 17.2.3 shows the perimeter superconducting magnets which have a maximum rated current of 4 MA (for MHD boost) and nominal current of 3 MA under other, less energetic propulsion modes. If for example, 1 MA of current is transferred from the lower magnet (reducing it from 3 MA to 2 MA) into the upper magnet (increasing it from 3 MA to 4 MA), then a "net" current of 2 MA will torque the lightcraft in the Earth's magnetic field (as in Figure 17.2.2). Typical pitch or roll rates for that "net" current of 2 MA are given in Table 17.2.1, along with the elapsed times to rotate through 30°.

Table 17.2.1: LTI-20 Geomagnetic torquing performance vs. Θ for 30° pitch/roll change.

Θ "Geomagnetic Alignment" (degrees)	Angular Acceleration In Pitch or Roll (degrees/sec²)	Time to Perform 30° Pitch or Roll Change (seconds)
30	8.64	1.08
60	15.0	0.82
90	17.3	0.76

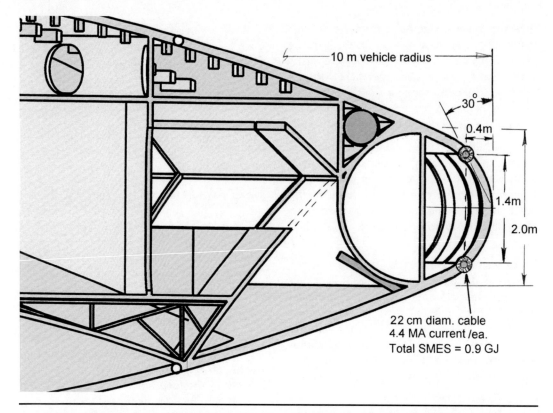

Figure 17.2.3: Location and geometry of rim magnets in LTI-20 lightcraft. *(Courtesy of RPI.)*

17.3 Supersonic PDE Flight Regime

When accelerating beyond 160 km/hr and through Mach 1 in lateral flight using the PDE thrusters, the Lightcraft hull must be rotating in order to have intrinsic stability. The aerodynamic pitching moment and lift on a lenticular disc are not affected as the spin rate is increased, but small Magnus rolling moments appear along with an increasing gyroscopic resistance to pitch and roll. This increased stability is essential for a secure and solid lock-on to the remote power beam.

The basic criterion for stability of a spinning disc is that its pitch and roll rates and velocity perturbations must return to equilibrium when disturbed. This condition does not, however, require that the position or angle of attack of the Lightcraft will be the same after a disturbance as it was before.

17.3.1 Stability and Control in PDE Flight

The LTI-20 Lightcraft's FMS computer models the forces and moments acting on the Lightcraft as those on a lenticular disc with a diameter-to-thickness ratio of 6:1. The physics of unpowered, Frisbee-mode flight is well understood (see Section A5).

The physics of external flow control and thrust vectoring for spin-stabilized disc Lightcraft cannot be revealed here (restricted information).

17.3.2 Pitch and Roll in PDE Flight

The theory, use, and subsystem structure of the Pulsed Detonation Engine (PDE) are discussed in detail in Section 6. A quick review of these concepts is needed to understand how the PDE thrusters are used for pitch, yaw, and rotation.

The PDE thrusters are powered by a high-intensity microwave beam transmitted from a power station or relay satellite. This beam must be precisely aligned with the Lightcraft antennas, and hence with the principal axis of the Lightcraft, for the PDE engine to receive adequate power to function properly. For stealth (maximum noise abatement), the microwave beam is pulsed at

FLIGHT DYNAMICS

either subsonic or ultrasonic pulse repetition frequencies. When this beam is received at the rectennas it is reflected to focus just outside of the Lightcraft rim, where it triggers electrical breakdown of the air. Microwave intensities at the focal line girdling the rim are strong enough to produce detonations with peak pressures up to 30 atmospheres and temperatures high enough to partially ionize the air into an electrically conductive plasma. This plasma is then vectored in the desired directions by applied electromagnetic fields that emanate from the rim superconducting magnets. In bright daylight, these thrust-vectored pulses are barely visible to the unaided eye, in part because each pulse lasts only 1 millisecond. A schematic of the PDE thruster cycle is given in Figure 17.3.1 (next page).

The PDE's main function as an air-breathing engine is to push the Lightcraft rapidly through Mach 1, jumping it to supersonic speeds (i. e., a hyperjump). As Figure 17.3.2 shows, to initiate an axial hyperjump the PDE thrusters fire sequentially all around the Lightcraft rim, creating a uniform exhaust flow that is vectored downward to produce a lifting thrust. The plasma exhaust can just as easily be vectored upward, producing a downward acceleration. In a maximum-performance maneuver the LTI-20 can jump a few kilometers in 1 to 2 seconds, using the airspike to reduce drag and heat transfer after breaking Mach 1.

Aside from rapid acceleration in the axial and lateral (transverse) directions, the PDE can be used to hover, pitch, roll, and spin the Lightcraft. A pitch-up or roll maneuver requires exerting a torque in the desired direction, as shown in Figure 17.3.3. The pulses are so quick and discrete that the Lightcraft can torque sharply to almost any attitude in a millisecond (when the crew is secured in the de-spun maglev landers).

The PDE exhaust is vectored by manipulating the electric current in the lower and upper rim magnets to torque the Lightcraft in the desired direction. These current changes in turn alter the intensity and direction of the magnetic field, whose effect is to act as a magnetic nozzle that controls the direction of the hot plasma exhaust

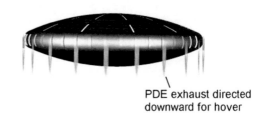

Figure 17.3.2: Downward accelerated PDE exhaust produces lifting force. *(Courtesy of RPI.)*

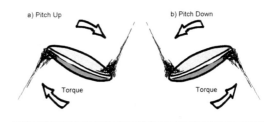

Figure 17.3.3: Pitch / roll maneuvers with PDE thrusters, showing plasma exhaust plumes and resultant applied torque. *(Courtesy of RPI.)*

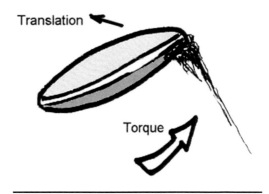

Figure 17.3.4: Counterclockwise pitch with PDE thruster; generates torque and translation. *(Courtesy of RPI.)*

stream. Holding the current in the upper superconducting magnet higher than that in the lower, as indicated in Figure 17.3.4 (see also Figure 6.3.2), vectors the plasma exhaust downward and pitches the Lightcraft in the counterclockwise sense (as indicated). Note that only the right side PDE thrusters are engaged (i.e., illuminated by the microwave beam) for this pitch maneuver.

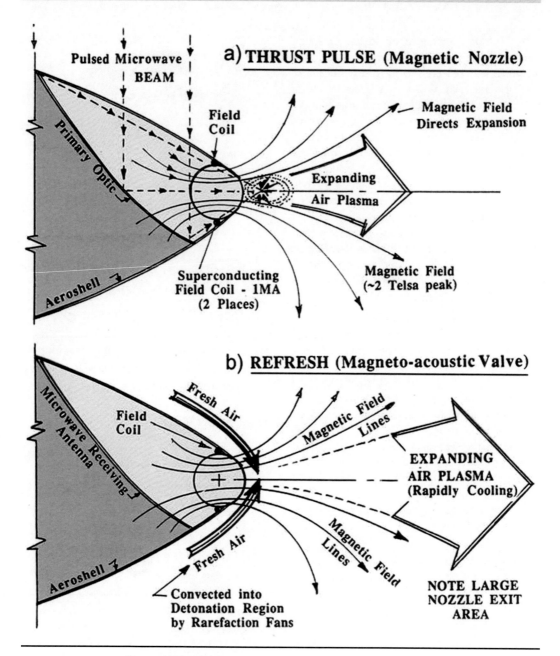

Figure 17.3.1: Air-breathing pulsed detonation engine cycle.

17.3.3 Spin Rate Control

Aided by the MHD system, the PDE thrusters can be used to rotate the Lightcraft about its axis of symmetry. By detonating the thrusters in sequence around the Lightcraft, oblique rotating detonation wavefronts form around the rim (see Figure 17.3.5), causing the vehicle to spin. This torque is greatly magnified by MHD forces as indicated in Figure 17.3.6. To spin up the Lightcraft under MHD power, the rectennas extract some portion of the incident microwave beam for conversion into electric pulses, which are delivered to the upper and lower rim electrodes. Electrical discharges, driven through the (microwave-detonated) air plasmas in the presence of a 2-Tesla magnetic field, induce the azimuthal acceleration through the action of powerful Lorentz forces.

17.4 Hypersonic Flight Regime

The combination of the airspike and the annular MHD slipstream accelerator makes the Lightcraft capable of transatmospheric flight into space. The core portion of the axially aligned microwave beam is reflected off the central rectenna and focused above the disc (about 1 vehicle diameter away) to create a very hot ionized region from which the plasma expands, forming a conical volume of hot, low-density air. The conical air plasma volume is called a Directed Energy AirSpike or DEAS. In the process of airspike formation, forebody drag and heat transfer to the hull are greatly reduced and the propulsive efficiency of the MHD accelerator is greatly increased. The airspike acts as a variable geometry inlet for the air-breathing annular MHD engine, delivering compressed air

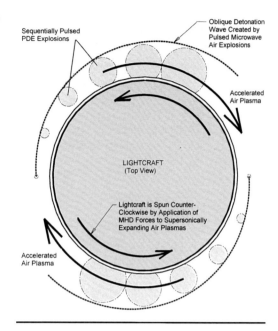

Figure 17.3.5: Top view of lightcraft being spun by combination of PDE thrusters and MHD applied Lorentz forces acting upon expanding air plasmas from rim. *(Courtesy of RPI.)*

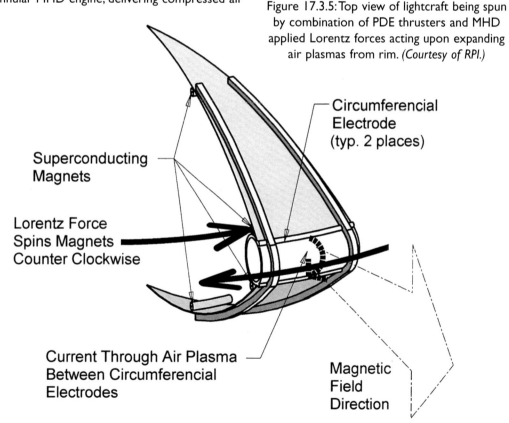

Figure 17.3.6: MHD induced "yaw" (or spin) maneuver to torque lightcraft hull about central axis. *(Courtesy of RPI.)*

just outside the rim of the Lightcraft, trapped between the bow shock wave and the plasma contact surface (see Figure 5.6.1).

The electrical energy required by the MHD engine comes from two rectifying antennas positioned just inside the top hull of the Lightcraft. These 35 GHz rectenna arrays can deliver up to 9 GW of electrical power to the engine. The biggest advantage of the airspike technology is that the Lightcraft can be streamlined without a mass penalty, which is an extremely valuable feature for a trans-atmospheric vehicle.

Flight stability throughout the hypersonic regime is accomplished by active thrust vectoring of the MHD slipstream accelerator, and by precise positioning of the airspike over the disc. By moving the airspike slightly off-axis, precise pitch and roll moments can be exerted on the forward hull to help steer the Lightcraft on the desired flight path.

17.5 Reentry and Aerobraking Methods

The key mission parameter that defines the geometry of the LTI-20 Lightcraft is the requirement of a capability for atmospheric and space flight. The vehicle is designed to transport its crew safely around Earth as well as between Earth and the Moon. Return to Earth from any mission at speeds above Earth's escape velocity requires dissipation of excess kinetic energy to permit capture by Earth's gravity. This use of aerobraking is called aerocapture.

For successful aerobraking, the Lightcraft must orient itself to achieve the desired angle of incidence as it approaches contact with the upper atmosphere, at approximately 150 km altitude. Because the Lightcraft typically enters the atmosphere at a speed of 8 km/s for return from low orbit, or 11 km/sec for return from the Moon, deceleration during atmospheric reentry will require high g loads unless some means is sought to alleviate the deceleration. The LTI-20 uses three different methods of handling the reentry deceleration problem.

17.5.1 Aerobraking

The principle of aerobraking was first put to practical use in the early days of space exploration. Most space capsules of that era, notably those in the Mercury, Gemini, and Apollo programs, had blunt parabolic bases (similar to that of the Lightcraft), and were heavily insulated to act as heat shields. Upon reentry the capsule was oriented shield-first to create a strong normal shock wave on the curved surface (Figure 17.5.1). This shock wave generated a tremendous amount of drag, decelerated the vehicle rapidly, and imposed high g loads (in direct proportion to the reentry incidence angle, of course).

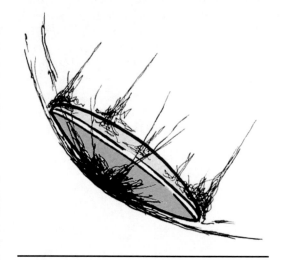

Figure 17.5.1: LTI-20 under passive aerobraking during reentry. *(Courtesy of RPI.)*

However, the reentry velocity of an LTI-20 returning from lunar or interplanetary missions can be as high as 20 to 25 km/s, twice the entry speed of an Apollo command module returning from the Moon. Aerobraking alone does not allow the Lightcraft sufficient flexibility for safe direct entry and recovery. Alternatives to such direct entry include adopting a different flight profile, using magnetohydrodynamics in flight, or resorting to powered deceleration during reentry.

17.5.2 Skip Trajectory

Entry deceleration can be alleviated by a skip-glide trajectory (Figure 17.5.2). A reentry vehicle that generates lift may use multiple shallow passes through the upper atmosphere to reduce its original kinetic energy. After each increment of velocity loss, the vehicle may use its lift to kick it back out of the atmosphere where aerodynamic heating is negligible. Emerging from the atmosphere, the vehicle loses lift, cools off, and drops back into the atmosphere. By repeating such a maneuver the craft may reenter the atmosphere safely without incurring the penalty of high thermal or g loads.

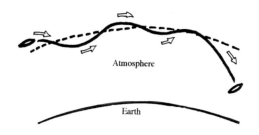

Figure 17.5.2: Hypersonic "atmosphere skipping" flight mode. *(Courtesy of RPI.)*

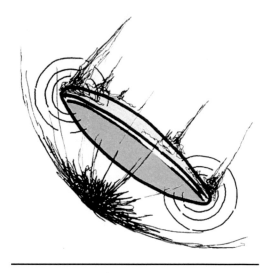

Figure 17.5.3: Magnetic aerobraking enhancement with increased bow shock standoff, reducing heat transfer. *(Courtesy of RPI.)*

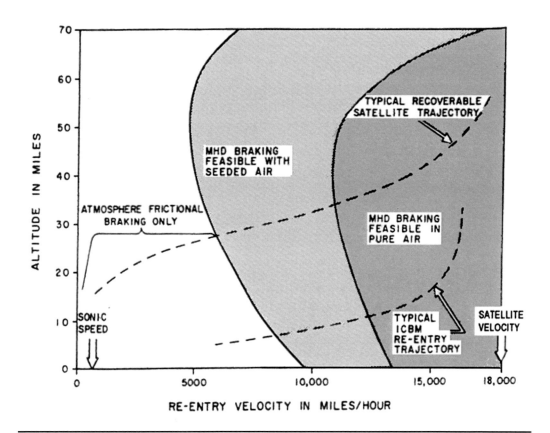

Figure 17.5.4: Flight regime for MHD aerobraking. *(Courtesy of A. Kantrowitz and AERL.)*

17.5.3 Magneto-Aerobraking

One of the drawbacks associated with traditional aerobraking is the rapid heat buildup on the reentry shields. Hypervelocity air slams into the outer hull, heating the air and the hull's heat shield to incandescence. Aerobraking heat loads are so high that a refractory hull, a massive heat sink, or a heat-dissipating mechanism is required. All three of these approaches incur serious weight penalties.

The concept of flight magneto-aerobraking was first demonstrated in the 1950s at the AVCO Everett Research Laboratory by Arthur Kantrowitz, who discovered that a strong magnetic field increased the stand-off distance of the bow shock from the hull of a reentry vehicle (see Section 22.8.5.3). The LTI-20 Lightcraft takes full advantage of this principle upon reentry. By charging the two superconducting magnets along its rim, a powerful 2-Tesla magnetic field forms around the vehicle. This field pushes the bow shock away from the hull (Figure 17.5.3), thus preventing the hypersonic air stream from coming into close contact with the hull and greatly alleviating the reentry heat load due to aerothermodynamic heating. Radiant heating of the hull by the incandescent shock wave is also hindered by the bright mirror-like surface of the Lightcraft. Figure 17.5.4 shows the hypersonic flight regime (i.e., altitude vs. velocity) where magneto-aerobraking is feasible.

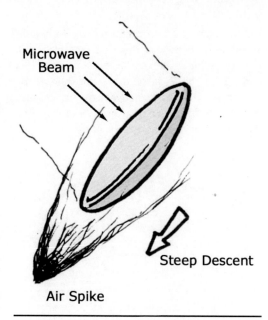

Figure 17.5.5: Beam powered reentry for maximum performance descent. *(Courtesy of RPI.)*

17.5.4 Powered Reentry

In suborbital or fractional orbital flights around Earth for the surveillance and rescue missions, the LTI-20 may require a maximum-performance descent to low-altitude hover (Figure 17.5.5). In this powered reentry mode, the Lightcraft enters the atmosphere with lateral flight orientation, with the airspike and the MHD slipstream accelerators energized.

Lightcraft (lower left) in MHD propulsion mode, powered by 1 km diameter SPS-01 (upper right) in low Earth orbit. *(Bob Sauls, Frassanito and Associates; Courtesy of NASA.)*

Upper view of lightcraft in MHD propulsion mode, showing electric discharges from rim electrodes. *(Bob Sauls, Frassanito and Associates; Courtesy of NASA.)*

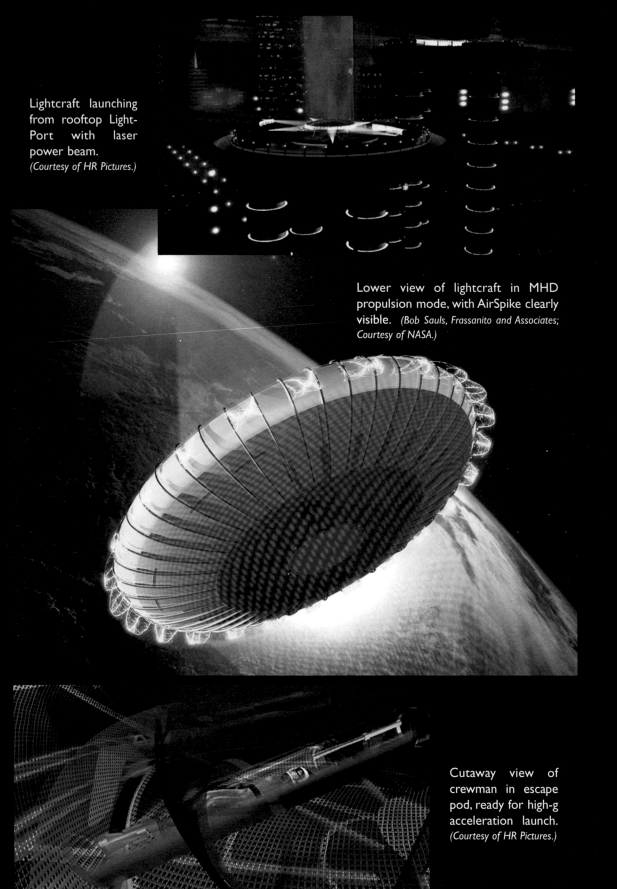

Lightcraft launching from rooftop Light-Port with laser power beam. *(Courtesy of HR Pictures.)*

Lower view of lightcraft in MHD propulsion mode, with AirSpike clearly visible. *(Bob Sauls, Frassanito and Associates; Courtesy of NASA.)*

Cutaway view of crewman in escape pod, ready for high-g acceleration launch. *(Courtesy of HR Pictures.)*

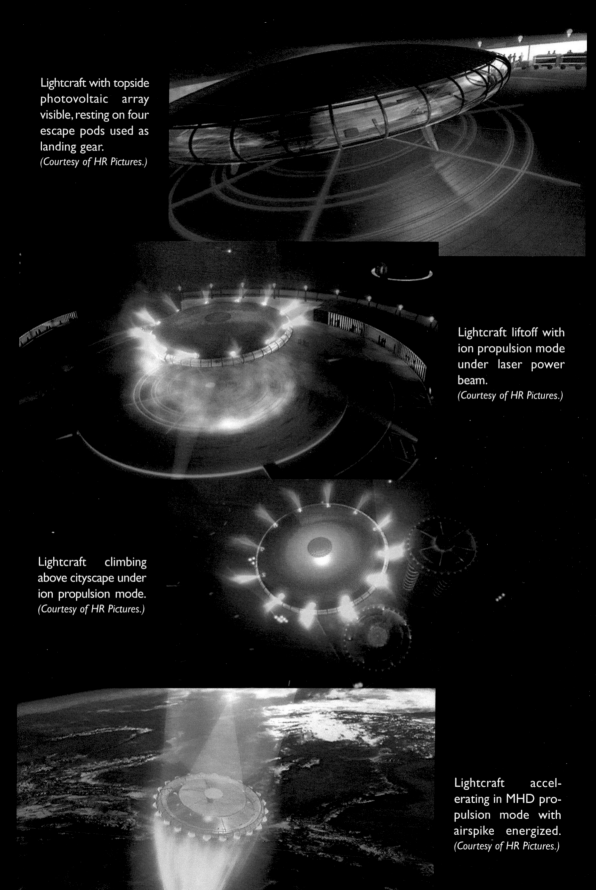

Lightcraft with topside photovoltaic array visible, resting on four escape pods used as landing gear.
(Courtesy of HR Pictures.)

Lightcraft liftoff with ion propulsion mode under laser power beam.
(Courtesy of HR Pictures.)

Lightcraft climbing above cityscape under ion propulsion mode.
(Courtesy of HR Pictures.)

Lightcraft accelerating in MHD propulsion mode with airspike energized.
(Courtesy of HR Pictures.)

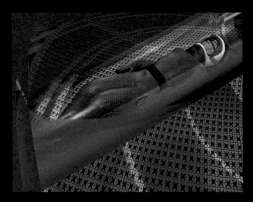
Cutaway view of escape pod with crew-member in ultra-g suit. *(Courtesy of HR Pictures.)*

Cutaway of lightcraft hull showing two crew-members in escape pods. *(Courtesy of HR Pictures.)*

Lightcraft reentry from low Earth orbit. *(Courtesy of HR Pictures.)*

Lightcraft resting on rooftop LightPort. *(Courtesy of HR Pictures.)*

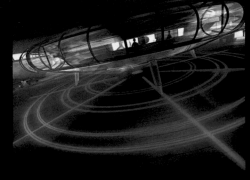
Crew preparing lightcraft for flight. *(Courtesy of HR Pictures.)*

Lightcraft liftoff with ion propulsion mode. *(Courtesy of HR Pictures.)*

Transparent view of lightcraft hull revealing 6 of the 12 escape pods. *(Courtesy of HR Pictures.)*

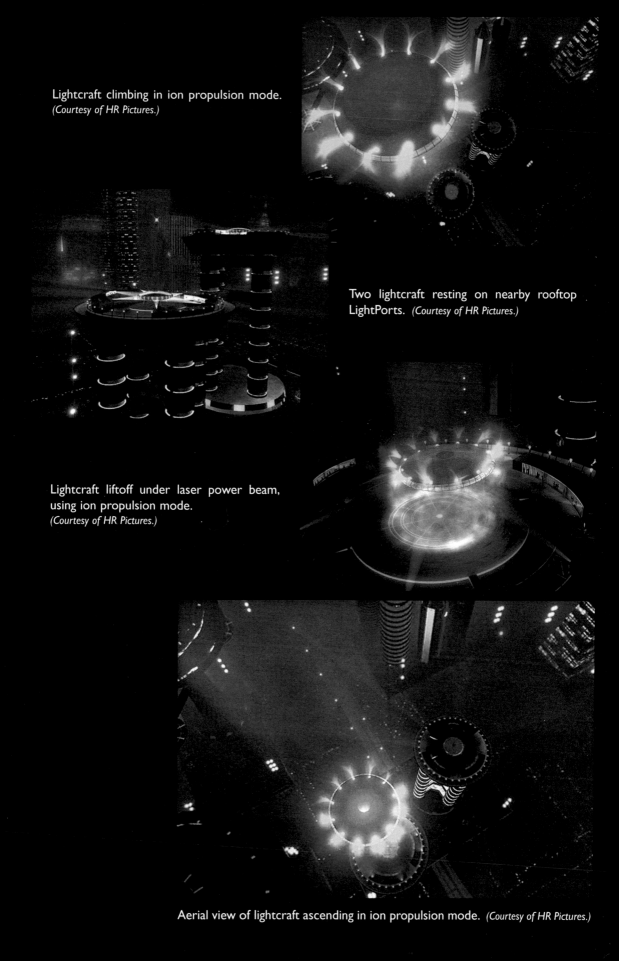

Lightcraft climbing in ion propulsion mode. *(Courtesy of HR Pictures.)*

Two lightcraft resting on nearby rooftop LightPorts. *(Courtesy of HR Pictures.)*

Lightcraft liftoff under laser power beam, using ion propulsion mode. *(Courtesy of HR Pictures.)*

Aerial view of lightcraft ascending in ion propulsion mode. *(Courtesy of HR Pictures.)*

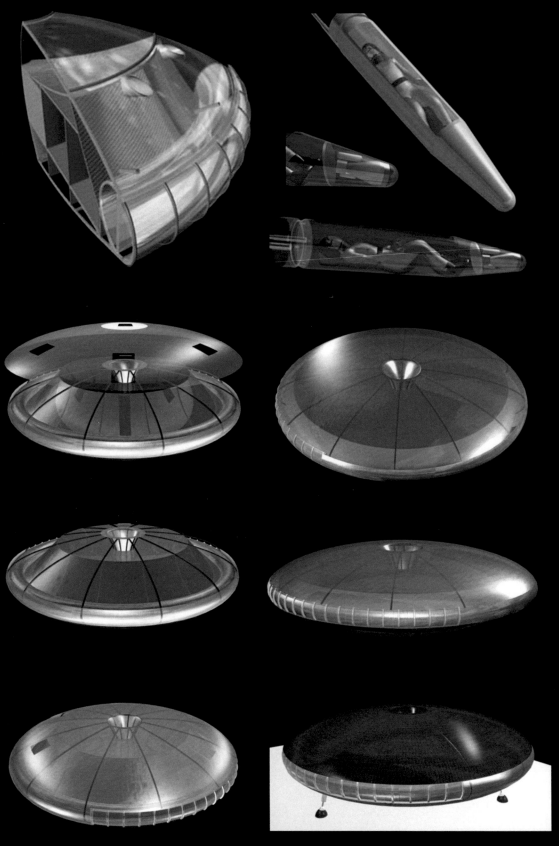

Composite of 6 lightcraft renderings *(Media Fusion; Courtesy of NASA)*, with 4 cutaway views of hull and escape pods. *(Courtesy of HR Pictures.)*

Lightcraft in MHD propulsion mode with AirSpike energized. *(Media Fusion; Courtesy of NASA.)*

Lightcraft in hypersonic MHD propulsion mode with AirSpike on. *(Media Fusion; Courtesy of NASA.)*

Lightcraft accelerating laterally in pulsed detonation mode. *(Media Fusion; Courtesy of NASA.)*

Crew disembarking from lightcraft in southwest USA desert. *(Lightcraft Website Team; Courtesy of RPI.)*

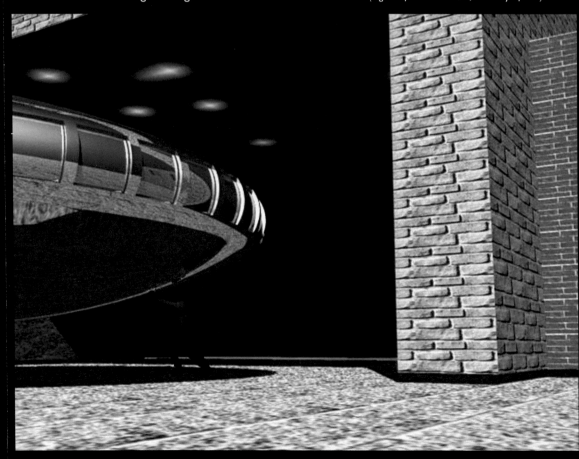

Neutrally buoyant lightcraft being pushed out from hanger by ground crew. *(Lightcraft Website Team; Courtesy of RPI.)*

Lightcraft hovering at Yosemite using Maglev lander to scavenge lake water. *(Lightcraft Website Team; Courtesy of RPI.)*

Lightcraft climbing under ion propulsion mode in humid day, with water vapor condensing in wake. *(Lightcraft Website Team; Courtesy of RPI.)*

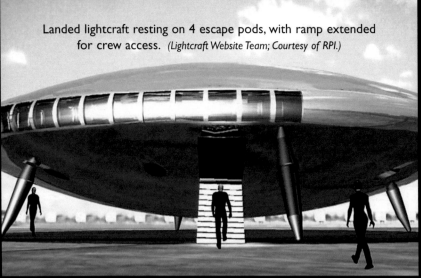

Landed lightcraft resting on 4 escape pods, with ramp extended for crew access. *(Lightcraft Website Team; Courtesy of RPI.)*

Above: Lightcraft bridge crew at flight control station, with rectenna panels raised for direct viewing of exterior airspace through transparent hull section. *(Lightcraft Website Team; Courtesy of RPI.)*

Below: Lightcraft hovering under ion propulsion mode in nighttime cityscape. *(Lightcraft Website Team; Courtesy of RPI.)*

Maglev lander decending with hatch open. *(Chuck Lindgren;. Courtesy of RPI.)*

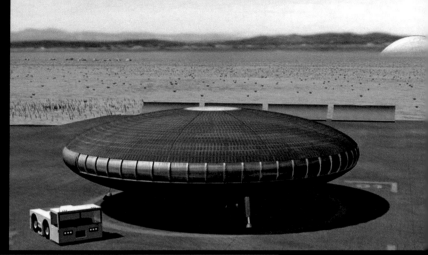

Lightcraft maintenance ramp at HEMSTF, with dome-covered PDS-01 (upper right) in background. *(Media Fusion, Courtesy of NASA.)*

Lightcraft flying in ion propulsion mode, with condensation in wake. *(Lightcraft Website Team; Courtesy of RPI.)*

Lightcraft crew embarking via landing ramp. *(Lightcraft Website Team; Courtesy of RPI.)*

PDS-03 power beaming station in charging mode, with liquid droplet radiators extended.
(Chuck Lindgren; Courtesy of RPI.)

Right: Lightcraft hovering in dense fog, with ion propulsion mode. *(Lightcraft Website Team; Courtesy of RPI.)*

PDS-03 power beaming station in "standby" mode, fully charged, with LDRs closed.
(Chuck Lindgren; Courtesy of RPI.)

Lightcraft "hovering" at high altitude in circular flight mode with crew in annular corridor. *(Lightcraft Website Team; Courtesy of RPI.)*

Maglev lander resting on tripod gear. *(Chuck Lindgren; Courtesy of RPI.)*

Landed lightcraft at remote desert site. *(Lightcraft Website Team; Courtesy of RPI.)*

Moonlit lightcraft hovering at street level, with condensation in wake.
(Lightcraft Website Team; Courtesy of RPI.)

Lightcraft accelerating in MHD propulsion mode with AirSpike energized.
(Lightcraft Website Team; Courtesy of RPI.)

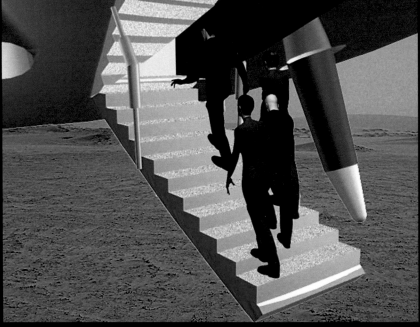

Crew entering lightcraft via landing ramp.
(Lightcraft Website Team; Courtesy of RPI.)

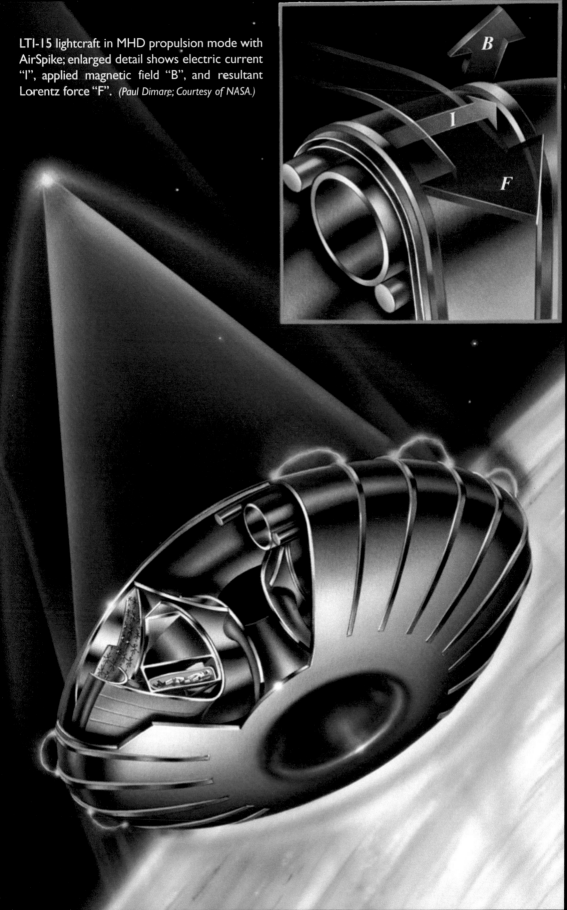

LTI-15 lightcraft in MHD propulsion mode with AirSpike; enlarged detail shows electric current "I", applied magnetic field "B", and resultant Lorentz force "F". *(Paul Dimare; Courtesy of NASA.)*

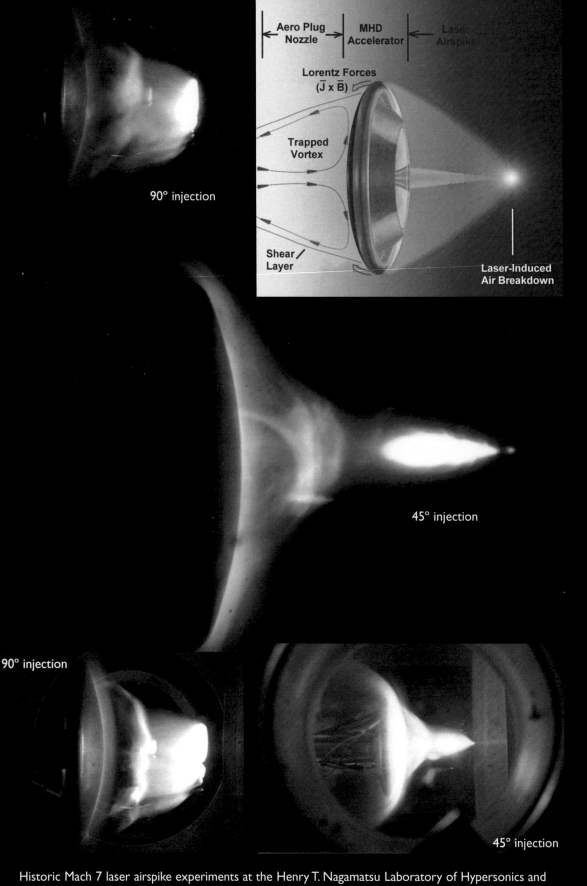

Historic Mach 7 laser airspike experiments at the Henry T. Nagamatsu Laboratory of Hypersonics and Aerothermodynamics. *(Photos by A.C. Oliveira and I.I. Salvador; Courtesy of IEAv-CTA, Sao Jose dos Campos, Brazil)*

18.0 EMERGENCY OPERATIONS

The LTI-20 is capable of various types of emergency operations. These operations are essential to the safety of the crew members and their vehicle. These features include ejection pods, a protective gas bag, and specific plans for rescue and evacuation operations.

18.1 Introduction to Emergency Operations

Safety is the single most important factor in the design of the Lightcraft. The central element and controlling intelligence for assuring the safety of the Lightcraft and crew is the FMS computer. Among other tasks, this advanced intelligent computer system monitors and controls all of the vehicle's safety features.

The exterior double hull of the Lightcraft is designed to function as a gas bag that protects the crew against crashes. The exterior silicon carbide hull is inflated with HeliOx at 1.5 atmospheres pressure (at sea level) so that it acts like a cushion should the Lightcraft come into contact with atmospheric detritus, trees, buildings, or the ground. The primary structural member of the hull is a toroidal pressure vessel, inflated by 10 atmospheres of HeliOx, that constitutes the rim of the Lightcraft disc. The lenticular double hull is attached directly to this toroidal tube and is stabilized by it. Under certain crash conditions the outer layer of the lenticular hull is designed to separate from the vehicle. The shed skin temporarily shields the remaining structure from catastrophic damage such as complete structural collapse or disintegration.

The Lightcraft is also equipped with plasma shields as a first line of defense against solar flare proton bombardment in space. In the event of a solar flare, the computer activates the space plasma shield (see Section 15.5) to protect the crew from protons with energies up to 200 MeV. During flight through space the FMS computer constantly monitors the space environment for orbital debris on collision paths, using the two hemispherical LIDAR sensors. Upon detecting an imminent collision, one of the three onboard 300 kW laser tracks and vaporizes debris large enough to cause damage to the Lightcraft. Often such debris can be "laser propelled" out of the Lightcraft's orbital path by rapidly evaporating (or ablating) the surface material. In the event the debris is too massive to be deflected in time, Lightcraft orbital maneuvering system (OMS) engines are fired up to move the vehicle out of harm's way.

Additionally, the Lightcraft is equipped with 12 internal crew ejection pods. In the unlikely event of catastrophic Lightcraft failure, the FMS computer instructs the crew members to evacuate the ship and then directs all subsequent ejection and rescue operations. Once the crew members are securely positioned in their escape pods the computer ejects the pods. Crew members are later retrieved by another Lightcraft.

The maglev landers provide an additional method of escape for the crew. The Lightcraft is equipped with two landers, one located on the top and the other located at the bottom of the Lightcraft. Each lander can safely carry six passengers through reentry.

The maglev landers are retrieved by means of the maglev coils located at the base of the rescuing Lightcraft. Upon approaching the lander, the Lightcraft's maglev coils lift the lander into position at the base of the central donut hole. Section 10.2.2 covers the maglev lander systems in more detail.

The Lightcraft must be in a stationary hover or landed to retrieve an individual escape pod. The Lightcraft uses a special maglev belt to lift the escape pod aboard.

18.2 Lightcraft Hull as Protective Gas Bag

The exterior double hull of the Lightcraft is a tough, thin-film, pressurized gas envelope designed to absorb enormous crash loads, and thus help protect the crew. This shock-absorbing gas bag acts as a cushion in the event that the Lightcraft comes into contact with any airborne or surface object.

The rim of the Lightcraft disc is girdled by a toroidal pressure vessel filled with HeliOx at a pressure of 10 atmospheres. At low velocities this "outer tube" acts as a bumper to cushion the force of any impact and to reduce greatly the peak acceleration loads on the crew. At very low subsonic speeds the Lightcraft will actually bounce off most of the objects it may collide with.

As an additional safety feature, the tough double outer hull enables the Lightcraft to graze a massive foreign object above or below it without causing catastrophic damage to the primary structure of the craft. The outer layer of the hull is considered sacrificial in such glancing collisions because its loss may avoid hull rupture upon impact while preserving the Lightcraft structure and crew in safety. Nonetheless, loss of the outer hull makes the Lightcraft unsafe to fly, and perhaps even impossible to repair.

The level of impact the Lightcraft can withstand in a collision with the ground depends on the impact velocity and angle of incidence with the ground. The higher the velocity, the lower the glide path angle must be for the Lightcraft to survive its first contact with the ground.

Specific strong or sharp surface obstacles such as tree trunks, utility poles, antennas, and guy wires greatly increase the hazard of contact. Also, when impact occurs at high velocities the pressurized hull may burst. The Lightcraft is therefore equipped with additional collision-avoidance detectors that actively search for collision hazards. Since the craft is capable of accelerations of 200 g in atmospheric flight, it can normally dodge any collision threat, but only if all crew members are safely protected in their escape pods.

18.3 Biconic Escape Pod Ejection

In the event of a catastrophic structural failure of the Lightcraft during high-g boost, the personnel in the crew escape pods must be removed as quickly and safely as possible. To do this, the pods must be ejected from their launch tube "bays" at a high enough speed to escape the debris cloud from the disintegrating Lightcraft.

There are two basic methods of achieving pod ejection. The first involves using the 1.5 atm of pressure already inside the craft to propel the pod away, as shown in Figure 18.3.1. At sea level, with a 1-atm back-pressure, the pressure differential of 50 kPa acting on a 60-cm diameter pod exerts 14.1 kN of force on the pod base. Assuming a total mass of 100 kg for the pod plus occupant, the acceleration of the pod will be about 15 g, more than sufficient to remove the pod from the damaged Lightcraft. Retention structures such as the sliding bar attached to the pod will break away under intense acceleration and aerodynamic forces, freeing the pod completely from the Lightcraft. At altitudes above about 30 km, where the back-pressure of the external atmosphere is much smaller than sea level, the ejection acceleration is still 15 g because the FMS computer maintains the gauge pressure at 0.5 atm.

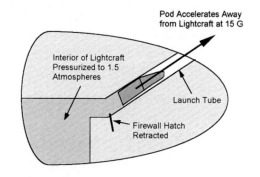

Figure 18.3.1: Pod ejection by internal lightcraft pressure. *(Courtesy of RPI.)*

The second method of ejecting the pods is used if the 1.5 atm of HeliOx pressure in the hull has been lost because the hull has ruptured. In this situation, the pod launch tube acts like a cannon, and the pod is ejected by firing a small rocket gas generator charge behind it (see Figure 18.3.2). First, the landing gear mechanism that extends the pods during normal landing will push the pod into the launch tube. Next, a small firewall door is closed behind the pod, completely isolating it from the Lightcraft

EMERGENCY OPERATIONS

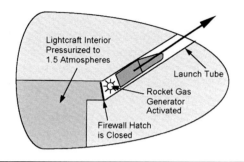

Figure 18.3.2: Pod ejection by rocket gas generator in launch tube. *(Courtesy of RPI.)*

interior. The propellant charge is then fired behind the pod, creating high-pressure exhaust gases that swiftly propel the pod out of the tube. Any gear-extension structure attached to the pod is sheared away in the process. If the propellant charge misfires, a pressure regulated backup system bleeds high pressure HeliOx from the perimeter toroid (15 atm reservoir) to carry out the ejection process.

Once the ejection pods have safely cleared the vicinity of the Lightcraft, they negotiate reentry and maneuver through the atmosphere to make a safe, controlled landing. There are several mechanisms on the pod to assist in landing.

In the nose of each pod is a small pressure-fed chemical rocket engine fueled by gaseous oxygen and hydrogen. The crew member uses this engine to de-orbit the escape pod for reentry into Earth's atmosphere. By firing in the direction of motion, the engine acts as a retrorocket to drop the speed of the pod by 100 m/s below circular orbital speed and hence dip into the upper atmosphere. The pod is equipped with a thermal protection system to protect it from the searing heat of reentry. This engine can also be used to further slow the pod in the final stages of landing.

Once in the lower atmosphere, the pod deploys an aerodynamic flap from its rear lower edge. The pod passenger can control the pod's descent path by adjusting the deflection of this flap, much in the manner of a military maneuverable reentry vehicle. Tiny reaction control thrusters on the rear of the pod provide pitch, yaw, and roll control throughout the flight as the velocity of the pod decreases. When the pod is within 1 km of the ground the passenger can deploy a ram-air mattress-type parachute, which slows the pod enough for a safe horizontal landing. In hostile territory the deployment of the parachute can be delayed until an extremely low altitude – about 100 m – is attained. The passenger's ultra-g suit will provide adequate protection against injury in the ensuing rough landing.

The pod passenger is fully aware and informed of the surroundings by numerous sensors throughout the entire reentry procedure. Optical images and altitude, attitude, velocity, and position data are fed to the pod's onboard computer system, which in turn displays the data on the passenger's virtual reality goggles. Two side windows are provided so that the pod passenger will have the backup option of direct viewing in the event of failure of the electronic systems. If the passenger is unconscious for any reason, the pod's computer is capable of scanning the immediate vicinity and choosing a suitable landing site.

18.4 Crew Retrieval Methods

During normal retrieval operations, crew members can be lifted by their maglev belts into access portals in the maglev landers. From there, they can enter the interior of the Lightcraft by using the lander as an air-lock in the manner described in Section 10. At other times, however, air-lock access through the maglev landers may be blocked by damage, or the lander may not even be docked with the Lightcraft. In these cases, the crew member can gain entrance to the LTI-20 through another ship portal. These portals include the pod launch tubes, the landing ramp, or the maglev lander portal in the central "donut hole" of the Lightcraft.

When the Lightcraft is flying with the rectenna arrays facing the ground, and the pods are deployed as landing gear, the pods are immediately available for access. Crew members simply step into an open pod (see Figure 14.8.1), the door closes, and the pods are swiftly retracted into the Lightcraft interior.

Open launch tubes may also be used by crew members to gain access to the Lightcraft. These launch tubes have auxiliary maglev equipment that is extended down the launch tube, and are fitted with hand grips recessed into the exterior hatch door of the launch tubes (see Figure 18.4.1). This special equipment consists of a magnetic ring that hugs the sides of the tube. The ring is moved into position by an extension assembly connected to the inner pressurization hatch. The auxiliary launch tube magnet assists in drawing the boarding crew member toward and into the launch tube. Once inside the launch tube, the crew member is held in place by the attraction of the maglev belt for the auxiliary launch tube magnet. After the crew member is secure the outer launch tube door is closed and the magnet actuator assembly retracts to draw the crew member into the innermost compartment of the vehicle.

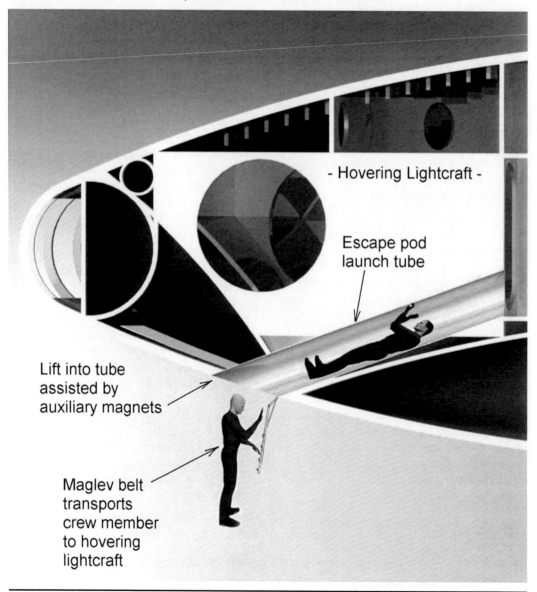

Figure 18.4.1: Emergency crew recovery by Maglev belt access into lightcraft interior through launch tube. *(Courtesy of RPI.)*

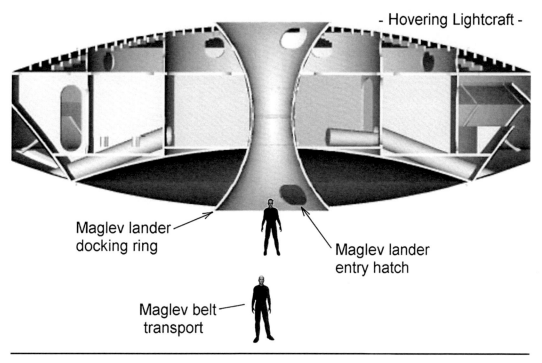

Figure 18.4.2: Maglev belt transport into open lander docking port. *(Courtesy of RPI.)*

When no maglev lander is docked on the dorsal side of the Lightcraft, entrance is possible through that lander's docking bay entrance hatches (see Figure 18.4.2). Two such hatches, both located in the central "donut-hole" of the Lightcraft, are normally used by crew members entering and exiting the maglev landers. The same superconducting magnets that normally retrieve maglev landers can also be used to lift crew members or cargo, via maglev belt, up into the empty docking bay. Once in this position the crew member is drawn into an internal air-lock compartment through a docking bay hatch, as described in Section 15.

Another boarding option is by way of a winch cable lowered from the docking bay hatch to the ground. When loading in this way the hull cannot remain fully charged for flight maneuvering in the ion-propulsion mode.

Another method of gaining entrance while the rectennas are facing the ground and the maglev systems are inoperative is by way of the landing ramps (Figure 18.4.3). Non-conducting rope ladders and cables can be lowered through the ramps to assist in recovering crew members or cargo. However, when the ramps are lowered a section of the hovering lighcraft must be depressurized, thereby temporarily weakening the structural integrity and lowering the flight worthiness of the Lightcraft. This method also decreases the amount of electrical charge that can be carried on the Lightcraft's hull because charge would be dissipated readily as corona discharges from the extended landing ramp. This method does, however, permit the recovery of large objects and multiple crew members who are not all wearing maglev belts. A winch operated at the ramp entrance can also reel in extended cables or ropes.

Flight propulsive power is severely restricted when any of the recovery options is used. For crew safety, only the ion-propulsion mode and SMES "battery" can be used, which prevents the use of megawatt-level beamed laser or microwave power. These power limitations in turn restrict Lightcraft speeds during recovery operations, as does the lessening of the structural strength of the Lightcraft due to the depressurization of parts of the hull required in certain recovery scenarios.

Chapter 18

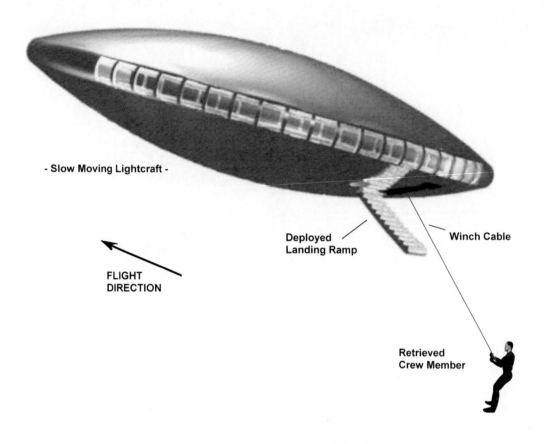

Figure 18.4.3: Crew retrieval by cable-winch into lowered landing ramp. *(Courtesy of RPI.)*

Open landing ramps place the most severe restrictions on flight operations.

All grounded crew members awaiting recovery by a hovering Lightcraft must be fully clothed in protective SAU suits that are laser- and microwave-reflective. The beam power must be closely monitored to ensure that crew members are not exposed to harmful levels of beamed energy. The potential for crew exposure can be minimized by switching to internal SMES power at the moment of actual recovery.

18.5 Fire Suppression

A fire onboard a partially buoyant Lightcraft is a serious threat to the lives of the crew. To minimize such risks, all materials within the crew compartment of the LTI-20 must adhere to strict standards of flammability mandated by Space Command. All equipment, furnishings, and crew's belongings must be pre-approved before boarding the Lightcraft to minimize the presence of unnecessarily flammable materials. The Lightcraft's interior is partitioned into small rooms designed to contain and prevent the spread of an accidentally ignited fire. Each room features airlock doors that help to isolate hazardous areas from the rest of the Lightcraft. Note that the high thermal conductivity of helium in the HeliOx mix already tends to inhibit fire propagation by robbing the flame of heat.

Atmospheric sensors, as described in Section 13.1, are positioned throughout the Lightcraft to detect sudden changes of temperature, pressure, and humidity. The FMS computer has a fire-suppression system that analyzes the sensor data and can instruct the HeliOx plenum system to discontinue the circulation of HeliOx gas through compartments that have fire burning. The FMS computer can also activate the air-lock doors to seal off a burning compartment to suffocate the fire. In addition, crew members can broadcast a fire alarm manually from any fire alert station located in every habitable compartment on board the LTI-20. By analyzing gaseous combustion products, the FMS computer system then determines the type and severity of the fire and classifies the fire into one of three categories:

- *Class A Fire*: A small, confinable fire, of no threat to vital support systems, hull integrity, or human life, such as burning waste materials in a trash receptacle.
- *Class B Fire*: A small fire, but one posing a threat to vital support systems, hull integrity, or human life, such as a fire in a critical electrical supply panel of the SMES unit.
- *Class C Fire*: A large fire, not easily contained, of extreme danger to vital support systems, hull integrity, or human life, such as ignition of any combustible or ultra-energetic material, or malfunctioning on-board beam weapon system.

After classification the FMS computer system takes appropriate action to fight the fire.

The Class A fire is controlled by sealing the air-lock hatches around the fire, thereby confining the fire to the smallest possible volume. The environmental control system then begins to starve the fire by cutting off the HeliOx supply, leaving the fire to burn itself out.

The Class B fire is considerably more dangerous to the Lightcraft. Again, the hatches are sealed to confine the fire in the minimal volume. The burning compartment is then flooded with pure helium gas that slightly overpressurizes the compartment, swiftly smothering the fire with inert gas.

The Class C fire is of grave danger to the Lightcraft. Again, the compartment containing the fire is sealed off. In addition, personnel are evacuated from all adjacent compartments and those compartments are sealed. The fire is sprayed with liquid helium, simultaneously cooling and smothering the flames. All adjacent compartments are flooded with pure helium gas to help contain the blaze.

To limit the risk of spontaneous re-ignition or flash-back, corrective actions are maintained until the combustible material has cooled to a level below the spontaneous combustion temperature.

If a fire reaches the point where it is uncontrollable and has already spread to more than 15 - 25% of the Lightcraft's interior, the emergency evacuation procedure is initiated, whereupon the crew ejects using their escape pods or maglev landers. Figure 18.5.1 visually summarizes the fire suppression system of the LTI-20.

18.6 Emergency Use of Maglev Landers

In an emergency operation, such as the rescue of ground forces from a hostile landing zone, the maglev landers may be used for evacuation. Well within the rated load limit of 2400 kg for the maglev lander, as many as seven adults can be rescued together (volume permitting).

In addition, the maglev landers are equipped for individual reentry and can be used to escape a crippled Lightcraft in orbit and reenter Earth's atmosphere.

18.7. Emergency Airspike Support

Emergency power is drawn from the LTI-20 SMES unit and beamed out from the central nested rectenna – operating in mm-wave phased array transmit mode – for emergency maintenance of the airspike, in the rare event that microwave power transmission from remote space or terrestrial stations is momentarily interrupted. Figure 18.7.1 (from Reference A144) illustrates the essential features of this system. Note that numerous solid-state microwave oscillators (monitored continuously) are positioned about the periphery of the central rectenna paraboloid; they are both frequency- and phase-locked to each other, and supply the source of injected RP power to every sub-array in Figure 18.7.1, with sufficient redundancy. When the remote power beam is lost, each sub-array switches instantly from its rectenna function to a transmit function. Roughly 900 MJ are stored in the LTI-20 SMES unit, which is sufficient to perform this emergency function for the 2 - 10 seconds necessary for rapid deceleration, without exceeding the LTI-20's aerothermostructural limits. Reference A144 covers this topic in substantially more detail.

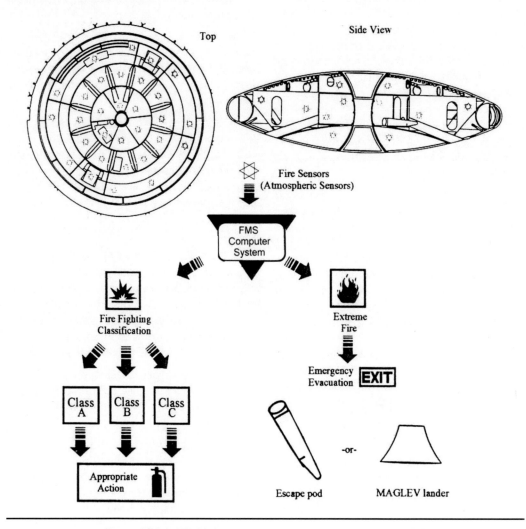

Figure 18.5.1: LTI-20 fire suppression system. *(Courtesy of RPI.)*

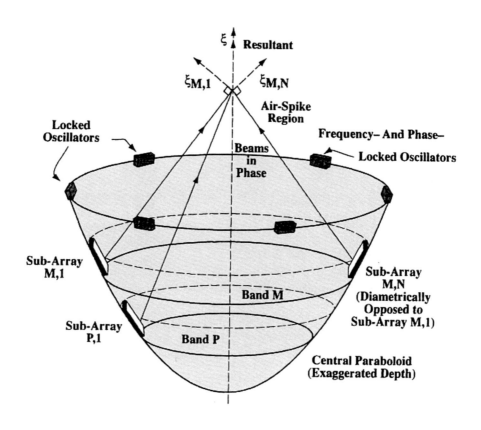

Figure 18.7.1: Paraboloidal phased array for emergency maintenance of airspike (depth of parabola is exaggerated). *(Courtesy of A. Alden.)*

19.0 POWER BEAMING INFRASTRUCTURE

The USSC power beaming infrastructure is configured to provide "highways of light" for the existing fleet of microwave Lightcraft that, depending on their mission, must perform hyper-jump maneuvers, suborbital boost-glide trajectories, and flights to low Earth orbit or escape velocity. By design, a Lightcraft must incorporate a vehicle structure, beam receiving device (optics, rectenna arrays, photovoltaics, etc.) and propulsion energy converters to produce thrust. All such elements are very tightly integrated into the vehicle design.

Lightcraft engines collectively demand an unusually high degree of flexibility from the power beaming infrastructure, whether ground-based or space-based. Hence the power-beaming station must efficiently deliver either continuous-wave (CW) or repetitively pulsed (RP) waveforms, and be throttleable over a wide power range.

The power-beaming infrastructure, as noted in Figure 19.0.1, encompasses the energy source, transmitter device, and transmission links (e.g., uplink, downlink, and cross-link) – essentially everything but the Lightcraft itself. Long range power transmission can be accomplished by laser, microwave, and millimeter-wave sources. Table 19.0.1 lists the *energy systems* options for primary sources as well as power-processing / conversion, energy storage, and transmission technologies.

Note that SMES systems permit "off-line" transmission of microwave power levels well in excess (e.g., 100x or more) of a dedicated primary energy source. The unique abilities of large SMES "batteries" to store terajoule level energies and discharge at gigawatt powers makes them the exclusive energy storage technology of choice in both ground- and space-based microwave power-beaming stations.[1,2]

19.1 SMES for Power Beaming Stations

Table 19.1.1 gives the SMES specifications for four microwave stations that comprise the existing power-beaming infrastructure. These

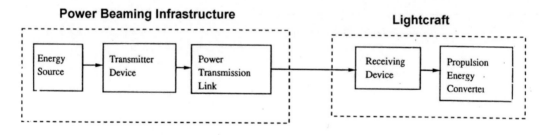

Figure 19.0.1: Architecture for beamed energy propulsion of lightcraft.

Table 19.0.1: Energy systems alternatives for beamed power transmission to lightcraft.

Energy Source	Electric Power Generation	Energy Storage	Transmission	Conversion (at Lightcraft)
• Nuclear • Solar • Chemical • Terrestrial electric grid	• Gasdynamic • Magneto-hydrodynamic • Thermoelectric • Thermionic • Photovoltaic (solar)	• SMES • Inertial (flywheels) • Electrochemical • Mechanical	• Laser • Microwave • Millimeter wave	• Rectenna • Photovoltaics • Direct absorption into propellant

POWER BEAMING INFRASTRUCTURE

Table 19.1.1: SMES specifications for microwave power beaming stations.[1]

Parameters	PDS-01	PDS-02	PDS-03	SPS-01
Year Built	2013	2017	2022	2024
Number of Rings	1	2	2	2
Capacity (GJ)	1250	2500	2500	2500
Fully Discharged, with 16% remaining (GJ)	200	400	400	400
Available Energy (GJ)	1050	2100	2100	2100
Maximum Discharge Rate (GWe)	18	30	33.3	31.7
SMES Ring Diameter (m)	500	250	250	1000
Total Ring Current (MA)	42.7	71	71	26.9
Ring Minor Diameter (m)	3.42	5.68	5.68	2.14
J_C – operating (GA/m^2)	0.146	0.164	0.183	0.201
J_C – maximum (GA/m^2)	4.00	4.50	5.00	5.50
Carbon fiber limiting stress (GPa)	1.5	3	5.5	5.5
Ring Self Inductance (mH)	1.37	0.496	0.496	3.47
MMC Belt Thickness (cm)	6.66	4.7	2.6	1.98
Superconductor/Cu Mass—2 rings (tonnes)	798	1178	1060	1458
Hoop Carbon Fiber Mass (tonnes)	2199	2510	1271	1395
Toroid pair Sub-Structure, Cryostat. Refrigeration, Power Supply (tonnes)	238	256	140	108
Total SMES Mass—2 rings (tonnes)	3235	3944	2471	2961
Specific Energy (kJ/kg)	386	634	1012	844
Max. Specific Power (kW/kg)	5.56	7.61	13.48	10.71

SMES "batteries" were constructed from commercially available, high-production-quantity Metal Matrix Composite Superconductors (MMC_Sc), comprised of MgCNi$_3$ microfibers in a copper matrix.

Microfiber superconductors (Sc) were first demonstrated in 2004[3] with an 80-nm thick annular shell of polycrystalline MgCNi$_3$ synthesized directly onto 7-µm diameter, high-strength carbon fibers. The MgCNi$_3$ microfibers were manufactured by reacting nickel-coated carbon fibers in excess magnesium vapor at 700° C, resulting in a 21.6:1 ratio of carbon to Sc material (cross-sectional area).

Figure 19.1.1 (not to scale) shows the modular pattern of the C/Sc composite unit in modern MMC_Sc superconductors. The metal matrix components (either Cu or Be inserts) are added as bulk and integrated metallurgically by means of electrolytic filling and bonding to the superconductor, which is prepped with a thin Ni coating for lattice matching and compatibility.[1]

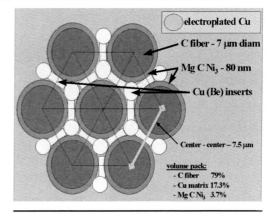

Figure 19.1.1: Geometry for metal matrix composite superconductors. *(Courtesy of D. Gross.)*

The electrolytic operation proceeds in a relatively weak ultrasonic field background, which improves oxide and particulate removal and dissolution. (The substitution of beryllium for copper has distinct advantages of lower mass and higher energy density, especially attractive for flight-weight superconducting magnets in the LTI-20.)

For example, consider the PDS-01 SMES unit in Table 19.1.1. The C/Sc fibers in the tight modular unit form an 85% fill fraction (hexagonal pack) with a carbon fiber limiting stress of 1.5 GPa. Strands are manufactured into belts of appropriate widths to facilitate spiral-wrapping of the 500- x 3.42 m toroidal loop substructure. This MMC_Sc material is designed to sustain the massive self-magnetic "pinch" pressures generated when the SMES "battery" is fully charged, carrying 42.7 MA in the MMC_Sc belt. The fully operational SMES unit has a specific energy of 0.386 MJ/kg, roughly equal to 2008-vintage LiPoly batteries (0.4 to 0.6 MJ/kg); however, the SMES specific power of 5.56 kW/kg is twice that of LiPoly cells (e.g., 2.8 kW/kg @ 20°C discharge rate) from that era.

19.2 Millimeter Wave Sources

Gyrotrons are the dominant technology for millimeter-wave power generation at atmospheric window wavelengths in the range of 35- to 220 GHz and beyond.[4,5] Table 19.2.1 summarizes the development history for gyrotron technology over the past 17 years, starting with the 1 MW baseline unit of 2008.

Although the mass of a gyrotron's superconducting magnet is the dominant factor limiting specific power, the vacuum space environment permits aggressive light-weighting of these devices for SBM stations.

19.3 PDS Basing Considerations

Power may be beamed to a flying Lightcraft either from below (i.e., "pusher" beam in Figure 19.3.1a), or above ("tractor" beam in Figure 19.3.1b or 19.3.1c), either as laser light or microwave power. Microwave Lightcraft, like the LTI-20, are configured for "tractor" beam propulsion but can also accelerate at right angles to the power beam in any direction. To reduce "uplink" atmospheric transmission losses, ground-based microwave (GBM) power stations must be located in the high desert (as with PDS-01) or 3 - 4 km mountain peaks (e.g., PDS-02). Space-Based Microwave (SBM) power beaming stations (Figure 19.3.1c) like the PDS03 and NASA's new SPS-01 provide the greatest transmission flexibility, most capable infrastructure, and lowest atmospheric transmission losses (because they use "downlink" only). Since Lightcraft quickly depart the dense atmosphere where water vapor can heavily attenuate the beam, space-based microwave stations have a clear advantage over their ground-based counterparts.

19.4 Microwave Transmission Through Atmosphere

One must recognize that the efficient "up-link" and "down-link" propagation (through the atmosphere) is a decisive factor in determining the feasibility of any desired Lightcraft maneuver (hyper-jump, suborbital boost trajectory, etc.). To accomplish this, the phase and intensity across the microwave transmitter must be very carefully controlled, aided by beacon information from the Lightcraft in flight.

19.4.1 Lightcraft Beacon System

All Lightcraft (e.g., LTI-20) are equipped with an on-board beacon that directs a signal beam back through the optical train towards the space-based relay (SBR) satellite and/or microwave source (Figures 19.3.1b and 19.3.1c). Lightcraft beacons transmit an initially undistorted

Table 19.2.1: Gyrotron 20-year development history.[6]

Year Built	Power Station	Basing	Gyrotron Power (MW)	Mass (kg)	Specific Power (kW/kg)
2008	(Baseline)	Ground	1	170	5.88
2013	PDS-01	Ground	2	240	8.33
2015	PDS-01	Ground	3	295	10.2
2017	PDS-02	Ground	5	380	13.2
2022	PDS-03	Space	6.7	155	43.2

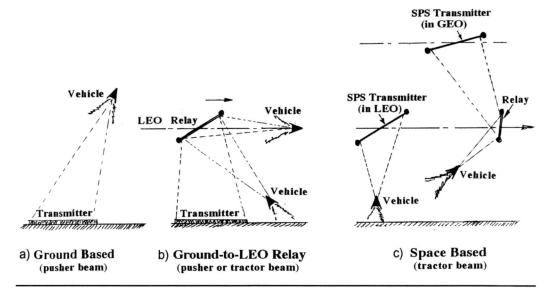

Figure 19.3.1: Power-beaming infrastructure – source siting options.

a) Ground Based (pusher beam)
b) Ground-to-LEO Relay (pusher or tractor beam)
c) Space Based (tractor beam)

spherical wavefront to the SBR mirror which, in turn, reflects the signal toward the microwave transmitter. The transmitter aperture is shared by a distributed sensing system that detects the incoming phase front of the low-power signal beam. This sensing system provides feedback correction signals for the adaptive (i.e., phase controlled) transmitter elements to pre-bias outgoing high-power wavefronts so that they arrive at the Lightcraft as a nearly diffraction-limited beam. Note that the Lightcraft beacon communicates not only phase information, but also vehicle position (exact GPS coordinates) and the updated power-demand schedule needed for a given flight maneuver.

19.4.2 Safety Protocols

Safety protocols require that beacon-derived data are updated every millisecond, and when the beacon signal is not promptly received (for whatever reason), the power beam is instantly defocused and dimmed to extinction. In the event that any "intruder" is about to fly into the power beam, the same shut-down procedure is applied.

Every USSC power beaming station has its transmitter(s) interlocked with a safety surveillance radar subsystem that protects against the accidental irradiation of friendly or unidentified spacecraft, aircraft, and/or unprotected astronauts. The importance of illuminating the face (i.e., receptive optics) of the Lightcraft precisely cannot be overstated. Both microwave and laser beam intensities can be sufficient to ignite fires and injure or kill exposed plant and animal life. Absorbing surfaces exposed to the unattenuated power beam are quickly heated to temperatures in excess of 3000 K. (Normally, when unprotected personnel are operating in the vicinity of a Lightcraft, the vehicle must be on internal power.)

19.4.3 Atmospheric Models

Atmospheric extinction (i.e., attenuation) of microwave and millimeter-wave transmission through the atmosphere is well understood and eminently predictable. Various microwave-propagation atmospheric models (Mid-Latitude Summer, Tropical, Polar Winter, etc.) are stored in USSC beam-director computers, and updated frequently with real-time meteorological data in the flight region of active interest.

19.4.4 Updated Meteorological Conditions

Local meteorological conditions ultimately determine the real-time water vapor content of the atmosphere. Water vapor is by far the strongest absorbing molecule in the millimeter regime. Radiation fog, clouds, and precipitation in the form of drizzle, light rain, moderate rain, and heavy rain can present transmission difficulties. The scattering effects of suspended particulates (e.g., pollution, dust, haze, ice crystals, snow) must be included in extinction (attenuation) estimates.

Of course, with 10- to 20 GW at its disposal a microwave power station can, if needed, readily evaporate a 0.5- to 1 km diameter hole through clouds, fog, or light rain to deliver and/or extract a Lightcraft. Such drastic measures are reserved for emergencies and other conditions that demand immediate action.

19.4.5 Microwave Power Transmission Limits
The physics of propagating high-power microwave radiation through the atmosphere constrains the operating regime within specific limits. See Appendix A10 for details.

19.5 Ground-Based Power Stations

The USSC ground-based stations, designated PDS-01 (initially 6 GW at 35 GHz) and PDS-02 (9 GW at 94 GHz), represent the earliest deployment of a microwave power-beaming infrastructure for full-sized Lightcraft flight demonstrations. Note that GBM installations present no inherent limitations on power system mass, volume, or overall dimensions – nor must its components survive the high-vibration launch environment. GBM subsystems are engineered for 20- to 30-year operational lifetimes (much like terrestrial electric power plants) and efficient maintainability. To "tractor" a Lightcraft into orbit, GBM power transmissions must be uplinked to a microwave relay mirror in low Earth orbit, which in turn directs a downlink to the Lightcraft as shown in Figure 19.3.1b. These relay mirrors have precise spherical contours with adaptable focal range, and can aim the reflected beam accurately at a Lightcraft in flight.

19.5.1 High Desert Installation: PDS-01
Direct orbital launch of a Lightcraft by a single power-beaming station on Earth's surface (Figure 19.3.1a), is feasible only for line-of-sight transmission through relatively modest air masses, such as with a direct "pusher" beam boost into orbit.[4] However, "pusher" beam infrastructures heavily restrict the Lightcraft design process, making DEAS airspike and MHD slipstream accelerator technology difficult to accommodate. Note that the family of LTI-10, 15 and -20 Lightcraft requires a "tractor" beam infrastructure.

The first 500 m diameter power beaming station, called PDS-01, was built over a period of 3 years at the High Energy Microwave Systems Test Facility (HEMSTF) at White Sands Missile Range (WSMR), NM and finally came on-line in Dec. 2013. The 35 GHz transmission frequency was selected for its excellent transmittance and the mature SOA rectenna technology at the time.

PDS-01 charges up its single-loop, 1250 gigajoule SMES "battery" overnight to exploit the decreased nighttime demand on the terrestrial electric power grid, in preparation for the next day's Lightcraft flight test program. The 500 m diameter SMES "battery" was constructed from commercially available, high-production-quantity Metal Matrix Composite (MMC) Superconductors, using exotic $MgCNi_3$ microfibers in a copper matrix. Basic specifications for the PDS-01 SMES unit are provided in Table 19.1.1.

Figure 19.5.1 gives the layout of the PDS-01 GBM boost station, showing the geometry of the large antenna array measuring 500 m in diameter. To provide adequate transmission and beam focusing, the array is divided into 3,000 segments (each measuring 9 meters in diameter) over which the phase is controlled. A single microwave source (i.e., gyrotron) cannot easily feed more than one 9 m optical segment if the phase is to be independently controllable, so a total of 3,000 water-cooled gyrotron sources with 2 MW time-average output power (per source) were needed for the large array. Figure 19.5.2 shows a close-up view of the array segments. To protect the array from contamination (dust and debris) blown about in the dry desert environment, a frequently-cleaned microwave-transparent inflatable dome was installed over the entire transmitter aperture (see aerial view in Figure 19.5.2). Figure 19.5.3 shows a landed LTI-20 with a partial view of the dome-covered PDS-01 in the remote distance (upper right).

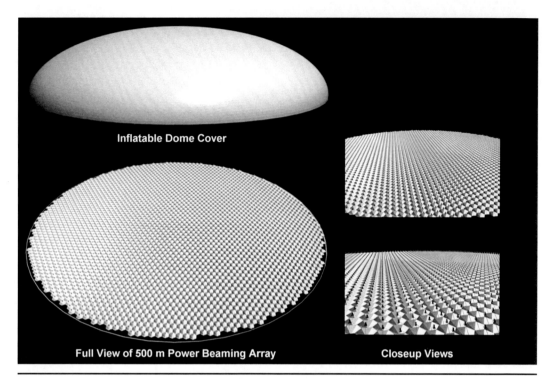

Figure 19.5.1: Layout of PDS-01 microwave power station at HEMSTF. *(Courtesy of Media Fusion.)*

Figure 19.5.2: Aerial view of protective dome over PDS-01 at HEMSTF. *(Courtesy of Media Fusion.)*

Figure 19.5.3: LTI-20 with partial view of dome-covered PDS-01 in the distance. *(Courtesy of Media Fusion.)*

Pictured in Figure 19.5.4, the standardized array module consists of a 2 MW gyrotron with 45% "wall-plug" efficiency, and beam waveguide (BWG) parabolic antenna of ~9 m diameter.[5] The 9 m, 2 MW module specification resulted from an economic analysis[9] that considered cost tradeoffs between the individual antenna segments (including cost of pointing, acquisition and tracking) and the cost of gyrotron sources (including raw power, power supply, cooling, and exciter). As indicated in Figure 19.5.4, the gyrotron source and Low-Noise Amplifier (LNA) are installed in the pedestal mount for the 9-m antenna, and are duplexed to the BWG feed system by means of a flip-mirror arrangement.[5]

Each 9-m parabolic antenna can be accurately pointed (in unison with all other array segments) to zenith angles of 45° in any direction. Commanded phase steering of the full array power beam is accomplished by providing by an open-cycle high pressure water cooling system that collects and vents steam from a plume stack, well positioned for the prevailing wind patterns at the WSMR site. The water must be ultra-purified and de-ionized to prevent contamination of gyrotron micro-channel heat exchangers.

The microwave beam from PDS-01 can directly accelerate the LTI-10 to supersonic lateral speeds in the PDE mode before transitioning to the MHD propulsion mode – in brief high altitude excursions – aided by the DEAS "wedge" to cut aerodynamic drag.

In 2016, the upgraded 9 GW station successfully "tractored" the LTI-10 into its first suborbital flight over the WSMR base, aided by the new 550 m microwave relay satellite (MRS-01) in LEO (see Section 19.6).

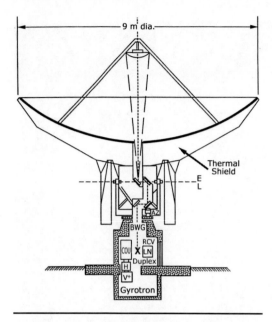

Figure 19.5.4: Microwave gyrotron phased array module. *(Courtesy of J. Benford.)*

each transmitter segment with a common frequency reference, via fiber optic lines from the GBM master oscillator. The entire 550 m array is periodically "phased up" using a reference beacon signal sent from the Lightcraft and/or LEO relay satellite, simultaneously received at all 3000 elements. Once the phase calibration conjugate values have been determined by the array receiving equipment, the computed beam steering commands insure that the high-power microwave beam is pointed in precisely the desired direction with the proper focal distance.

By 2015 improvements in 35 GHz gyrotron technology permitted an upgrade of PDS-01 radiated power to 9 GW. All 3000 gyrotrons were replaced with more powerful (3 MW) and efficient (now 50% "wall plug") units that were easily accommodated by the original, very robust (and over-designed) power supplies. Although pushed to its operational limit, the single-loop SMES "battery" was successfully "tweaked" to deliver 18 GWe, for a specific power of 4.46 kw/kg (max. is 5.56 kW/kg – see Table 19.1.1).

As with the earlier PDS-01 gyrotrons, waste heat (now ~9 GW^{th} at max. power) is removed

19.5.2 Mountain Top Installation: PDS-02

For maximum atmospheric transmittance, ground-based microwave (GBM) installations like the PDS-02 must be placed upon 3 or 4 km mountain tops in order to minimize beam distortion and extinction effects of the atmosphere. Late in 2017 after a 3½ year construction project that began in 2014, the 94 GHz power-beaming station called PDS-02 finally became operational with a personnel staff of 98. Shortly thereafter, the total complement grew to 150 (15 officers, 135 enlisted) or 2 MW_e/person – roughly the same "staff loading" as with commercial nuclear power plants on the terrestrial electric-grid.

Pictured in Figure 19.5.5 is the 250 m diameter transmitter that dominates the secure 3-km alpine PDS-02 site in the southwest USA sunbelt. The aperture comprises 3000 segments, each with its 4 m reflector dish and 5 MW water-cooled gyrotron (~50% "wall-plug" efficiency). Later, in 2020, these units were replaced with 6.67 MW gyrotrons operating at 140 GHz with 60% "wall-plug" efficiency, thereby upgrading the total radiated power to 20 GW.

PDS-02 is also equipped with three 40 MW free-electron lasers (FEL), tunable through the infrared wavelength range of 650 nm to 860 nm (i.e., 0.65 μm to 0.86 μm), each linked with its

POWER BEAMING INFRASTRUCTURE

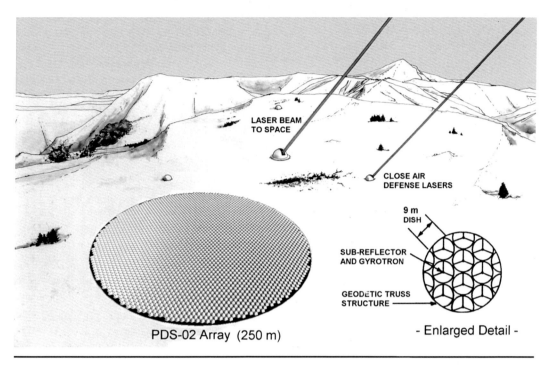

Figure 19.5.5: Secure 3 km alpine site of PDS-02 power-beaming station (cooling towers not shown).

own 10 m beam director (i.e., telescope). These water-cooled FELs have a "wall-plug" efficiency of 40%, so each laser can run continuously on electric power supplied by a single reactor, if necessary. Under emergency conditions, the FELs can reliably deliver up to 50 MW (i.e., 125% rated power level). Being a fenced-off military facility, PDS-02 has a secure perimeter that can be defended against most any conceivable airborne or ground threat.

The USSC holds itself accountable for every joule of transmitted microwave and laser power from PBS-02, and the station's Directed-Energy Transmission (DET) records are scrutinized at the highest command levels on a regular basis.

19.5.2.1 Open-Cycle Gyrotron Cooling System

When transmitting at maximum rated power, these new gyrotrons reject a total of 13.3 GW_{th} into their open-cycle, high-pressure water cooling system – similar to that deployed with PDS-01. Within micro-channel coolant passageways at each gyrotron's base, elevated water pressures suppress boiling until after the heated water exits the heat exchanger. Later, at a point farther downstream when this pressure is suddenly relaxed, the water flashes into steam and expands into a network of much larger steam-collection ducts. As with PDS-01, the plume stack is positioned with respect to prevailing wind patterns at the site so as not to interfere with microwave power transmissions. During a typical 2-minute Lightcraft boost, roughly 8900 kg (19,600 lb) of ultra-purified water is converted to steam, while draining roughly 2350 US gallons or 8.9 m^3 (314 ft^3) from the PDS-02 water reservoir.

19.5.2.2 Military Training Center

Along with the PDS-02 upgrade, a new military "training center" was added for educational purposes, in preparation for deployment of the next-generation space-based PDS-03 in 2022. The PDS Training Center facilities are a maze of 4 - 5 m diameter cylindrical modules resembling submarine and International Space Station (ISS) accommodations (command center, laboratory, cooking, R&R, habitation, etc.); all modular compartments are linked by air-locks and utility nodes. The overall layout is identical to that planned for deployment in PDS-03 with a total crew complement of 63.

19.5.2.3 PWR Prime Power

Prime power for the PDS-02 is supplied by three independent 300 MW_{th} GE-derivative pressurized water reactors (PWR), each driving a pair of ~50 MW (70,000 shp) turbine / generator units for a combined electrical power output of just over 300 MW_e. The automated nuclear-powered machinery produces pressurized steam to drive the twin turbines with a thermodynamic cycle efficiency of 33%. The primary coolant circuit operates at 560° K (550° F) and 2500 psi, a sufficiently high pressure that water never boils in the main coolant loop. In the secondary circuit, the steam generator delivers 500° F steam at 1000 psi to the pair of main turbine / generator sets. With the reactor at full throttle (300 MW_{th}), nearly all of the turbine's shaft power (140,000 shp) is converted into electrical power (~100 MW_e). When the throttle is closed, a much lower flow rate of steam is diverted to a small turbo-generator sized to provide station-keeping power needs of the facility.

Note that the three 300 MW_{th} naval reactors deliver high power from a small volume, and therefore run on highly enriched uranium (>20% U-235), normally in the form of uranium-zirconium or uranium-aluminum alloy. An internal neutron shield insures the long-term integrity of the compact reactor pressure vessel. The reactor core has a long core life (~50 years) and refueling is necessary only after 10 years or more.

Note that large reactor power plants designed for the terrestrial electric grid can swiftly be idled back to 10% of their rated power level, but normally take 12 hours (at the shortest) to recover from idle power back up to 100% capacity. In sharp contrast, naval reactors can be immediately shut down to nearly zero power output (or to low station-keeping levels), and when the command is given for a "hot start," they will respond in seconds. Naval reactors are extremely agile and can be cycled quickly, making them ideal for PDS-02 functions.

Reactors used in PDS-02 are identical to those powering the tenth and last Nimitz Class nuclear aircraft carrier, the USS George H.W. Bush (CVN77), which entered service in 2011. To be factual, a majority of components in the PDS-02 reactor power system were "appropriated" from stockpiled spare parts for CVN77 (and other nuclear naval vessels) made available as surplus equipment just one year before PDS-02 construction began in 2013.

Waste heat from the triple-reactor system is rejected through three cooling towers located far enough downwind from the transmitter aperture that water vapor plumes do not interfere with microwave or laser beam propagation. The cooling towers operate at their rated capacity only during the SMES charging-cycle, each rejecting about 13.5 MW_{th}.

19.5.2.4 SMES "Battery" Details for PDS-02

The station's 250 m diameter, twin-loop SMES "battery" can store up to 2500 gigajoules for long periods of time with negligible resistive losses. As with PDS-01, the SMES employs Metal Matrix Composite Superconductors (MMC_Sc) based on $MgCNi_3$ microfibers in a copper matrix, maintained at 4.25° K (-268.9° C) by liquid helium circulated through refrigerant passageways. The SMES battery can be charged from zero to its maximum capacity in 2.4 hours using the combined 300 MW_e output from all three reactors, although it is never depleted below 16% capacity. Since each reactor is independent of the others, the SMES recharging system is fully triply redundant

An aggressive design hoop stress of 3.0 GPa (ultimate of 3.7 GPa) was mandated for the imbedded carbon fibers since this SMES was to serve as a technology demonstrator for the space-based unit planned for PDS-03. This enabled an overall SMES specific energy of 0.634 MJ/kg; the maximum specific power is 7.61 kW/kg. Table 19.1.1 provides more detail on this twin-loop SMES "battery."

19.6 Orbital Power Relays

Space-based mirrors with adaptive optics can redirect a propulsive power beam originating at a surface or LEO installation to a Lightcraft operating at any altitude, especially in a location that is over the horizon from the beam source. Free-flying lightweight laser mirrors (e.g., 3 x 7 m) with highly precise aiming and adaptive optics can be inserted into any of a number of different orbits: e.g., as with the *Monocle* constellation deployed in 2023.

For microwave power, the infrastructure requires huge passive reflectors. Designed for 35 GHz, the 500 m MRS-01 became operational in 2016 (circular orbit = 476 km altitude; inclination = 28.5°; daily repeating orbit over WSMR). Anticipating the retirement of the 12-year old MRS-01, the new 250 m MRS-02 (designed for higher frequencies) was finally brought on line in mid 2025.

19.6.1 Microwave Relay Stations (MRS)

For a GBM station like PDS-01 or PDS-02, the simplest and most efficient orbital boost system is to uplink to a single satellite reflector in LEO (e.g., MRS-01 or MRS-01), and to use its 8 km/sec velocity to tractor the Lightcraft into orbit, thereby minimizing the distance over which the beam is transmitted. Atmospheric attenuation constrains the GBM station to beam upward at zenith angles of ±45°, so passive relay satellites are a necessity, especially for the final stage of LEO insertion when the Lightcraft is downrange from the launch site and accelerating to orbital velocity. The uplink is accomplished at small angles to the vertical to avoid excessive atmospheric attenuation of the power beam. When employing a passive satellite reflector to relay power received directly from the ground station, little attenuation is suffered in the cross-link or downlink. A typical launch scenario gives an engagement distance varying between 500 and 1200 km during the final boost phase.

19.6.1.1 Passive MRS Reflectors

The concept of using space reflectors for relaying Earth-generated microwave power was first suggested by Kraft Ehrike. The gyro-stabilized MRS-01 and MRS-02 microwave relay satellites have evolved directly from the orbiting flat reflector designs of Ehsani et al.,[10] following that of Hedgepeth.[11] The manufacturing, launch, and assembly system was inspired by the work of Flint.[12]

For MRS-01, the 500 m diameter passive relay reflector (Figure 19.6.1) is stretched across a bicycle wheel-type perimeter structure, having an area density of 0.1 kg/m^2 and total mass of 27,000 kg – basically one Shuttle or Delta IV load. The bicycle rim structure is a lightweight truss that separates two superconducting magnets that initially are used to magnetically inflate the reflector,[28] and later provide centrifugal tensioning of the reflector by gyroscopic forces. Also, the electric current in these "bucking" magnets can be varied in orbit to engage the Earth's geomagnetic field of 4.5 gauss and thereby torque the relay for precise active pointing at the Lightcraft. On-board ion thrusters are needed to torque the relay satellite about the other axis, to control the axis-symmetric spin rate, and for aerodynamic drag makeup; hence, the satellite requires yearly servicing for propellant resupply. Station-keeping power is provided by an 800 kW solar photovoltaic array.

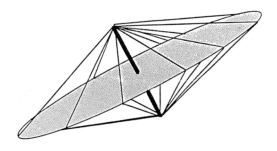

Figure 19.6.1: MRS-01 "bicycle wheel" passive microwave relay satellite (500 m diameter).

The relay's flat reflector figure contour is actively monitored with a small laser mounted to a central "bicycle hub" mast. To assure surface flatness to within acceptable limits, this figure is then corrected by using smart structures technology with actuators imbedded in the reflector's surface and perimeter structure.

19.6.1.2 Beam Transmission Efficiency

The launch flexibility of a given ground-based HPM boost station can be greatly increased through the aid of a passive microwave relay station in low Earth orbit. But for any GBM / MRS system, several factors (see Section 19.4) directly affect the beam transmission performance. To deliver a prescribed "brightness" to the relay mirror, the ground station must increase the transmitted power as a direct function of zenith angle to make up for losses caused by the atmosphere. Since the GBM station faces practical as well as physical upper limits to the microwave power, the required number of relay satellites (i.e., for global coverage) also becomes a function of zenith angle because one or more satellites must always be visible within the accessible cone of zenith angles. With the present microwave power-beaming infrastructure, global coverage is an impossibility. Within limits imposed by other mission constraints, the required number of relay satellites can be reduced by placing them in higher orbits (as with the Monocle laser relays) where each satellite moves slower and views more of the Earth; but this drives up the mirror size (and hence the mass) of each relay device, as well as system inertia and dynamic response complexity. For the foreseeable future, this approach has proven practical only at laser frequencies.

19.6.2 Monocle Laser Relay Constellation

The constellation of six 3- x 7 m *Monocle* laser mirrors (Figure 19.6.2) was launched in 2023 into equally spaced 3000 to 5000 km orbits to enable global delivery of laser power to Lightcraft in flight. The single-surface *Monocle* mirror concept owes its origins to early "Star Wars" studies in the late 1970's by Lockheed.[13,14]

19.7 Space Based Power Beaming Stations

Space-Based Microwave (SBM) power stations can provide increased operational flexibility over their GBM counterparts, in part because uplink atmospheric transmission problems (Figure 19.3.1c) are avoided altogether. Furthermore, for those positioned at lower altitudes, orbital mechanics enables direct downlinks to Lightcraft over much of the planet: i.e., whenever and wherever both the SBM and vehicle are within power-beaming range.

Figure 19.6.2: Monocle laser relay mirror in 3000 to 7000 km orbits (3 m x 7 m, elliptical). *(Courtesy of SDIO.)*

Microwave relay stations, located in favorable orbits, can greatly increase the range of a single orbiting power station. Aided by those same relay satellites, a few well-positioned SBM stations in LEO can provide complete, continuous coverage for Lightcraft power over the entire Earth.

However, SBM stations face very restrictive limits on overall system mass and dimensions. Lightweight engineering design of the energy source, microwave device, transmitter and integrating spacecraft structure is essential. Efficient packaging for the transport boost to orbit, as well as efficient erection (once in space), is also mandatory. The rejection of waste heat from both the energy source and power conversion elements is of paramount importance. Heavy lift launch vehicles are mandatory for boosting the most massive SBM components into orbit at an affordable price. Finally, to keep the sizes of transmitting apertures reasonable (e.g., 250 to 1000 m) for microwave frequencies between 35 and 220 GHz and Lightcraft diameters between 10 and 20 m, SBM stations are deployed in LEO.

19.7.1 Orbital Power Station Vulnerability

Power stations in low Earth orbits are less expensive to launch and service than geostationary systems, but have much shorter visibility times and more severe beam-aiming (pointing and tracking) requirements because of their high speed relative to any point on the ground and subsequent rapid change in powersat incidence angle. Stations in lower orbits are also much more vulnerable to countermeasures such as lasers and kinetic energy weapons launched from the ground. Many nations which lack the technical ability to threaten GEO assets could readily reach LEO assets with relatively primitive "shotgun" payloads launched by vehicles as small as a Scud.

The nuclear power source itself is much smaller than a solar power station (SPS), and hence is a harder target to hit. The surface area of a nuclear-powered station such as PDS-03 is dominated by the radiators that dump waste heat from the power-generation process. Damage to conventional radiators will result in explosive loss of coolant which might easily destroy the station or set it to tumbling in an uncontrollable manner; complex mechanical systems with many valves are required to isolate punctured radiators and minimize coolant losses. For this reason, among others, PDS-03 employs liquid droplet radiators (LDR) which are highly reliable, allowing projectiles to pass through the coolant stream with negligible damage.

19.7.2 Orbital Power Station Failure Modes and Repairs

LEO solar power stations are vulnerable to orbital debris which, at the turn of the millennium, would have impacted these large structures at a prohibitive rate. Fortunately, the USSC orbital debris cleanup program was largely complete before SPS-01 entered operational service in 2025 – courtesy of pulsed FEL lasers aboard the PDS-03.

Damage to an SPS is far easier to inflict than with a hardened nuclear station (i.e., LDR equipped), but also not likely to result in catastrophic failure. Damage to a solar panel may be contained by simple, highly reliable solid-state electrical switches so that overall system degradation is predictably incremental and minor. In either case, damage to the power-transmitting antenna has similar consequences for both nuclear and photovoltaic systems.

On-orbit repair of the two types of power systems is another important factor. Photovoltaics operated at low temperature will degrade due to radiation damage over a period of a few decades. Nuclear systems require refueling every 10 years or so. Human operations near a live or decommissioned nuclear reactor are out of the question because of the high radiation levels. Shadow-type reactor shielding, as on PDS-03, is insufficient to protect a human repair crew operating in close proximity to the reactor. All servicing and repair must be by radiation-hardened robotic equipment which itself becomes radioactive due to its neutron exposure over time.

19.8 Orbital Nuclear Power Station: PDS-03

The PDS-03 microwave power station, officially commissioned in Dec. 2022 (nearly five years after project initiation in Feb. 2018), initially maintained a near circular orbit at 476 km altitude and inclination of 28.5° but later raised itself (electric thrusters) into a 500 year orbit. Note that the MRS-01 relay satellite shares the same orbit as PDS-03, and resides within power beaming range. The 433 m triangular truss "keel" of PDS-03 was laid in 2019.

19.8.1 Triangular Architecture with LDRs

The triangular architecture and overall dimensions of the PDS-03 station (see Figure 19.8.1) were dictated by five overriding considerations:

a) 250 m diameter transmitter aperture radiating 20 GW at 140 GHz;
b) triple-redundant nuclear reactors with minimum "shadow-type" radiation shields, producing 100 MW_e each;
c) three deployable 433 m triangle-shaped liquid droplet radiators (LDR) to cool the three reactors;
d) 2 RPM station rotation rate for 1 g artificial gravity at the reactor location (i.e., tips of triangle); and

e) 2500 gigajoule (i.e., 695 MW-h) SMES "battery" comprised of two superconducting cable assemblies (i.e., "loops') that encircle the 250 m transmitter, having a specific energy of 1.01 MJ/kg and maximum specific power of 7.61 kW/kg.

LDRs were specified for two technical reasons:

1) it is the only feasible space radiator technology suitable for a "hardened" military facility (projectiles can pass right through without causing damage); and
2) stored liquid coolant (Dow 705 oil) can serve dual-use to augment "storm cellar" shielding for the crew in times of heightened solar flare activity.

The storm cellar is located at the geometric center of PDS-03 and is equipped with a complete duplicate control center for a skeleton crew of 21. During severe solar flares, all non-essential personnel are deorbited in Lightcraft or crew return vehicles (CRVs). Augmented with LDR oil, the storm cellar's shield has a stopping power against energetic 200 MeV protons equivalent to a 1 m lead plate. During these solar events, the SMES "battery" is maintained in the fully charged condition, ready for duty.

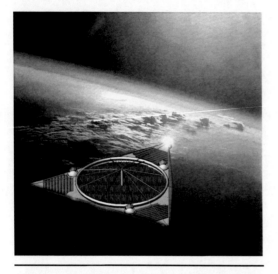

Figure 19.8.1: PDS-03 power beaming station in "fully charged," station keeping configuration. *(Courtesy of RPI.)*

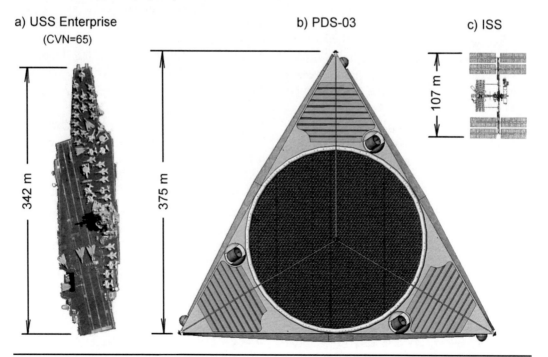

Figure 19.8.2: Size of PDS-03 relative to ISS and USS Enterprise aircraft carrier. *(Courtesy of RPI.)*

Figure 19.8.2 shows the scale of PDS-03 relative to the International Space Station (ISS), and the USS Enterprise (CVN-65) which was retired in 2013. The total pressurized habitable volume is approximately 3600 m^3 (~3x that of the ISS) for a crew complement of 63 (i.e., about 57 m^3 per person). The lightweight habitat modules are inflatables covered with CMC laser armor, whereas the cylindrical 4 to 5 m diameter command and laboratory modules are configured like their ISS prototypes.

With the LDRs closed, PDS-03 forms an equilateral triangle with a base of 433 m, a "height" of 375 m (~10% longer than the USS Enterprise), and a surface area of 81,190 m^2. Each leg of the PDS-03 triangular keel has exactly four times the major truss length of the ISS (i.e., 108.4 m). When the three LDRs are fully extended (see Figure 19.8.3), the overall dimensions of PDS-03 double and the total surface area increases by four times. Each LDR is deployed by a pair of 100-m long electromagnetic actuators, derived from early EMALS technology (ElectroMagnetic Aircraft Launch System). EMALS was developed in 2005 by the Navy for next-generation nuclear-powered aircraft carriers: e.g., CVN-78 (displacement of ~92,500 tonnes) which replaced the 53-year-old USS Enterprise in 2014. Note that CVN-78 is equipped with a 104 MWe power generation facility required for EMALS, Electromagnetic Aircraft Recovery System (EARS), electromagnetic (EM) weapons, pulsed energy, and laser weapons. CVN-78 employs four 91.4 m long EMALS catapults for launching aircraft, whereas PDS-03 has six that are 8% longer.

19.8.2 SMES Rechargeable "Battery" Details
The geometry for this rechargeable 2.5 terajoule "battery" is nearly identical to its sister PDS-02 unit. Like the PDS-02 unit, the two superconducting loops are arranged in a bucking configuration (i.e., two coils that repel each other), positioned sufficiently far apart (~25 m) that inductive coupling is minimal. Ramping of the current in one loop does not affect the current in the other. Each loop stores 1.25 terajoules in a 2.6 cm thick MMC_Sc belt applied over the 5.68 m (minor diameter) toroidal housing with a major diameter of 250

Figure 19.8.3: PDS-03 power beaming station in "charging" mode, with liquid droplet radiators deployed. *(Courtesy of RPI.)*

m. Fully charged, each loop carries 71 MA.

Table 19.1.1 summarizes the twin-loop 2.5 terajoule SMES "battery" specifications for PDS-03. A complete discharge of the SMES battery dissipates about 1 MJ of unrecoverable energy per loop, generated by hysteresis and eddy currents in the superconducting material. This waste heat must be removed by the cryocoolant refrigeration system prior to recharging for the next power-beaming cycle.[1,2] Figure 19.8.4 shows the SMES battery system schematic.

19.8.3 PDS-03 Mass and Power Loading
It is instructive to contrast the total mass and electric power loading of PDS-03 with the ISS and CVN-78. The fully assembled ISS has a mass of 419 tonnes and electric power of 0.11 MW$_e$; hence a power loading of 3810 tonnes/MW$_e$. In contrast, with 100 MW$_e$ available the CVN-78 aircraft carrier has a power loading of 925 tonnes/MW$_e$. At 4030 tonnes (i.e., 9.6x the mass of ISS) and 300 MW$_e$, PDS-03 has a power loading of 13.4 tonnes/MW$_e$. Table 19.8.1 gives the mass breakdown for the orbiting station.

Note that the SMES structure and transmitter array web adds significant stiffness to the station, well beyond that of the triangular truss superstructure itself. The 3000 fixed, non-

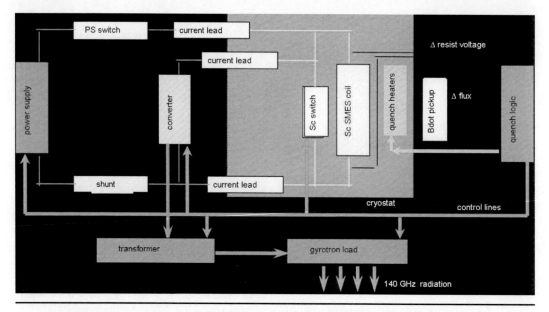

Figure 19.8.4: SMES battery system schematic for PDS-03. *(Courtesy of D. Gross.)*

articulating "telescope" dishes that comprise the 250 m transmitter array are tightly held within this a hexagonal web.

19.8.4 Launch and Orbital Assembly of PDS-03
Breaking the 4030 tonne PDS-03 station down into 4 to 5 tonne, or even 20 tonne packages for the boost into orbit and then assembling these pieces in LEO was deemed too complex. In particular, the SMES unit, which carries sizable hoop stresses, would be unable to meet its aggressive specific energy objective of 1.01 MJ/kg if assembled from numerous small segments.

Fortunately, NASA and the USSC had already developed an advanced version of Ares V, a shuttle-derived, heavy-lift launch vehicle (HLLV) designed to hoist 130+ tonne payloads into 400 to 500 km orbits, at a cost of only 2,000 dollars per kilogram. NASA's original Ares V proposal (see Figure 19.8.5) was to employ two reusable solid rocket boosters, and five commercial RS68 engines fueled by a 10 m diameter tank, close in size to the Saturn V's structure. *(In comparison, remember that the combined mass of the reusable Space Shuttle orbiter and its 30,000 kg payload totaled 100 tonnes – an Apollo-class moon rocket "throw weight.")*

Hence, 31 launches were required to boost the complete PDS-03 hardware kit into low Earth orbit for final assembly. After the 433-m

Table 19.8.1: Mass breakdown for PDS-03 (w/o personnel & accessories).

Component	Mass (tonnes)
Structure (triangular keel w/ reinforcing members)	350
Reactor Power System with 3 LDRs (1 kg/ kW$_e$)	300
SMES "Battery" (2.5 terajoules)	2330
Liquid Helium SMES Refrigeration System	140
Microwave Transmitter Array (3000 elements)	465
3 FEL Lasers (10 MW/ea.) with 3 Telescopes	150
Command, Habitation, and Laboratory Modules	295
TOTAL MASS	4030

triangular truss keel was laid on 2019 and the three LDR arms were attached, station construction proceeded at a frenzied rate to meet the 2022 deadline. Progress was greatly aided by the three LDR arms that were designed to function as construction cranes (for both initial station assembly and ongoing maintenance) – much like the remote manipulator arm on the Space Shuttle orbiter.

19.8.5 Reactor Power System Selection
Three nuclear power source options were seriously considered for PDS-03:

Option #1 was to retain the same 33% efficient, 300 MW_{th} PWR power systems used in PDS-02. (Note that pressurized water reactors require a 1 g environment to function properly.) However, this option would have required each LDR to radiate ~200 MW_{th} to space, or roughly twice the capacity of a 433 m equilateral triangle radiator using Dow 705 silicone oil as coolant.

Option #2 specified a 49.5% efficient helium gas-cooled reactor power plant with a high-temperature Brayton topping cycle. Note that the simple Brayton cycle, by itself, would have been only 40% efficient, so a low temperature "bottoming" cycle was clearly mandated. (Exact thermodynamic cycle details are classified.)

Option #3 considered a dual-mode gas-cooled reactor power plant. The plant would generate electric power in closed-cycle mode using gaseous helium as the working fluid (as in Option #2), but occasionally would also run the reactors (only) in an open-cycle rocket mode to produce thrust with on-board hydrogen propellant using converging-diverging nozzles. In principle, this auxiliary mode (i.e., a nuclear thermal rocket) could provide three valuable functions for PDS-03:

a) aggressively spin-up (or slow down) the station's rotation rate;
b) thrust the station into higher orbits to counteract aerodynamic drag forces that constantly degrade the station's orbit; and
c) quickly adjust the attitude (pitch or roll) of the station in emergencies.

Figure 19.8.5: Ares V, Apollo Saturn V, and Space Shuttle – drawn to scale. *(Courtesy of NASA.)*

Due to its complexity, Option #3 faced several formidable problems including hydrogen embrittlement of structural materials at elevated temperatures, and the potential of leakage from the dual-mode valves.

Ultimately the engineering team selected Option #3 for PDS-03, but retained the 1 g specification (at the triangle tips) from Option #1 to provide artificial gravity for the crew and eliminate the calcium depletion problem plaguing extended stays at zero g. Although spin stabilization of the station was, of course, mandatory, the 1 g specification "stuck" even though a lower artificial gravity could have sufficed. The PDS-03 reactor design benefited from nuclear rocket technology developed under the international manned Mars mission, launched in 2018.

19.8.6 Liquid Droplet Radiator Details
With 200 MW_{th} of reactor thermal power driving a 49.5% efficient helium-cooled power cycle, each 433 m LDR in Option #3 must radiate 102 MW_{th}. This dictated an average Dow 705 oil temperature of 330 K (340 K is the max. allowable) and LDR sheet emissivity of 0.90. (See References 15 and 16 for technical details on liquid droplet radiators.) Table 19.8.2 gives

Table 19.8.2: Properties of liquid droplet radiator sheet — (Dow 705).[17]

LDR Property	Quantity
Specific Gravity (Dow 705 at 298 K)	1.09
Molecular Weight (Dow 705)	546
Boiling Point (Dow 705)	518. K
Droplet Diameter (μm)	100. μm
Droplet Emissivity (Dow 705 coolant)	0.7
LDR Sheet Emissivity	0.9
Sheet Thickness	1.0 m
Coolant Mass Per Unit of Radiator Area	0.2 kg/m²
Optical Depth of Radiator Sheet	3.0
Droplet Spacing (~ 14x droplet diameter)	1.4 mm
Radiated Power at Average Temperature of 330 K	102. MW$_{th}$
Coolant Mass in 433 m Triangular LDR Sheet	16.2 tonnes
Total LDR Coolant Mass (~ 2x that of droplet sheet)	32.4 tonnes

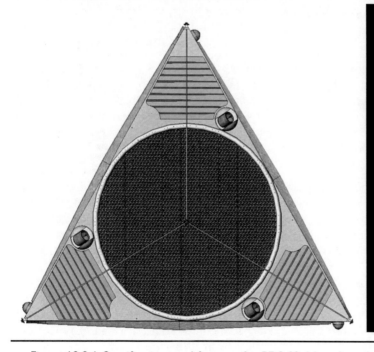

DETAILS:
- 433 m triangular truss structure, cable braced
- 250 m diameter phased array transmitter (20GW at 140 GHz)
- 2.5 TJ twin-coil SMES battery
- Three 202 MWth nuclear reactors, each with its own liquid droplet radiator
- Three 10 MW tunable FEL lasers
- Crew habitation modules (inflatable) at 1 gravity
- On-axis, de-spun docking port

Figure 19.8.6: Specifications and features for PDS-03 (closed mode shown). *(Courtesy of RPI.)*

the LDR properties[17] for a typical 81,200 m² triangular sheet radiator – assuming Dow 705 as the coolant.

19.8.7 Specifications and Features of PDS-03
The principal specifications and design features for the PDS-03 are highlighted in Figures 19.8.6 and 19.8.7, and more fully described below:

- Triple redundant, 49.5% efficient 100 MW$_e$ reactor power system. Helium gas-cooled, 202 MW$_{th}$ reactors with high-temperature Brayton topping cycle; Compact reactor with minimum mass "shadow-type" radiation shield at each point of triangle station (Figure 19.8.8); Power system density of 1 kWe/kg, inclusive of LDR heat-rejection system; Total

POWER BEAMING INFRASTRUCTURE

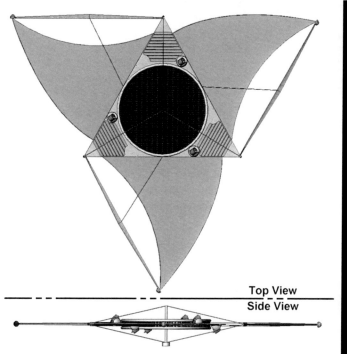

DETAILS:
- Retractable liquid droplet radiator arms
- SMES unit takes 2 hours to fully recharge, drawing 300 MWe from the triple-redundant reactor power plants
- Reactors shut down (to idle power level) and LDR arms retract when SMES unit is fully charged
- Coarse pointing of station during microwave boost is accomplished by geomagnetic torquing
- Fine pointing of BEP beam accomplished by electronic steering of phased array
- Six 100m linear electromagnetic actuators (2 per arm) open & close LDRs

Figure 19.8.7: Design details for PDS-03 (LDRs open; charging mode shown). *(Courtesy of RPI.)*

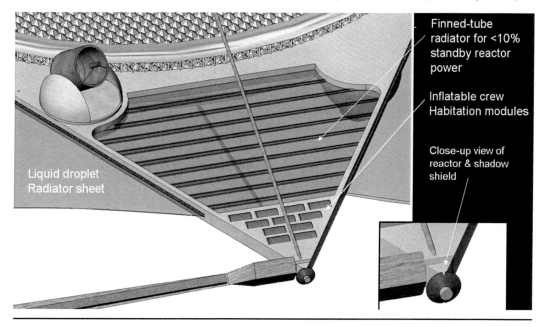

Figure 19.8.8: Reactor power system with low power radiator and shadow shield. *(Courtesy of RPI.)*

electric power output of 300 MWe for the station. Each reactor is provided with a finned tube radiator sized for handling the waste heat generated at the small "idle" or "stand-by" power rating (Figure 19.8.8).
- Three 433 m triangular liquid droplet

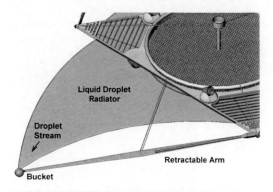

Figure 19.8.9: Closeup view of operating liquid droplet radiator (1 of 3 LDRs shown). *(Courtesy of RPI.)*

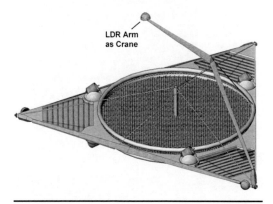

Figure 19.8.10: Dual use of LDR arm as remote manipulator crane. *(Courtesy of RPI.)*

radiators running on Dow 705 silicone oil coolant, with droplet collectors (at the end of each LDR arm) to retrieve and recirculate coolant (Figure 19.8.9).

- Dual use of all three LDR arms as remote manipulator cranes in constructing, servicing, and maintaining the PDS-03 station (Figure 19.8.10).
- Lightweight laser armor (i.e., "skin") over the LDR outer arm's truss. When radiators are shut down, the arms close over their linear "shower heads" designed to spray Dow 705 coolant into the radiator, thereby protecting them against damage from micro-meteorites or laser fire. Each LDR arm is equipped with an articulated finger (at the far end, near the droplet collection "bucket") that can grab and latch the arm securely against the station, providing a positive lockdown against centrifugal forces that would tend to dislodge it.
- Six 100-m linear electromagnetic actuators (two per arm) to open and close the LDR radiators. When the LDRs must be opened (and the station is spinning, as usual), the actuators must slow the arm deployment rate. After the SMES units are fully charged and the LDRs must again be closed, the actuators work against the centrifugal forces due to station rotation. (Note the station rotation rate increases slightly as arms are retracted.)
- Maximum radiated power of 20 GW at 140 GHz from 250 m diameter phased-array transmitter composed of 3000 reflector dishes, all 4 m in diameter (see Figure 19.8.11). Each of the 60 % efficient gyrotrons produces 6.67 MW of beam power. The space-rated gyrotrons were evolved from their ground-based predecessors, but stripped of the vacuum "tube" enclosure / jacket and refined for absolute minimum mass.
- SMES battery sized to store 2500 GJ (695 MW-h) with a specific energy of 1.01 MJ/kg, and maximum specific power of 13.5 kW/kg (see Table 19.1.1).
- Normal station rotation rate of 2 RPM to provide a 1 g environment for crew at triangle tips (countering calcium depletion problem with zero g life in space).
- Geomagnetic torquing of station in Earth's magnetic field to provide coarse pointing of transmitter (and entire station) at the Lightcraft / target. The desired gyroscopic precess rate is accomplished by transferring a precise amount of current (a few thousands of amperes) between opposing SMES loops, to create the required magnetic moment.
- Three tunable 10 MW Free-Electron Lasers (FEL) operating in the wavelength range of 650 nm to 860 nm (i.e., 0.65 to 0.86 μm), linked to six 15-m diameter beam directors (Figure 19.8.12). Each RF FEL has access to two telescopes – split by the plane of the triangular station (one looks up, the other looks down) – for full spherical coverage. The FEL resonator optics are separated by 430 m to reduce flux levels incident upon those mirrors to sustainable levels.

POWER BEAMING INFRASTRUCTURE

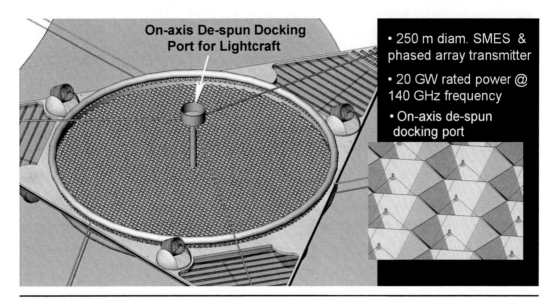

Figure 19.8.11 Rear view of 250 m diameter transmitter array (encircled by SMES unit) showing on-axis, de-spun docking port. *(Courtesy of RPI.)*

- 250 m diam. SMES & phased array transmitter
- 20 GW rated power @ 140 GHz frequency
- On-axis de-spun docking port

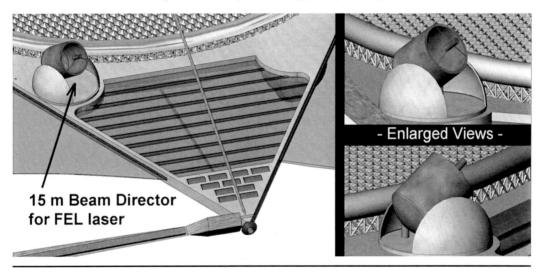

Figure 19.8.12: Tunable 10 MW free-electron laser with 15 m beam director. *(Courtesy of RPI.)*

- Total crew complement of 63 with "submarine-like" accommodations at 1 g for 21 members near each triangle tip (Figure 19.8.8).
- Seven large escape pods (3-person, Apollo Command Module size) stored at each triangle tip. triangle tip. These Crew Return Vehicles (CRV) are identical to the maglev landers carried on the LTI-20.
- Station "elevator" system to transport maglev landers from one triangle tip to any of the others.
- On-axis, de-spun magnetic docking port provided for up to six visiting Lightcraft (Figure 19.8.11). Elevator transport of crew in pressurized maglev landers to and from command modules at each triangle tip. Lightcraft must not approach the vicinity of PDS-03 when the station is in SMES charging mode and the LDRs are operational.)
- Central Storm Cellar for 21-member skeleton crew; dual use of Dow 705 coolant from LDRs

to augment radiation protection against 200 MeV solar proton storms; approximately 99 tonnes of coolant (33 metric tons from each LDR) is pumped into the double-walled metal hull of the cylindrical storm cellar, giving the equivalent "stopping power" of 1 m thickness of lead.

- Inflatable crew habitation modules are covered with lightweight laser armor that also serves as a micrometeorite "bumper" (Figure 19.8.8)
- Efficient electric propulsion (ion engine or electrodynamic station-keeping system) to counter the drag force acting on the power station at 550 km orbital altitude.
- The PDS-03 has the means to defend itself against any conceivable military threat.
- The entire station must pitch to coarsely align the transmitter axis with the Lightcraft / target under boost, using the SMES magnets (interacting with the Earth's magnetic field) to apply a geomagnetic torque to the rotating station. Fine pointing of the microwave beam is provided by adaptive optics on the array's sub-reflectors, and phase control is adjusted using information extracted from the incoming Lightcraft beacon signal.

Other PDS-03 details: Note that the nuclear fuel is extremely stable and doesn't contaminate the helium coolant of the closed-cycle conversion system. The twin turbo-generators on each reactor power system must counter-rotate to cancel torque inputs to the PBS-03 station. Every 4 m diameter reflector dish in the 3000 unit transmitter array also serves as a thermal radiator for its own 63% efficient gyrotron, linked with heat pipes. At the maximum transmitted microwave power of 20 GW (or 6.67 MW per gyrotron), every SiC reflector dish / radiator reaches a steady state temperature of 1,244 K (971°C): i.e., "glowing and still going." Note that thermionic cathodes inside the gyrotrons run at an even higher temperature of 1273 K (1000°C). The hexagonal structure around each 4 m reflector also participates in the thermal radiator function.

19.8.8 Requests for Information on the PDS-03 Requests for more detailed data on the PDS-03 (beyond that revealed above) should be submitted to: William Q. Storer, Secretary of the U.S. Space Command, Pentagon, Wash., D.C. The classified nature of PDS-03 activities prohibits release of technical and operational details.

19.9 Orbital Solar Power Station: SPS-01

This collaborative NASA / aerospace industry venture to construct the world's first rechargeable LEO Solar Power Satellite, SPS-01, was brought to fruition in mid-October 2024, about 6½ years after project initiation in March 2018. The 1-km diameter, rotating disc microwave power station is covered with photovoltaics on one side, a 35 GHz phased-array on the other, and a 2.5 terajoule energy storage system ("battery") wrapped about the perimeter. The SPS-01 has double the radiated power of the original SPS concept (Figure 19.9.1; continuous power of 9 GW) proposed by Peter Glaser in the mid 1970s for geostationary orbit. The advanced version of the Ares V launch vehicle lofted the entire structure to LEO in 43 launches at a cost of $6.6 billion. The government's role in the SPS-01 space development and commercialization venture

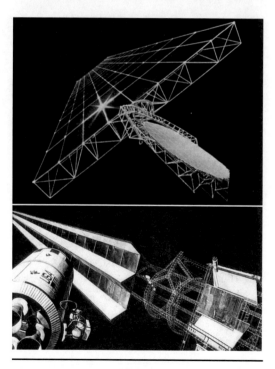

Figure 19.9.1: Two 1970's concepts for ~9 GW solar power satellite in GEO. *(Courtesy of NASA.)*

POWER BEAMING INFRASTRUCTURE

Figure 19.9.2: Lightcraft (lower left) in MHD propulsion mode, powered by 1 km diameter SPS-01 (upper right) in low Earth orbit. *(Bob Sauls, Frassanito & Assoc.; Courtesy of NASA)*

has emphasized R&D support for risk reduction, whereas most actual development of operational systems has been left to the commercial sector.

19.9.1 Dedicated Missions for SPS-01 The expressed intent of this government / industry construction project was to establish a fledgling 35 GHz energy-beaming infrastructure for *commercial* "Highways of Light." As the launch rate for commercial Lightcraft flights grows into the future, investors expect that more SPS stations will be financed, built, and linked into the space power grid to service the increasing microwave transmission load.

Sized to boost LTI-20 Lightcraft, this fully operational prototype for a 20 GW station maintains a circular orbit at 476 km altitude and inclination of 28.5°. Figure 19.9.2 shows the orbiting SPS-01 solar-power station (upper right) beaming microwave energy to an ascending Lightcraft (lower left). Although MRS-01 and PDS-03 share the same orbit as this new station, SPS-01 was purposefully deployed one-half orbit away (i.e., diametrically opposed, on the far side of the planet) to facilitate global coverage.

NASA has committed this prototype 35 GHz station for a series of international power beaming experiments with ground-based rectennas (~1 km diameter) linked into terrestrial electric grids. Over the next five years, SPS-01 will beam intermittent power to rectenna sites in 15 countries located within 900 km of the equator. The duration of each transmission test is limited to 100 seconds or less. The objective is to generate technical data of relevance for future 1 to 5 GWe geo-stationary SPS construction projects, possibly earmarked for developing countries. However, an economic analysis of powering Lightcraft flights for the commercial space tourism market has revealed a 100-fold greater return on

Table 19.9.1: Mass breakdown for SPS-01 (w/o personnel & accessories).

Component	Mass (tonnes)
320 MWe Photovoltaic Array (1 kWe/kg)	320
20 GW Solid State Transmitter Array (40.8 kW/kg)	490
SiC/SiC Disc Structure & tensioning spokes	503
SMES "Battery" (2.5 terajoules)	2853
Liquid Helium SMES Refrigeration System	108
Maintenance/ Habitation Modules (skeleton crew)	23
TOTAL MASS	4297

investment than that derived from energizing the terrestrial electric grid.

19.9.2 Powersat Specifics

The one-kilometer diameter structure of SPS01 resembles a giant, bicycle wheel (see Figure 19.9.2) orbiting at an altitude of 476 kilometers. The orbital period is ~94 minutes. The station rotates at 1.34 RPM to gain gyroscopic stability while providing the skeleton crew with a 1 g artificial gravity. Table 19.9.1 gives the mass breakdown for this 4,297 tonne station.

Besides the tensioning spokes, the wheel's thin-film disk structure is comprised of several hundred large, pie-slice segments of 0.32 millimeter thick silicon-carbide (SiC/SiC) film. Completely covering one side of this silicon-carbide web are 30% efficient, thin-film solar photovoltaic cells (1 kW$_e$/kg) that supply 320 megawatts of electricity (at normal incidence) to charge the twin-loop, 2.5 terajoule SMES "battery" that forms the rim of the wheel. On the other side of the web/disc are 13.3 billion miniature solid-state transmitters (each just 8.5 mm across and delivering 1.5 watts) to give a maximum radiated power of 20 GW at 35 GHz. The specific power for this solid-state, phased-array transmitter is 40.8 kW/kg, which is comparable to the gyrotron-based PDS-03 array (43.2 kW/kg). The SiC/SiC web structure of SPS-01 provides a dual use as the semiconductor substrate for the photovoltaic and transmitter electronics.

Micrometeoroid impacts are expected to degrade 1% of the SPS-01 photovoltaics area over the projected 30-year operational lifetime. Large meteoroid impacts are anticipated, but with a very low probability. A skeleton crew onboard the SPS-01 station continuously monitors its systems and conducts repairs, as necessary, with robotic aids. Fortunately for SPS-01, the USSC orbital debris cleanup program was completed in 2024 just prior to station activation.

19.9.3 SMES Battery Specifications

The SPS-01 station is ringed by an energy storage device (the SMES unit) consisting of two superconducting loops, each with a mass of 1426 metric tons, and charged up with opposing electric currents. This arrangement eliminates the titanic magnetic torque that would be produced by a single loop in Earth's geomagnetic field. Table 19.1.1 provides specifications for the 2.5 terajoule SMES "battery" of SPS-01. As with the earlier SMES units, the SPS-01 "battery" may not be discharged below 16% of its maximum capacity while boosting a Lightcraft. Once discharged, the station must complete two orbits (or more) before its 320 MWe solar panels can fully recharge the SMES with 2100 gigajoules of energy, to reach the maximum 2500 GJ capacity. As mentioned earlier, this SMES unit was constructed from metal matrix composite superconductors using $MgCNi_3$ microfibers in a copper matrix.

19.9.4 Orbital Mechanics and Power-Beaming Mission Analysis

As shown in Figure 19.9.3, the SPS-01 configuration is a 1 km diameter bicycle wheel design with its web paved with 30% efficient solar photovoltaics. The station spins for stability in an orbit with 28.5 deg. inclination, and

POWER BEAMING INFRASTRUCTURE

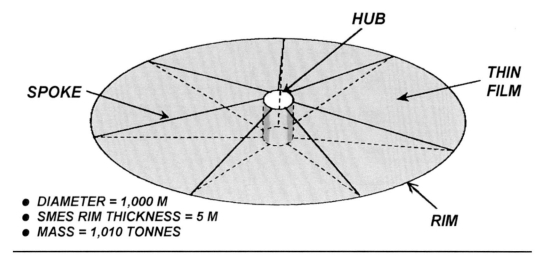

- DIAMETER = 1,000 M
- SMES RIM THICKNESS = 5 M
- MASS = 1,010 TONNES

Figure 19.9.3: Bicycle wheel geometry of 1-km diameter SPS-01 powerplant.

has an effective thickness of 5 m. Each of the SMES rings has a minor diameter of 2.14 m (see Table 19.1.1); additional "frontal" areas are contributed by the SMES interconnecting structure and tensioning cables (i.e., wheel-bracing spokes). When not beaming microwave power, the powersat normally orbits "edge on" to minimize drag force, so the projected area used for orbital lifetime calculations is 5000 m².

Reference 18 contains an analysis of SPS-01 orbital mechanics and power-beaming missions, from which the following details are extracted. Figure 19.9.4 shows the calculated aerodynamic drag vs. altitude for the SPS-01 assuming a +2 σ Earth atmosphere density. Drag force declines from a value of ~0.58 N for the daily repeating orbit of 476 km, down to approximately 0.2 N at an orbital altitude of 550 km. Ultimately NASA mandated the daily repeating orbit since the powersat would pass over one designated Lightcraft launch site each day. To maintain the daily repeating orbit feature, a continuous thrust opposing atmospheric drag must be applied by the powersat's electrodynamic station-keeping engine (and/or backup ion propulsion system).

If stationkeeping engines were to fail, Figure 19.9.5 reveals the rate at which the powersat will lose altitude until it reenters the atmosphere in approximately 3 years. Although these calculations were performed for an earlier 1010 tonne powersat design (i.e., ~4.25x less mass than the SPS-01) of identical dimensions, the predicted results highlight the extreme criticality of reliable stationkeeping propulsion.

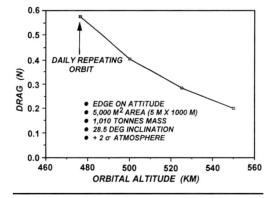

Figure 19.9.4: SPS-01 aerodynamic drag vs. altitude. *(Courtesy of NASA.)*

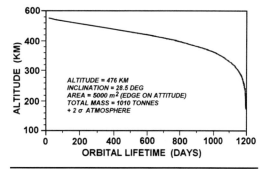

Figure 19.9.5: Powersat orbital lifetime. *(Courtesy of NASA.)*

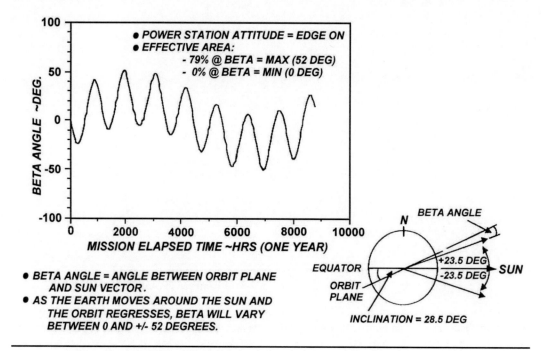

Figure 19.9.6: Powersat beta angle vs. mission elapsed time. *(Courtesy of NASA.)*

Figure 19.9.6 (see lower right) presents the coordinate system for describing powersat orbital mechanics about the Earth; note the North Pole, equator, and powersat orbit plane inclination of 28.5 degrees. Beta (β) is defined as the angle formed between the powersat orbit plane and a line from the Earth's center to the sun. For a 28.5° inclined orbit, β will vary between 0° and ±52° over a 12 month period as shown in Figure 19.9.6. The combined effects of the Earth's motion around the sun, and regression of the orbit's "line of nodes" produces the β angle excursions. With the powersat orbiting in "edge-on" attitude around the Earth, the effective solar-illuminated area will vary from 0% for a β of 0°, to a maximum of 79% when β is at ±52°. Clearly the combination of "edge-on" attitude and β angle ultimately dictates the powersat rate-of-charge, with the maximum solar charge rate occurring at the ±52° beta angle. In short, maximum solar illumination time occurs when β is also at its maximum.

Figure 19.9.7 gives the percentage of time that the powersat spends in the sun; i.e., that portion of the orbit period (about 92 minutes) that the powersat is exposed to solar illumination. Over a 12 month period, this percentage varies from 62% for a β of 0°, up to 70% for a β of 52°.

The rate at which the powersat SMES "battery" can be charged is directly related to the β angle because the station is orbiting in "edge-on" attitude. Therefore the maximum charge rate occurs when β reaches its maximum, simply because the "percent-time-in-the-sun" and projected photovoltaic area are also at their

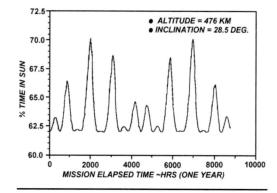

Figure 19.9.7: Powersat percent time in sun vs. time. *(Courtesy of NASA.)*

maximums. Figure 19.9.8 shows the time required to charge the SMES with 1265 GJ of solar-derived energy. The calculations assume:

a) the 785,400 m² solar array is perpendicular to the sun line;
b) solar cell efficiency of 30%;
c) maximum time in the sun of 1.1 hours per orbit; and
d) solar flux constant of 1353 W/m².

The maximum charge necessary for a LTI-20 boost to LEO is 2100 gigajoules. Hence for a maximum β of 52°, just over 3 hours (2+ orbital revolutions) are necessary for the full charge, whereas ~16 hours (10.3+ revolutions) are needed when β is 10°.

The force acting on the powersat due to solar photonic pressure also varies with β because of the edge-on operating attitude. The minimum powersat "sail" area is projected for a β of 0°, which results in the minimum force. With the maximum β angle, the projected sail area is also at maximum, with the resulting force being the

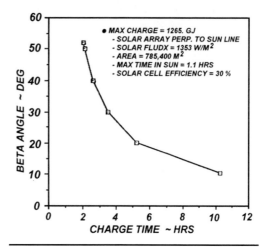

Figure 19.9.8: Powersat beta angle vs. charge time for edge-on attitude. *(Courtesy of NASA.)*

largest for this powersat orientation. Since the absorptivity and reflectivity of powersat web surfaces (i.e., photovoltaics on one side vs. microwave transmitter array on the other) and SMES housing also affect this force, the results in Table 19.9.2 are displayed for total reflection,

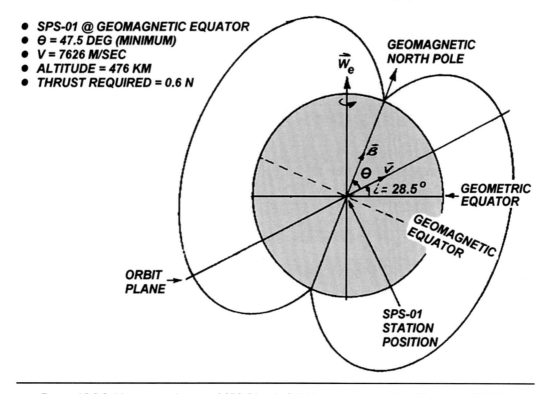

Figure 19.9.9: Magnetic reboost of SPS-01 with 0.6 N to maintain orbit. *(Courtesy of NASA.)*

total absorption, and the average. Note that the average force due to solar photonic pressure is 7 times that of atmospheric drag at the 476 km orbit. However, solar pressure acts as both a propulsive and retarding force as the powersat travels around the orbit. For the SPS-01, the effect of solar photonic pressure is the major consideration in sizing the orbit-keeping propulsion system, which is a combination of electrodynamic engines and electric rocket thrusters (see Figure 19.9.9 – magnetic reboost with 0.6 N to maintain orbit).

19.9.5 Ascent to Orbit and Power-Beaming Relationships

When fully charged, SPS-01 can beam down 20 gigawatts of microwave power onto a Lightcraft at a maximum range of 1170 kilometers. Coarse pointing of the power station transmitter is accomplished by geomagnetic torquing (by shifting small amounts of electric current from one SMES cable to the other), but fine control comes from a beacon mounted on the Lightcraft. This beacon signal continuously coordinates the individual solid-state transmitters on the SPS-01 to create a spot 10- to 20 meters in diameter at the Lightcraft launch site. Under power, the LTI-20 can reach orbit in less than five minutes, subjecting its occupants to approximately 3 g of acceleration. Or the powersat can unload all its energy in a 54-second burst that offers a nearly vertical 20 g boost to geostationary orbit or even to escape velocity.

An analysis of the LTI-20 Lightcraft ascent to orbit and SPS-01 power-beaming relationships is contained in Reference 19, from which the principal results are summarized below. The basic ground rules and assumptions of the analysis are:

a) LTI-20 acceleration history is a modified version of the BEP launch trajectory reported in Reference 20 – see Figure 19.9.10;
b) powersat and Lightcraft major axes are constrained to be parallel throughout power-beaming;
c) Lightcraft launch is from 28.5° latitude;
d) first 10 seconds of flight are provided by internal LTI-20 power;
e) powersat power-beaming begins 10 seconds after liftoff;
f) power-beaming range between powersat and Lightcraft is constrained to less than 1170 km;
g) launch trajectory targets an apogee altitude of 211 km; and,
h) autonomous Lightcraft RCS propulsion is employed for orbit circularization and deboost for reentry.

Figure 19.9.11 is a schematic of the LTI-20 ascent mission profile, for which the sequence of events is annotated in numerically:

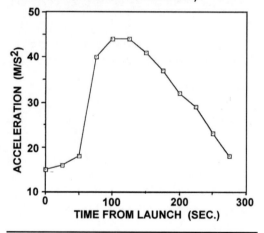

Figure 19.9.10: Lightcraft acceleration profile.
(Courtesy of RPI.)

Table 19.9.2: Solar pressure force on 1-km powersat with "edge-on" attitude.[18]

Reflective Properties of Powersat (web and SMES housing)	Solar Photonic Force (N)	
	Beta (ß) Angle = 10°	Beta (ß) Angle = 52°
Total Reflection	0.045	5.6
Total Absorption	0.023	2.8
Average	0.034	4.2

POWER BEAMING INFRASTRUCTURE

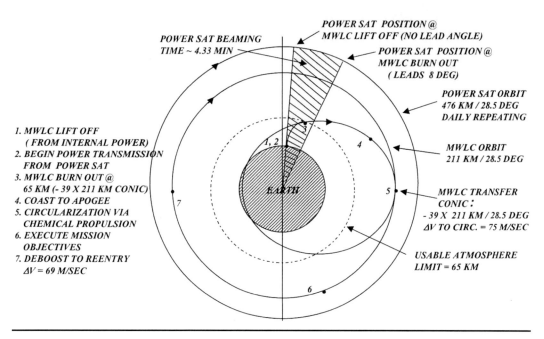

Figure 19.9.11: Schematic of lightcraft ascent mission profile. *(Courtesy of NASA.)*

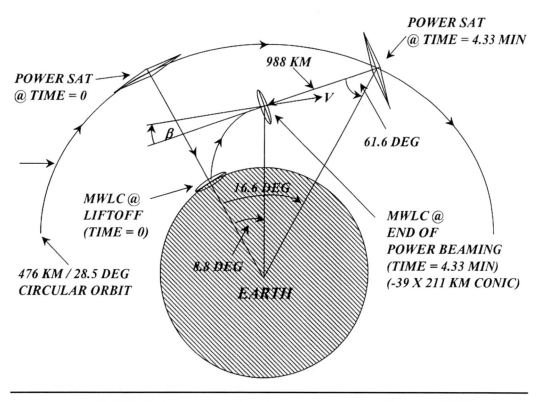

Figure 19.9.12: Mission geometry for lightcraft and powersat. *(Courtesy of NASA.)*

1) Lightcraft liftoff (from internal power);
2) begin power transmission from powersat;
3) Lightcraft "burnout" at 65 km (-39 x 211 km conic) – end of MHD slipstream accelerator mode;
4) coast to apogee of 211 km;
5) circularization via autonomous RCS propulsion;
6) execute mission objectives; and
7) deboost to reentry ($\Delta V = 69$ m/s).

Note the powersat position in Figure 19.9.11 at the moment of Lightcraft liftoff (no lead angle); at Lightcraft "burnout", the powersat leads by 8 degrees. The maximum extent of "usable" atmosphere for air-breathing MHD slipstream accelerators is 65 km. The total powersat beaming time is 4.33 minutes for the launch.

Figure 19.9.12 gives the time-dependent geometry for the LEO launch mission, assuming the major axes of both Lightcraft and powersat are maintained parallel during the power-beaming. Note that engine thrust is directed along the Lightcraft velocity vector (V), and that β (in Figure 19.9.12) is defined as the engine thrust angle measured from the Lightcraft minor axis. At time = 0 the Lightcraft lifts off when the powersat is directly above in a **476 km -28.5°** circular orbit. At power-beaming termination (i.e., time = 4.33 minutes), the Lightcraft has advanced 8.8° from the launch point, whereas the powersat has moved to 16.6° along its orbit – increasing the maximum transmission range to 988 km. Note the average powersat pitch rate is ~12 deg./min, and the powersat pitch angle is 61.6 deg.

Figure 19.9.13 shows how the Lightcraft engine thrust angle (β) varies with time during the launch. The Lightcraft thrust vector angle varies from 73 to 25 deg. during the first minute of flight, and holds near 20 deg. for the last 150 seconds of the boost. The LTI-20's combined cycle propulsion system is able to accommodate large variations in the thrust direction "β" required during the ascent.

Figure 19.9.14 presents the powersat pitch rate throughout the power-beaming boost of 260 seconds. Note that the pitch rate falls rapidly from an initial requirement of 1 deg./s down to near zero at time = 200 seconds. To satisfy the initial pitch rate, the powersat must already be pitching at 60 deg./min. at time = zero; thereafter its pitch rate must decelerate at about 0.01 deg./s². However, the average pitch rate is only 12 degrees/minute, and the powersat can easily accommodate the highest pitch rates "artificially" by using the transmitter's phased array for pointing control – to as much as ±20° Pitching the spinning powersat improperly can result in significant control, pointing, and structural impacts.

Construction Chronology

This section presents the construction chronology for the SPS-01, a rotating "bicycle wheel" powersat designed and engineered for

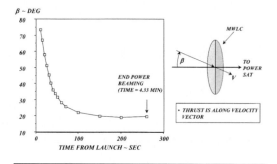

Figure 19.9.13: Lightcraft thrust angle β vs. time. (Courtesy of NASA.)

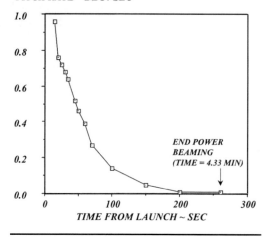

Figure 19.9.14: Powersat pitch rate during power-beaming boost. (Courtesy of NASA.)

telerobotic assembly in low earth orbit, from terrestrial manufactured parts (see Figure 19.9.15). This innovative construction and assembly method was first proposed by Flint.[12] The modular SPS-01 elements include a central mast, prefabricated thin film array (TFA) wedges (Figure 19.9.16), radial stabilizing guy wires, "bicycle" rim structure comprised of multiple tube segments, and the twin-loop perimeter SMES unit. These "dual use" rim-tubes serve five important functions in the construction process:

a) array wedges are rolled onto the tubes after terrestrial manufacture;
b) tubes stabilize rolled array wedges during boost to orbit (Figure 19.9.17);
c) tubes aid in unfurling thin film wedges in LEO;
d) tubes are joined together (with rim couplings) to complete the "bicycle" rim for a high stiffness perimeter structure; and finally,
e) SMES loops are then mounted onto opposite sides of the structural rim.

The size and mass of powersat modules is dictated by the ARES V booster's payload capacity.

Table 19.9.3 from Reference 12 lays out the essential 6-step construction process, for which the "story board" is portrayed in Figures 19.9.18 to 19.9.21. The "living" construction base in Figure 19.9.18 expedites telerobotic assembly through the use of temporary spars and submasts (i.e., ultralight trusses, recycled later) to anchor and stabilize deployment guy wires rigged for unfurling the thin-film wedges (Figures 19.9.19 and 19.9.20). The final SPS-01 configuration is captured in Figure 19.9.21, with central mast, despun section, stiffening guy wires, and solar TFA wedges with phased array microwave antenna on the back side.

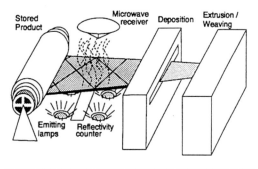

Figure 19.9.16: Manufacturing process for thin film arrays.[12] *(Courtesy of E. Flint and SSI.)*

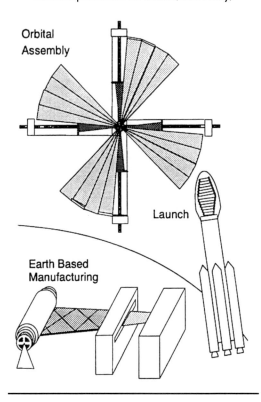

Figure 19.9.15: Overview of SPS-01 construction method showing terrestrial fabrication of thin film arrays, subsequently launched in compact form and then unfurled.[12] *(Courtesy of E. Flint and SSI.)*

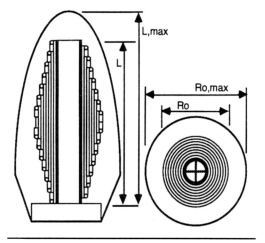

Figure 19.9.17: Launcher payload bay considerations.[12] *(Courtesy of E. Flint and SSI.)*

Table 19.9.3: Construction schedule.[12] *(Courtesy of E. Flint & SSI.)*

Step	Construction Activity
1	Deployment of central mast
2	Attachment of mounting rings
3	Unfurl thin film array
3a	Move wedge from storage
3b	Attach wedge and upper and lower guys
3c	Roll out wedges
3d	Attach rim to its neighbors; attach guys
3e	Check out wedges (add to power net)
3f	Rotate partially completed disc satellite so next wedge can be deployed
3g	Repeat until done
4	Separate construction base
5	Attach "despun" unit; top off propellant tanks
6	Spin up, lock out rim, and boost for orbit

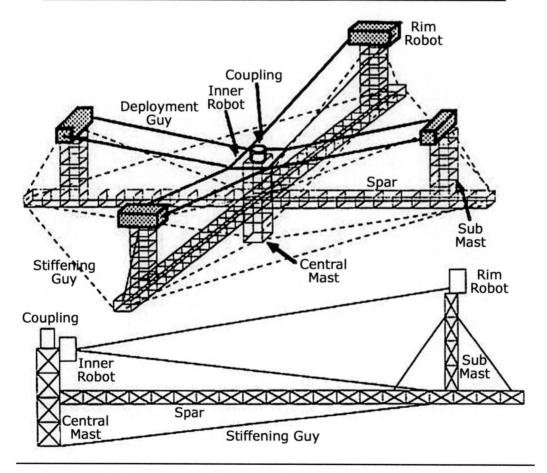

Figure 19.9.18: Construction base (upper image) and details of one arm (lower image).[12] *(Courtesy of E. Flint and SSI.)*

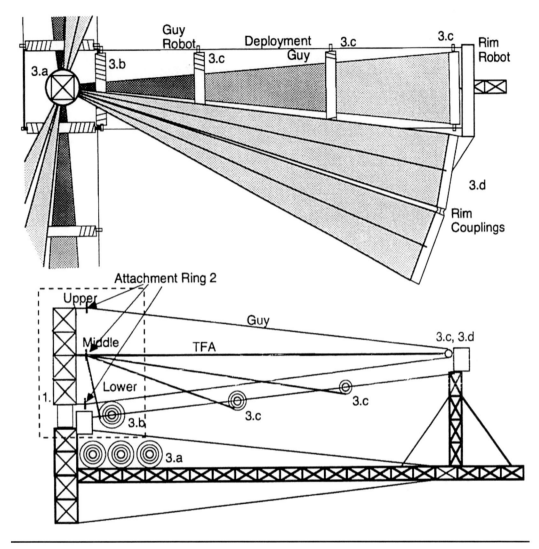

Figure 19.9.19: Sequential "snapshots" of the unfurling of thin film wedge.[12]
(Courtesy of E. Flint and SSI.)

Historic First Flight to LEO on SPS-01 Power Next Generation rectennas on the newest LTI-20s can now function equally well at 35 GHz and 140 GHz. Anticipating this upgrade, the USSC had planned and discretely carried out a rigorous 10 month flight test program on the SPS-01 (with NASA cooperation) commencing in January 2025. The historic flight on 23 October 2025 proved conclusively that SPS-01 can easily "tractor" the LTI-20A with its 6-person crew right into orbit (Figure 19.9.22). In securing this NASA objective, SPS-01 has effectively paved the way for a whole industry of orbital solar stations that, in the near future, could spawn a robust power-beaming infrastructure for the planet. Given such alternative energy infrastructure to "fuel" the emerging commercial Lightcraft fleet, rapid and low-cost travel around the globe, into orbit, to the moon and beyond should soon be feasible and affordable *for the civilian population*. Industry observers now predict the price of civilian Lightcraft flights to LEO will fall by a factor of 100x to 1000x below that of chemical-rocket powered spacecraft within a decade.

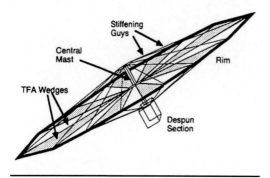

Figure 19.9.20: Thin film wedge deployment and guy wire stowage details.[12] *(Courtesy of E. Flint and SSI.)*

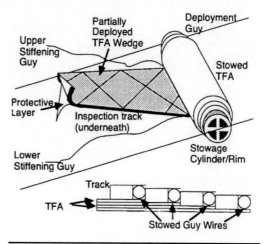

Figure 19.9.21: Final SPS-01 configuration with central mast, solar collector array (i.e., wedges), outer rim, guy wires, and despun section.[12] *(Courtesy of E. Flint and SSI.)*

Figure 19.9.22: Historic flight of LTI-20A to LEO, powered by SPS-01 (23 October 2025). *(Courtesy of Media Fusion and NASA.)*

20.0 HUMAN FACTORS AND g-TOLERANCE

Human tolerance to g forces depends not only on physiological and psychological variables, but also on the structure of the body. The fact that the body is not a solid and will internally deform must be considered. The physical extent of such deformations, exacerbated by density differences between muscle, organs, and bone, ultimately sets the limits of g tolerance. Although the physiological responses to the effects of low-g forces (0 to 30 g) have been in the open literature for decades, the extensive DoD database on the hyper-g range (beyond 30 g) has only been recently been declassified.

20.1 Introduction to Human Factors

A quick review of human structural anatomy will demonstrate the specific areas of concern.

20.1.1 Head

The skull encases and protects the brain. Shifts in blood flow which result in over- or under-pressurization of the blood vessels produce corresponding pressure changes in the brain. Brain damage could result if pressures exceed known limits. Rapid blood pressure changes to the brain are one of the causes of g-induced loss of consciousness (GLOC or GLC). In extreme cases, aneurysms or blood vessel rupture (stroke) may also occur. The eyeballs are recessed in sockets in the skull and surrounded by pads of fat. They are naturally protected from low g forces, but may deform slightly under high g forces, resulting in vision changes. The inner ear, although encased in a part of the skull, is also susceptible to g forces. Overstimulation by the otoliths in the semicircular canals of the vestibular system may result in balance and disorientation pathologies.[1]

20.1.2 Thorax

The most prominent structure occupying the anterior part of the chest is the heart. Under high $+g_X$ accelerations (i.e., positive g; subject in prone position), when the chest is subjected to anterior compression, the heart may deform against the spinal column. This fact is used during cardiopulmonary resuscitation (CPR) to create or enhance blood flow when the heart has stopped or has malfunctioned. The exact amount of deformation that the heart may tolerate and still function is not known. The application of g forces from posterior to anterior (i.e., $-g_X$; prone position, "eyeballs out") can deform the heart and descending aorta less than if the g forces are applied anterior to posterior ($+g_X$; prone position, "eyeballs in").

The cardiovascular effects of high g forces are well known. The most important of these is that a reduction of blood flow to the head effects vision and diminishes the visual field.

The majority of the chest is occupied by the lungs. The thicker portions of the lungs lie towards the posterior and bases. Simpura[2] demonstrated that accelerations up to 13 g did not cause significant shifts in ventilation and gas exchange. Above 13 g, there was a marked decrease in effectiveness. Hlastala, et al.[3] showed that pulmonary blood flow (PBF) is remarkably similar in each region of the lung. When g force is tripled, only 19% of the variance of PBF is due to gravitational force. This means that the pulmonary blood flow is largely resilient to increasing levels of acceleration.

20.1.3 Abdomen

The upper abdomen contains a number of solid organs including the liver, kidneys, and spleen. The kidneys are quite well protected on the posterior wall. The spleen and liver, although less well protected, are supported by ligaments and their position against the abdominal walls. The lower abdomen, containing the large and small intestines, is structurally the most compressible section of the body. Special supports are mandated for this area, but should not be so restrictive as to prevent movement of the diaphragm during breathing.

The discussion above clearly demonstrates that structural deformation must be considered when designing equipment and systems to support the body under high-g loads. Additional factors which can further improve human g tolerance are as follows.

20.2 Improving Human g Tolerance

Three opportunities for improving human g tolerance are in evidence: by physiological, psychological, and/or mechanical means, as described below. Figure 20.2.1 indicates the standard convention for accelerations relative to human body orientations:

a) transverse supine acceleration (+g_X or "eyeballs in");
b) transverse prone acceleration (-g_X or "eyeballs out');
c) positive acceleration (+g_Z); and,
d) negative acceleration (-g_Z).

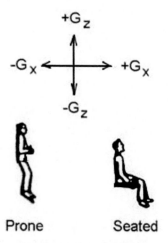

Figure 20.2.1: Standard coordinate system convention for accelerations relative to human body orientation.

20.2.1 Physiological Means

The g-tolerance standard of a 15-second, and 1 g/s onset loading to a +7 g_Z was first adopted in 1977 by the USAF School of Aerospace Medicine for evaluation of prospective crew members.[4] Air and space forces customarily train fighter aircrews to increase their tolerance to high-g environments using human centrifuges.[5] The Naval Air Development Center (NADC) initiated high-g centrifuge training for fighter pilots and developed the world's first full system, total g-force environment, called the Dynamic Flight Simulator.[6] The NADC subsequently adapted its centrifuge as a total g-force environment to simulate spins and recoveries up to minus 5 g.[7] A high-g training curriculum including GLOC detection and recovery was also adopted for fighter pilots assigned to Hollomon AFB.[8] Throughout history, human g tolerance has been increased by selecting pilots and crew that have had:

a) a naturally high tolerance to g stress,
b) a high degree of physical conditioning, and
c) frequent exposure to g stress.[9]

20.2.2 Psychological Means

Human tolerance to g force also depends on psychological factors, of which the most crucial is lack of motivation.[10] Hence, Lightcraft crew candidates are self-selected, highly motivated, and preconditioned to the effects of acceleration. Presented in Figure 20.2.2 are data on the upper limits of voluntary endurance by highly trained pilots in withstanding accelerations. Throughout the times indicated, they were required to operate a side-arm control device and satisfactorily perform a tracking task.[11] Under +g_X accelerations, note that 1 g can be sustained indefinitely, 6 g for 550 seconds, 10 g for 220 s, and 14 g for 120 s. At 14 g the pilot's velocity is increasing by 137 m/s every second, so after 10 seconds the pilot is

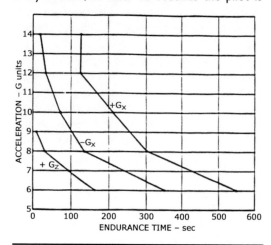

Figure 20.2.2: Upper limits of voluntary endurance.[11]

HUMAN FACTORS AND g-TOLERANCE

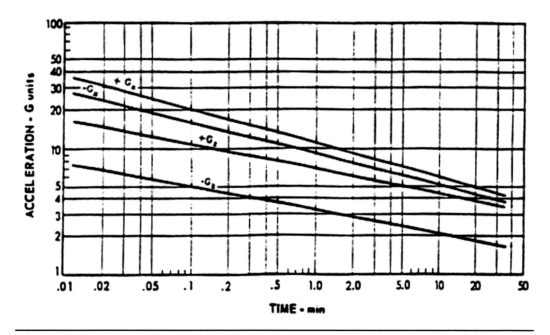

Figure 20.2.3: Average tolerance for positive (+gz), and transverse prone (-gx) acceleration.[12]

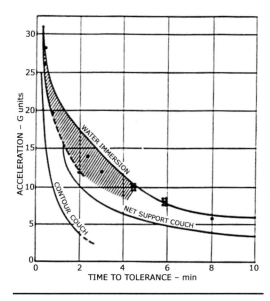

Figure 20.2.4: Enhanced g-tolerance afforded to subjects immersed in small water tank; g-tolerance curves also shown for subjects protected by optimal support and positioning.[12]

going 1.37 km/s (3000 miles/hr) and has traveled a total distance of 6.86 km (4.26 miles). In just under one minute, the pilot will reach orbital velocity, and 21.6 seconds later exceed escape velocity. In contrast to the above data, Figure 20.2.3 shows the average acceleration tolerances,[12] which are somewhat lower than the upper limits to voluntary endurance.

Note that Apollo astronauts were routinely trained to withstand 8 to 9 g, which the second stage booster reached near the end of its burn; this significantly exceeded the Space Shuttle's 3 g peak acceleration. Furthermore, in the event of fire or explosions, the emergency escape rocket was designed to separate the Apollo CM from the Saturn V, subjecting the crew to 13 g for the brief boost period.

20.2.3 Mechanical Means

A perfectly supported human body, as in the pressurized Ultra-g protection suit and escape pod, can withstand very large g forces. The rigid, individualized and custom-fitted shell with liquid immersion prevents excessive deformation of the human body and increases the g limit under high g_X loads. Figure 20.2.4 from Reference 12 clearly shows the enhanced g-tolerance afforded to a subject when immersed in a small tank of water; e.g., up to 2.7 minutes at 15 g, 75 seconds at 20 g, and ~5 seconds (or more) at 30 g.

Chapter 20

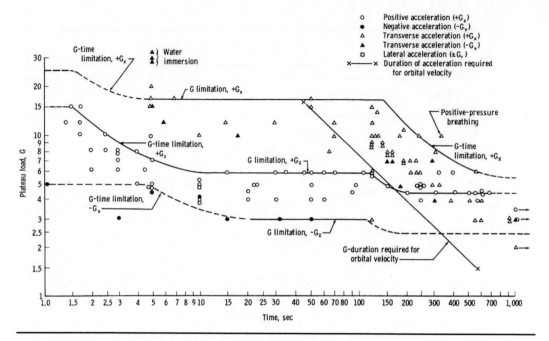

Figure 20.2.5: Human experience of sustained acceleration; data from many sources.[13]

Accelerating at 30 g, one reaches speed of sound in just over one second, or conversely, comes to a sudden stop after cruising at Mach 1 – without sustaining damage. Figure 20.2.5 summarizes the human g-tolerance data for sustained acceleration, from open literature sources.[13] Note the three diagonal lines (right side) showing the g-duration required for Earth orbital velocity, escape velocity, and twice escape velocity. All fall within the demonstrated limits of human tolerance. Figure 20.2.5 also includes data on positive-pressure breathing, which indicates a measurable improvement in g-tolerance.

20.3 Improving High-g Ventilation

It is apparent from the previous discussion that breathing must be augmented while in high-g (beyond 30 g) environments, if g-tolerance is to be maximized. Several methods and systems have been adopted for LTI-20 flight OPS:

a) HeliOx and special mixture gases, and,
b) total or partial liquid ventilation.

20.3.1 Basics of HeliOx and Special Mixture Gases

Normal breathing air is 21% oxygen and 79% nitrogen. When breathing air in the diving environment, nitrogen narcosis may be noticed at depths of around 30 m. Nitrogen becomes a narcotic and should not be used beyond depths of 50 m.

Nitrox is any combination of nitrogen and oxygen, usually with a greater proportion of oxygen than compressed air (i.e., Nitrox 21). For example, Nitrox 32 is 32% oxygen and 68% nitrogen; the recommended diving depth limit is 40 m (130 ft.), but using it may reduce decompression time.

Trimix is a mixture of oxygen, helium and nitrogen. Some still consider it the "poor man's HeliOx." HeliOx is a special mixture of helium and oxygen customarily used by professional and technical divers when diving to greater than 60 m (200 ft). Beyond 150 m (500 ft), the use of HeliOx may cause dizziness, nausea, vomiting, and marked tremors of the hands, arms, and torso called High Pressure Nervous Syndrome (HPNS). The major advantage to HeliOx is the avoidance of nitrogen narcosis, the "rapture of the deep;" however, the decompression time is the same as when breathing air. One of the disadvantages of HeliOx is that it has a high heat conductivity, so divers tend to get cold quickly.

This problem can be corrected by warming the gas and diver or by using a recirculating rebreather. Another minor inconvenience is that HeliOx has a lower density than air, which causes the vocal cords to vibrate faster. This voice pitch increase, the "Donald Duck" phenomenon, is overcome simply by using a voice unscrambler.

Any discussion of special mixture gases in high-g environments (i.e., akin to high pressure at depth) must include facts on the relative narcotic effects of its components (see Table 20.3.1). Helium and neon show virtually no narcotic effects. Hydrogen is less narcotic than nitrogen, but more so than helium or neon,[14] but of course hydrogen-oxygen mixtures are violently explosive. One must not forget the dangerous effects of oxygen itself, for it can be toxic at a partial pressure greater than 2 bars, although that value varies from individual to individual.

For low g maneuvers in LTI-20 missions, HeliOx is clearly the mixture of choice. HeliOx is used to inflate and pressurize the LTI-20 Lightcraft, and hence the normal breathing mixture on the bridge and in the pods. Pulsed pressurized HeliOx is applied in the pods for short duration accelerations (20 seconds or less), generally up to 30 g; hence, crewmembers do not receive saturation exposure to helium and therefore do not need to decompress. Beyond 30 g, liquid ventilation is the only feasible alternative.

20.3.2 Basics of Liquid Ventilation

Liquid ventilation is not a new concept. Groundbraking experiments date from the mid-1960s, when researchers used various liquids to support and facilitate ventilation. In 1966, Clark and Gollan were the first to apply an oxygenated perfluorocarbon (PFC) liquid to support the respiration of several small animals that were totally immersed in the liquid.[15] The concept of liquid breathing received national attention in the mid 1990s when the science fiction film "The Abyss" was first released. A demonstration in "The Abyss" movie showed a pet white rat being immersed in a pink liquid. (The pink coloration was a Hollywood addition.) Gollan and Clark[16] as well as Kylstra[17] demonstrated that PFC-breathing mice could be decompressed without suffering from decompression sickness. The liquid-filled lung prevented excessive inert gas from dissolving in the blood and tissues when exposed to large changes in partial pressure.

PFC liquids have physical properties that make them ideally suited for high-g lightcraft applications. These include a high oxygen and carbon dioxide solubility, a low surface tension (14 - 20 dyn/cm) and a viscosity less than water. They are chemically and biologically inert, insoluble in water, nonabsorbable, and essentially nontoxic. Most importantly, PFC's are immiscible with either exogenous or native surfactants, such as those produced by the Type II alveolar cells in the lungs. Surfactant reduces surface tension as the radius of the small airways and alveoli decreases. Without it, surface tension would collapse the alveoli and small airways, reduce their availability for gas exchange, and promote instability and atelectasis. The boiling points of PFC's are quite low, which allows vaporization at body temperature and enhances their suitability for instillation into the tracheobronchial tree of the lungs.

Table 20.3.1: Relative narcotic effects of special mixture gases.

Gas		Relative Narcotic Potency		Molecular Weight
He	Helium	4.26	least narcotic	4
Ne	Neon	3.58		20
H_2	Hydrogen	1.83		2
N_2	Nitrogen	1		28
A	Argon	0.43		40
Kr	Krypton	0.14		83.7
Xe	Xenon	0.039	most narcotic	131.3

A review of the literature documents the use of a number of perfluorocarbon liquids, usually with specific gravities between 1.77 and 1.95, or nearly twice the density of water. The most recent work has been done with PFOB (perfluorooctylbromide) or "Perflubron," Liquivent, and a sterile perflubron called AFO 141 (manufactured by Alliance Pharmaceutical Corporation).

Typically, liquid breathing (LV) is separated into two categories: Total Liquid Ventilation (TLV), and Partial Liquid Ventilation (PLV). In both cases, the lung is filled with PFC's to the Functional Residual Capacity (FRC). FRC is defined as that volume of air left in the thoracic cage at the resting expiratory level. Total liquid ventilation requires a special ventilator to circulate the PFC in and out of the lungs. In contrast to TLV, partial liquid ventilation requires a conventional ventilator to deliver breathing gas to the liquid-filled lungs. Gas-exchange oxygen delivery and carbon dioxide removal occurs between the liquid in the lungs and the gas delivered by the ventilator.

20.3.3 Total Liquid Ventilation
The significant historical contributions by Moskowitz and Shaffer[18-20] spurred the development of a demand-regulated liquid ventilator that allowed the subject to control the cycling of the ventilator as it circulated PFC liquid in and out of the lungs. They demonstrated for the first time that effective delivery of oxygen and adequate removal of carbon dioxide in the liquid-filled lung was possible. BioPulmonics Incorporated, a subsidiary of Alliance Pharmaceutical Corporation, developed the very first "advanced" liquid ventilator.

Subsequent studies by Modell[21] and Shaffer[22] proved that experimental subjects could breathe PFC for extended periods without undesirable side effects.

20.3.4 Partial Liquid Ventilation
Once the terminal alveoli are filled with liquid, PLV (or TLV) eliminates the air-liquid interface, recruits collapsed regions of the lung, improves compliance, and eliminates high surface tension and collapse. Because of the reduction of surface tension, lower inflation pressures are required when using conventional ventilators. The net result is an effective distribution of gas exchange and pulmonary stability.

20.3.5 Liquid Breathing for High-g Environments
LTI-20 crewmembers facing ultra high-g missions (>30 g) must be trained to breathe perfluorocarbon liquids (Figures 20.3.1 and 20.3.2). Clearly, all humans developed in a liquid-filled womb environment, so there is little reason to believe that the skill cannot be relearned. In fact, a decade of lightcraft flight experience has proven conclusively that indeed it can ... but not without some assistance. Breathing a liquid requires more work than breathing air because liquid is denser and more viscous than air. Assistance from a liquid or

Figure 20.3.1: Headgear for partial liquid ventilation. *(Courtesy of RPI.)*

HUMAN FACTORS AND g-TOLERANCE

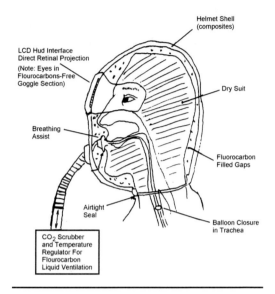

Figure 20.3.2: Schematic of liquid ventilation system headgear. *(Courtesy of RPI.)*

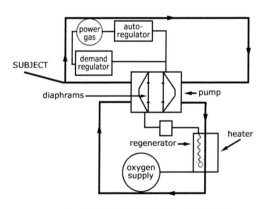

Figure 20.3.3: Liquid ventilation respirator system.

had been used, at least on an experimental basis, in deep ocean oil drilling operations and other recovery missions requiring levels of manual dexterity that could not be matched by existing robotic vessels. Since the limiting pressure (without damage) on the human central nervous system is about 50 to 80 bar (735 to 1176 psi), working depths beyond 500 to 800 meters were generally assumed "off limits."

Once the psychological aversion is overcome, loading the PFC is achieved by directly inhaling the liquid over a period of time. The swallow reflex must be diminished (or absent) to prevent PFC liquid from entering the stomach. Since the liquid is nontoxic, swallowing some of it is not of any consequence. As portrayed in "The Abyss," the PFC may also be inhaled and circulated in a "fishbowl" helmet that engulfs the head. However, with this method, separate protection **must** be provided for the corneas of the eyes, because current formulations of PFC's may possibly damage them.

Another "loading" option is through a series of computer controlled "smart tubes" that are advanced through the nostrils and placed at the entrance to the trachea for direct instillation or inhalation. The current LTI-20 apparatus is a further refinement of the Laryngeal Mask Airway (LMA) in Figure 20.3.4. The device is advanced orally to the area below the epiglottis, the cuff is inflated, and ventilation performed, or,

traditional gas ventilator is necessary so that the work of breathing can be reduced (Figure 20.3.3). As dramatized in "The Abyss," there is also a significant psychological aversion to breathing a liquid that must be overcome. Only recently have unclassified articles been published in the open literature that involve conscious, voluntary breathing of PFC's by a human subject, or undergoing high-g tolerance tests with PFC's. Prior to such disclosures, informed sources have speculated that PFC's

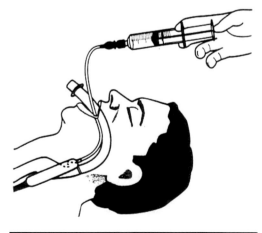

Figure 20.3.4: Laryngeal mask airway. *(Courtesy of Gensia, Inc.)*

in the LTI-20 case, liquid instilled. The inflated cuff helps prevent regurgitation and aspiration by isolating the esophagus and stomach contents from the lungs.

Once PFC is in the lungs, it can also be used to heat and cool the body because of its high heat capacity. Since the lungs have a surface area about the size of a tennis court, they can act as internal heat exchangers or even be used to administer pharmaceuticals.

Eliminating the PFC liquid from the lungs at the conclusion of the mission is accomplished by placing the body in a head down position and draining and coughing in a 1-g gravity environment. Also, computer controlled evacuation of the PFC from the "smart tubes" can precede and assist the voluntary expulsion. The residual PFC is simply eliminated by natural evaporation, assisted by warm-air flushing of the lungs to compensate for the absorbed heat of vaporization of the PFC liquid.

The fact that PFC's evaporate is deliberately exploited in the TLV / PLV ventilator system's design. Under PLV, active cooling of the exhaled gas can recover and conserve the exhaled PFC vapors. The condensed PFC liquid can then be recycled. Eliminating carbon dioxide is a critical feature in the design of the liquid ventilator system. Under PLV, carbon dioxide is eliminated by passing the exhaled gas through a CO_2 scrubber. Under TLV, carbon dioxide is dissolved in the PFC liquid and is more difficult to eliminate. Hence, the PFC must be passed by a selectively permeable osmotic membrane so that the dissolved CO_2 can diffuse through it; the large pressure differential between the PFC containing a high level of CO_2 on one side of the membrane, and a gas or liquid without CO_2 on the other side of the membrane, makes this all possible. Under active development is an alternative nano-membrane that is selectively permeable only to the gases and not the PFC liquid.

20.4 Improving Communication

The previous discussions clearly demonstrate that LTI-20 liquid ventilation systems provide superior distribution of gas exchange and pulmonary stability, thus enabling the human body to withstand accelerations far beyond 30 g. However, crew members must also perform perceptual, motor, and cognitive functions under these high-g conditions.[23] Significant developments in LTI-20 man-machine interfaces have now overcome the deficits of pre-existing flight equipment.

20.4.1 Wearable Man-Machine Interfaces

Smart interfaces should easily sense what an individual user wants, yet be so small or unobtrusive that they can be embedded in wearable objects. Historic examples of wearable interfaces are the "Evil Eye" and "Headhunter" tracking systems manufactured by ISCAN for the DoD. These systems, which exploited a video-based eye-tracking technique to monitor the position of the eye, were applied to target detection and tracking; it was subsequently used in upscale cameras to help the operator select the depth of field. In the LTI-20, an eye-tracking device can direct manipulate a virtual panel linked to the aerospace command, control, and communication systems.

Even seasoned pilots, when presented with conflicting sensory information concerning the direction of gravity, can often experience episodes of disorientation. For the LTI-20, a tactile interface was developed that extracts vehicle roll and pitch information from a gyro-stabilized attitude indicator, then maps it onto the pilot's torso by using a matrix of vibrotactors. The system enables the pilot to maintain an awareness of the vehicle's attitude without any visual cues.[24] This is another example of a wearable interface integrated into the LTI-20's flight systems.

The standard issue LTI-20 unitard (Figures 20.4.1 through 20.4.4) incorporates wearable, "full suit" non-invasive monitoring of the entire body, using microwave signals to penetrate and remotely diagnose human tissue. Tissue motion is sensed by a change in the phase between the microwave signal and that part which is reflected back to the distributed transmitter array, which is imbedded in the suit. Hence the microwave sensor-net can non-invasively monitor circulation, physiological and muscle motion, as well as many body functions.[25]

HUMAN FACTORS AND g-TOLERANCE

Figure 20.4.1: Space activity unitard (SAU). *(Courtesy of RPI.)*

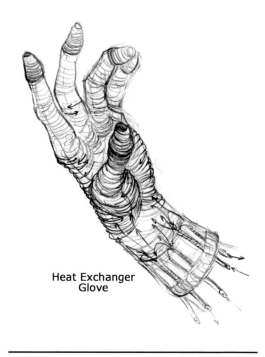

Figure 20.4.2: Interactive SAU glove for interface with FMS computer and virtual reality environment. *(Courtesy of RPI.)*

Lightcraft flight systems exploit Virtual Reality (VR) as an intelligence amplification technology that accomodates a variety of VR configurations ranging from a "wearable cockpit in a helmet," to "virtual vision" (Figures 14.5.1 and 14.5.2). The experience of being on the "inside looking out" might be called *VR immersion* and is the new wave of intelligence amplification technology exploited in Lightcraft flight systems. LTI-20 computer systems rely on artificial intelligence protocols to manage necessary functions and information so that both pilots and crew are rarely overwhelmed with *too much data*. The exponential explosion of information now available to Lightcraft crewmembers has mandated the evolution of superior VR systems and interfaces for the LTI-20.

VR technology is intimately integrated into LTI20 ultra-g suits to the extent that the flight crew can experience, process, and control their individual environments with enhanced efficiency.

Since visual disturbances may occur under high g loads, the latest VR system employs direct retinal projection for receiving and processing information. Custom-fitted contact lenses maintain the contour of the eye, cornea, and visual acuity, as well as directly project images and information to the retina. These contact lenses are sufficiently large to provide an eyeball retention function for high $-g_X$, posterior-to-anterior accelerations. Tom Furman (Director of the Human Interface Technology Laboratory of the Washington Technology Center) is credited with inventing the Virtual Retinal Display (VRD) which in its final embodiment projected an

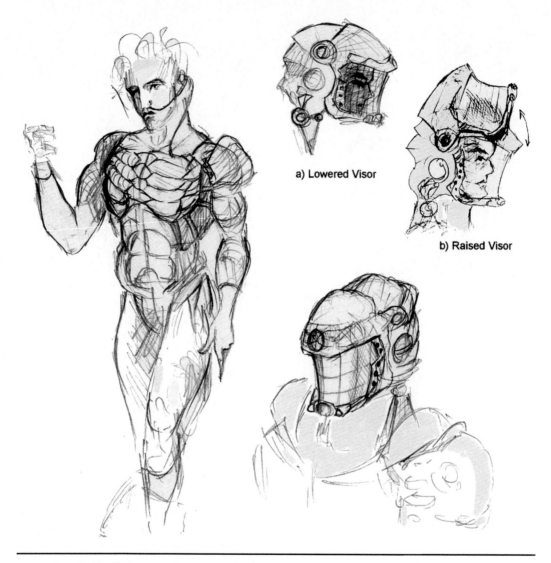

Figure 20.4.3: Space activity unitard and helmet, with active g-compensation features. *(Courtesy of RPI.)*

image that spans the eye's entire 120 degree field of view. True 3-D images are conveyed by providing each eye's retina with a different video image, using red, blue, and green laser diodes.

Although complete brain-computer interaction is still decades away, human brain waves of the alpha, beta, or mu type have successfully controlled a variety of small peripheral devices, including cursor and joy sticks. Further advances in such technology may allow future lightcraft crews to completely control their craft.

20.4.2 Implantable Interfaces

Implants, used as control or monitoring devices, continue to present inherent dangers of infection and rejection, as well as difficulties in placement, repair, and retrieval. However, the new 2-way personal communicator (see Section 9.4, Figures 9.4.1 and 9.4.2) is one fine example of recent successful implants for lightcraft use. Historically speaking, perhaps the most *benign* implant ever invented was the TongueTouch Keypad (TTK) by NewAbilities Systems (Figure 20.4.5). The interface, resembling an

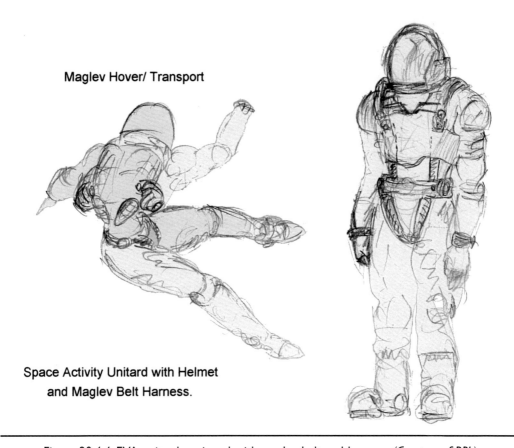

Figure 20.4.4: EVA unitard equipped with maglev belt and harness. *(Courtesy of RPI.)*

orthodontic retainer fitted to the roof of the mouth, contained nine tongue-activated microswitches that transmitted signals wirelessly to a remote controller of several electronic devices. Such interfaces are under consideration for ultra-g suit upgrades.

20.5 Super-Human Survivablility Levels

The successful integration of liquid-immersion g-suits (Figure 20.5.1), total or partial liquid ventilation, and individualized escape pods enables super-human levels of crew survivability in the LTI-20, a hypothesis prophesied in Reference 26. Historical advances in liquid filled g-suits such as Boeing's "Atlantis Warrior,"[27] linked with positive-pressure gaseous ventilation systems, helped yesterday's fighter pilots cope with the elevated combat g-forces in past aerial wars. Those experimental garments "intelligently" compressed the pilot's torso and lower body to squeeze blood towards the head, while simultaneously pumping oxygen into the lungs at elevated pressures. The pressurized vest kept the pilot's rib cage from expanding, while computer-controlled valves ensured quick system response, delivering the proper flow of oxygen – but not so quickly as to injure the pilot.

LTI-20 custom-fitted escape pods are expertly crafted from advanced composites and equipped with a 1 cm thick, liquid-filled bladder that virtually transforms the occupant into an exoskeleton "artificial lifeform" with superhuman survivablity. Pressurized HeliOx ventilation is sufficient for up to 30 g for several seconds, but ultra-high accelerations (beyond 30 g) necessitate total liquid ventilation. Prior to performing a hyper-jump, the pod's fluid pressure increases dramatically, "tensing" in anticipation of the energetic maneuver. This "tensing" feature enables much higher "g-onset" rates (g's per second) than with former systems.

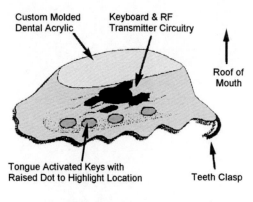

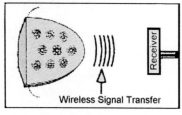

Figure 20.4.5: TongueTouch Keypad linked to wireless communications.
(Courtesy of NewAbilities Systems, Inc.)

For $+g_X$ loads (i.e., "eyeballs in"), evidence suggests that if all voids in the human body are completely filled with liquid, basically making it incompressible, then short-term accelerations of up to 200 g (or more) are sustainable, as long as the body is perfectly supported against deformations. Clearly, blood pressure levels in the head (and throughout the body) must be carefully monitored and intelligently controlled to avoid aneurysms and other medical injuries.

Secure in custom-fit pods, LTI-20 crewmembers perform perceptual, motor, and cognitive functions with high efficiency throughout a given mission. Standard procedure requires that high-G maneuvers be carried out by the Lightcraft FMS computers, pre-programmed just prior to negotiating an autonomous boost, hyperjump, or other ultra-energetic maneuver, requiring no real-time human intervention during the event. Crewmembers often just hold their breath for a few seconds under pressurized HeliOx, or several minutes with oxygen-rich PFC liquids.

20.6 Advancing BioElectronics Technology

Recent advances in bioelectronics technology are being purposefully tailored for the ultra high-g Lightcraft flight environment. Rapid developments in ultra-compact, low-power electronics and neural interfaces, often "spin-offs" from global Lightcraft support activities, make implantation of sensory, memory, and communications devices ever easier. Crews of future generations of Lightcraft will likely benefit from high-speed "intuitive" communications, command, and control interfaces with their Lightcraft vehicle – including the ability to make fully hands-off links with FMS computers, the entire multi-spectral sensor suite, and image-processing capabilities of the LTI-20. In the past decade, such bioelectronics and power systems advances have finally brought the concept of force enhancement through robotic exoskeletons into hardware reality. Figure 20.6.1 shows the new Space Command exoskeleton, which doubles as body armor.

20.7 Space Activity Unitard Origins

The standard issue USSC space activity unitard (SAU) was inspired by the pioneering work of Paul Webb.[28-30] His novel Space Activity Suit (SAS) consisted of a powerful elastic net leotard combined with a helmet (and trunk bladders) to provide oxygen under pressure for ventilation. The SAS suit featured maximum astronaut mobility at small metabolic cost relative to the motion-restricting, full-pressure suits of the day. Webb's suit was envisioned for active astronauts in extravehicular activity (EVA), working for 1 to 4 hours in the vacuum space environment.

20.8 Biconic Escape Pod Origins

The LTI-20 biconic escape pods are fully independent spacecraft once ejected from the lightcraft, and are capable of autonomous de-orbit and reentry from low Earth orbit. Each pod's interior can be "morphed" to custom-fit its crewmember. The pod's nose is equipped with a small rocket sized to give the 100 m/s delta V required for "de-orbit" from LEO, and a residual delta V for "retro-braking" on touchdown. Entry into the pod is through a retractable door that rotates around the

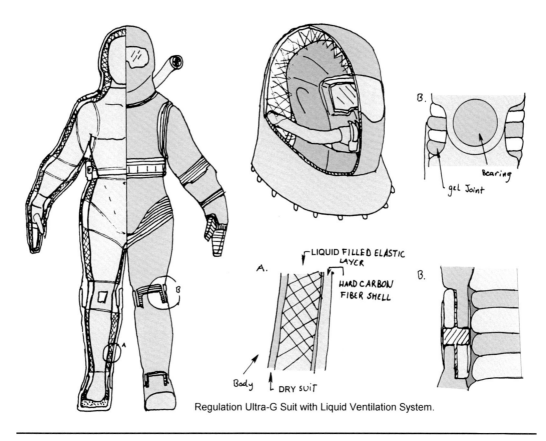

Regulation Ultra-G Suit with Liquid Ventilation System.

Figure 20.5.1: Liquid-immersion g-suit for total or partial liquid ventilation, for custom fit LTI-20 escape pods. *(Courtesy of RPI.)*

occupant, just inside the capsule liner. That way, when the door is rotated into position and the pod is pressurized to 1 bar, the door makes a vacuum tight seal as it is pressed against the tapered door opening. For more details on LTI-20 escape pod technology see Sections 3.4, 4.2, 14.7, 14.9, 16.11.2, and 18.3.

The historical foundations of the LTI-20 pod trace directly from Werhner von Braun's escape capsule design, as captured by artist Fred Freemann in the historic 1950's Colliers Magazine series on space (see Figure 20.8.1). In von Braun's vision, the escape capsules, guided by rails, were to be ejected by explosive power charges activated by the occupant. The capsule was equipped with a small retro-rocket in the nose, with a small enough exhaust diameter to avoid the clam-shell nose required in the Grumman concept (a notable liability during reentry). Apparently, the von Braun capsule's nose invoked a small "blow-out" plug (or one that rotated open), to expose the rocket exhaust nozzle at a time of need.

The LTI-20 Lightcraft's pod exploits a high L/D biconic aerocapture shape introduced in the 1980's[31-35] by Grumman, JPL, and others (e.g., note the manned Mars lander concept in Figure 20.8.2 from "The Case for Mars" by Carter Emmart). The external aeroshell geometry is closest to the robotic Grumman aerobraking biconic in Figure 20.8.3, and the interior layout retains a strong resemblance to von Braun's man-rated capsule. The LTI-20's pod retains the aft attitude control thrusters and retractable aft flap from the Grumman design, along with the retro rocket in the nose (like in the von Braun concept). The rear aerodynamic flap can be deployed (on demand) to adjust the lightcraft

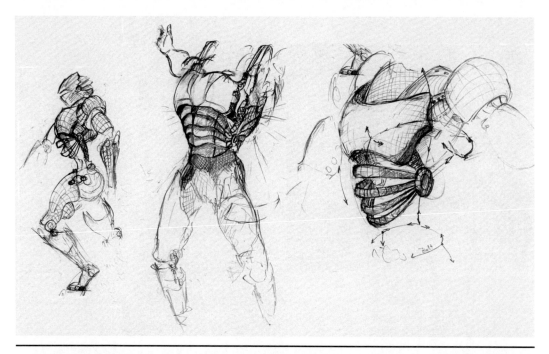

Figure 20.6.1: Robotic exoskeleton for force enhancement and body armor. *(Courtesy of RPI.)*

Figure 20.8.2: Mars lander (manned) with biconic aeroshell. *(Courtesy of Carter Emmart, Case for Mars)*

pod's angle-of-attack and rate of descent through an aerobraking maneuver. A modern "ram-air" parachute design (Figure 20.8.4) gently sets the LTI-20 pod upon the ground in a horizontal attitude, rather than the vertical water touchdown envisioned by Wernher von Braun.

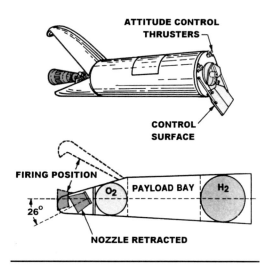

Figure 20.8.3: Grumman biconic aerobraking vehicle. *(Courtesy of JPL / NASA.)*

Figure 20.8.1: Wernher von Braun's escape pod concept. (Courtesy of Fred Freeman)

Figure 20.8.4: Modern ram-air parachute transports LTI-20 pod gently to ground. *(Courtesy of RPI.)*

21.0 PLANETARY DEFENSE MISSIONS (CLASSIFIED – Section Removed)

21.1 Monitoring Cislunar Space
21.2 Orbital Debris Mapping and Removal
21.3 Ballistic- and Cruise-Missile Defense
21.4 NEO Asteroid Imaging
21.5 BEP Launch of Deep Space Probes
21.6 NEO Trajectory Modification Tests
21.7 Intercept, Escort, and Retrieval of Unidentified Spacecraft

APPENDIX: HISTORICAL ORIGINS OF LTI-15 TECHNOLOGY

This appendix describes the historical origins of enabling technology, vehicle concept evolution, and development chronology for the LTI-15 Lightcraft – the immediate predecessor to the LTI-20. This 15-meter diameter craft was configured for transporting a crew of three (LTI-15A), or five (LTI-15B; *LC jockeys only*) space commandos in USSC excursions throughout cislunar space, including global transatmospheric flights, LEO / GEO access, and ultra-fast lunar shuttle missions.

The LTI-15 development cycle began with an exhaustive search of historical archives and technical data bases as far back as the dawn of the Space Age. This ground-breaking endeavor unearthed a wealth of ideas, concepts, and technology, some of which qualified as bonafide "historical origins" since those features were incorporated into the LTI-15 prototype. The most significant archival findings are covered below, along with a description of how specific features were systematically integrated within the evolving LTI-15 concept.

A1.0 Lawrence Passenger Spaceship

The LTI-15 owes its dimensions, geometry, and functional layout to an imaginative but unrealized (never built) concept for a nuclear-powered passenger spaceship (Figure A1.1) created in 1956 by Lovell Lawrence, Jr. when he was an assistant chief engineer of Chrysler Missile Operations in Detroit, Michigan. [For the LTI-15, the nuclear power-plant would of course be removed from the spacecraft and placed on a mountain top (PDS-02) or in LEO (PDS-03), with the propulsive power beamed to the vehicle in flight.]

<u>Take note</u>: Just 15 years earlier in December 1941, Lawrence had teamed up with John Shesta, H. Franklin Pierce, and James H. Wyld to form America's first commercial liquid-fuel

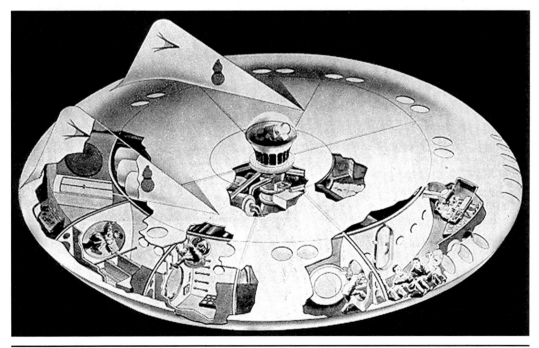

Figure A1.1: Cutaway view of passenger spaceship concept by Lovel Lawrence, Jr.
(Courtesy of Associated Press.)

APPENDIX

rocket company, Reaction Motors, Inc., which manufactured rockets to propel the Bell X-1 and the X-15 rocket planes.[1,2]

In 1956, Chrysler was managing the NASA installation that later assembled Saturn S-1 and S-1B boosters for the Apollo Moon program. Wernher von Braun's brother, Magnus von Braun (an electrical engineer), worked at the Chrysler Missile Division in Detroit during that period. Lawrence certainly had at his disposal the necessary aerospace engineering talent and resources to flesh out his imaginative concept.

The only known (surviving) archival record of Lawrence's creation is a brief Dec. 1956 newspaper article by the Associated Press,[3] which describes all the features that Lawrence considered essential for ambitious missions to the moon, Mars, and outer planets. The key features of the Lawrence spaceship were:

Lenticular geometry, 15 m (50 ft) in diameter and 1.83 m (6 ft) thick at the rim.
Crew complement of ~7 (noted from artist's sketch in Figure A1.1).
On-board nuclear powerplant, able to accelerate the vehicle to escape velocity, with high specific impulse performance (judging from small spacecraft internal volume).
Vectored thrust and flow control; hover capability (VTOL).
Take off like a conventional airplane (lenticular "flying wing" with twin vertical fins).
Artificial gravity in space (rotates at 3 RPM; walk on the walls under 1/5 Earth gravity.
Control room ("bridge" or "cockpit") has re-positionable chairs to accommodate either normal Earth or artificial gravity).
In space, pilot employs remote viewing view screen to see exterior environment, using a de-spun, gyro-stabilized TV camera.
Individually pressurized compartments, like a Navy submarine.
Double outer hull with self-sealing coatings to halt small meteoroids.
Retractable, multi-function work surfaces (tables, benches) that fit smoothly into walls, ceilings, and floors.
Closed-cycle environmental support system.

Temperature control in space by adjusting orientation of disc to sun.

The following excerpts from the AP article provide further insight: The power plant is described as an "atomic reactor" that can supply "prodigious thrust levels" once "the nuclear jets are opened wider" to speed the ship to escape velocity. [For what other, more energetic source for the propulsion system could have been imagined at the dawn of the Space Age?] The 15 meter (50-ft) diameter ship is 1.83 m (6 ft) thick at the outer edge and "somewhat thicker in the middle" where the power plant is housed. In the atmosphere, the "round and thin" craft functions as a flying wing, equipped with two vertical tail fins to provide directional stability as it takes off (or lands) much "like a conventional airplane"...as slowly as 15.6 m/s (35 mph). Apparently the propulsion system also employed thrust vectoring and flow control: "The jets of gas shoot out nozzles and through slots top and bottom, to control direction, yaw and pitch, even let this ship hover motionless in the air." After accelerating to 11 km/s (25,000 mph) and leaving Earth's gravity, the ship starts spinning at 3 RPM to create a centrifugal force equal to one-fifth g for the crew. In space, the crew would walk and sleep on the outer wall (Figures A1.2 and A1.3). At the center of the craft under a transparent dome is a "de-spun, gyro-stabilized optical telescope" connected to a remote video monitor watched by the pilot who keeps the target destination "always steadily in view."

Other details in the article relate to the meteoroid protection scheme: The outer toroidal living space is divided into a number of pressurized compartments, much like a Navy submarine. In case a large meteoroid punctures the hull "like an artillery shell," only one compartment would be decompressed. The ship also has a double outer hull with "self-sealing coatings" to halt micro-meteoroids that manage to puncture through the outermost wall. In penetrating the outer skin, small micro-meteoroids were assumed to disintegrate into finer particles before being stopped by the inner wall.

HISTORICAL ORIGINS OF LTI-15 TECHNOLOGY

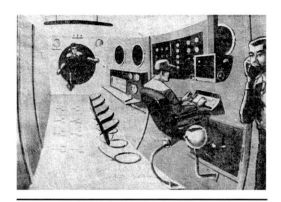

Figure A1.2: Interior view of control room ("bridge") in Lawrence spaceship under artificial gravity mode. *(Courtesy of Associated Press.)*

Figure A1.3: Recreational and dining area in Lawrence spaceship. *(Courtesy of Associated Press.)*

The Chrysler "cranked-arrow" logo (Figure A1.4) adorns both tail fins of the Lawrence spacecraft in Figure A1.1. Chrysler advertised this logo as "The Forward Look" – which emblazoned corporate reports and special products, including a very few Chrysler Imperial parade cars called "Thunderbolts" – in the 1955-56 era only. The companion logo on the tail fins appears to be the traditional "shell and flame" insignia (Figure A1.4) of the US Army Ordnance Corps.

Lawrence's original engineering report or proposal on the concept had apparently been destroyed, along with the aggregate corporate library of the (now defunct) Chrysler Missile Operations, according to a well-informed Chrysler librarian contacted in 1994. Many engineering design details remain unknown: e.g., initial launch mass, propellant load, structural concept, nuclear propulsion system specifics (air-breathing, rocket, or combined cycle?), nuclear radiation shielding, retractable landing ramp(s), reentry heat shield geometry (lower disc surface), atmospheric stability and control system, and exact crew complement. Nevertheless, clues to these details can perhaps be surmised from other archival technical and engineering reports from that *"dawn of the space age"* era.

A2.0 Nuclear Propulsion of Spacecraft

In May 1946, the U.S. Air Force established the Nuclear Energy for Propulsion of Aircraft (NEPA) project at Oak Ridge, TN. Secret work began in 1947 on nuclear rockets, originally as a part of NEPA. This project was continued until 1951 when it was disbanded in favor of a joint Atomic Energy Commission (AEC) and USAF program, called Aircraft Nuclear Propulsion (ANP) to pursue engineering development of

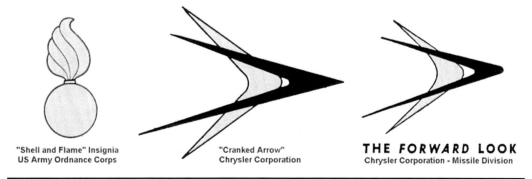

Figure A1.4: "Shell and flame" insignia (left) and Chrysler "cranked arrow logos" (right).

APPENDIX

aircraft reactors and engine systems. The director of the ANP program at Oak Ridge National Laboratory was R.C. Briant, who had maintained in 1954 that "manned nuclear aircraft pose the most difficult engineering development job yet attempted within this century."[4] And history has proved Briant correct, even though direct-cycle nuclear-turbojet and nuclear-ramjet engines were eventually demonstrated. For example under the Pluto program which required Mach 3 flight at low altitudes, the Tory 2-C nuclear-ramjet reactor reached 600 MW in 1963 in static tests. Before the ANP program was canceled in 1961, roughly a billion dollars was spent on the manned nuclear aircraft effort, split between the two major U.S. aircraft engine manufacturers.

At the height of the program in 1958, $100M/yr was going to General Electric for the direct-cycle program, and $25M/yr to Pratt & Whitney for development of an indirect-cycle that required a liquid-metal-to-air heat exchanger.

Clearly, the technological challenge presented by manned nuclear aircraft was far more difficult than initially assumed, but the program was also plagued with continuing reviews and changes of requirements. The mission kept changing from that of a high-subsonic cruise aircraft to one also requiring a supersonic dash mode.

Understandably, an orbital mission (i.e., the Lawrence spacecraft) was never considered under the ANP project, but was addressed in an ambitious analytical / design study of a hybrid nuclear propulsion system performed by Richard J. Rosa[5-7] in 1959 - 1960, while working for the AVCO Everett Research Laboratory in Everett, MA (now defunct). Rosa's launch vehicle was to be propelled by an air-breathing "MHD fanjet" supplied with electricity power from a closed-cycle nuclear reactor powerplant.

In the ANP manned nuclear aircraft adventure, substantial weight penalties came along with:

1) insuring adequate radiation shielding (up to 100 tonnes), and
2) satisfying nuclear safety requirements for crash protection of the reactor core.

Realistic engineering estimates forecast a flight platform gross mass of at least 454 metric tons (one million pounds) to make the manned nuclear aircraft concept practical.[8,9]

Clearly, this approach could not supply a suitable air-breathing propulsion system for the energetic spacecraft Lawrence had in mind. But then nuclear rockets don't pass the test either.

Nuclear rocket research was sponsored by the AEC, and NASA's Space Nuclear Propulsion Office under the KIWI and NERVA programs. Aerojet General, Westinghouse, and the Los Alamos Scientific Laboratory (LASL) were the major players, ultimately producing a family of nuclear rockets. The first nuclear-rocket reactor test (Kiwi-A) was conducted at the Nevada test Site of the AEC on 1 July 1959. In December 1967, Westinghouse Astronuclear Laboratory successfully carried out a test of the NRX A-6 reactor at a thermal power level of 1100 MW. Seven months later, LASL successfully operated the Phoebus 2B reactor at a power of 4000 MW. In general, the nuclear rocket programs enjoyed more technical success than their aircraft engine counterparts, while consuming only a few hundred million dollars – a mere fraction of that spent on the nuclear aircraft effort.

A nuclear rocket engine can indeed pack several thousand megawatts into a physically small volume, and generate truly impressive thrust-to-weight ratios with a "shadow-shielded" reactor. However, if installed in a Lawrence-type spacecraft, the local radiation environment would exceed the lethal dose for humans within a few seconds.

[The obvious solution to this dilemma was to remove the reactor from the spacecraft altogether, place it in a remote location, and use wireless power transmission to energize the engine from afar by beaming power from a mountain-top station like the PDS-02, or a space-based facility like the PDS-03. Incidentally, the three pressurized-water reactors (PWR) that "fuel" the PDS-02 are identical to those installed in Nimitz-class aircraft carriers. Each 300 MW_{th} reactor drives twin turbines, generating 50 MW_e (70,000 hp) per turbine.

These naval power plants are so reliable that the design life normally extends to 50 years. The exceptional safety record of these PWR reactors is a matter of record.]

A3.0 Artificial Gravity in Space

Artificial gravity is THE essential ingredient for countering the serious calcium-depletion problem associated with long periods at zero g, and, as such, strongly dominates the configuration of a compact, passenger-friendly spacecraft. The centrifuge feature of the toroidal cabin in the Lawrence "passenger spaceship" may have been inspired by rotating space station designs promoted by Wernher Von Braun in the mid-1950s.

However, the earliest toroidal spaceship design in science fiction lore was described by A. Train and R. Wood in the 1916 novel, "Moon Maker."[10] The interior of the 22.9-m (75-ft) diameter craft contained a number of cabins furnished with windows in the walls, floor and ceiling, a chartroom, along with twin gyroscopes and a power plant. Later, Herman Noordung in his 1929 book entitled "Problems of Space Flying," presented detailed cutaway views of a 30-m diameter "rotary house" that provided a 1 g environment in orbit. Still later, in 1948, H.E. Ross and R.A. Smith presented the British Interplanetary Society (BIS) with a detailed design for an "orbital base," a space station extrapolated from Noordung's proposal. The first known lenticular space station appeared in "Project Moon Base," a 1953 film based on a screenplay by Robert A. Heinlein.

A comprehensive study on the architecture of artificial gravity environments for long-duration space habitation was the subject of a Ph.D. thesis by T. Hall.[11] In this work, Hall had analyzed the specific "comfort zone" relationship between centrifuge diameter, rate of rotation, and average height of the occupants.

[Recent 2025 USSC studies have generally confirmed Hall's findings. For the tallest USSC crewmembers (statures of 196 cm) to avoid nausea, 30-m diameter Lightcraft will be mandatory and this craft is already "on the drawing boards." In contrast, the LCJs (statures of 130 to 162 cm) find the $1/6$ g artificial gravity of the LTI-20 quite comfortable. Small children (statures of 107 to 126 cm) delight in the LTI-15's $1/5$ g environment, and nausea is rarely reported.]

A4.0 Reentry Vehicle Options from Apollo Era

On Oct. 25, 1960, NASA selected three contractors to prepare feasibility studies for the Project Apollo spacecraft: General Electric, Martin Company, and Convair / Astronautics Division of General Dynamics. Convair subcontracted its reentry designs to AVCO Corp, and examined three options: a Mercury-type capsule, the M1 (a flat-faced cone and a half-cone), and a lens-shaped vehicle. Various lenticular reentry designs were created in this effort, and some were actually tested (see Figure A4.1). The winning reentry vehicle design for America's moon mission was, of course, the Apollo Command Module (CM).

[Note that the 3.9 m diameter, lower reentry heat shield of the Apollo CM is fitted with a spherical segment of 469.4 cm constant radius. If this heat shield is mirrored to make a lenticular disc geometry, its diameter-to-thickness ratio would be 3.39:1]

A4.1 Pyewacket Program

"Pyewacket"[12-24] was the code name for a formerly classified NASA program established in circa 1959-60 to generate experimental data (subsonic, transonic, and hypersonic) on the aerodynamics of lenticular disk reentry vehicles with diameter-to-thickness ratios ranging from 2.73 to 3.44. Much of this research was conducted at NASA Langley and NASA Ames Research Centers. Pyewacket was finally declassified in June 1967. Although Reference 25 investigates the aerodynamics of clean disks, most of the research explores disk geometries stabilized by extendible vertical and horizontal fins. Note that two Pyewacket configurations with twin vertical fins (see Figure A4.1) bear a striking resemblance to the Lawrence spaceship.

A4.2 Stability and Control of Landable Disks

Reference 26 summarizes an investigation into the stability and control of "landable" disk reentry vehicles. The work was performed by

APPENDIX

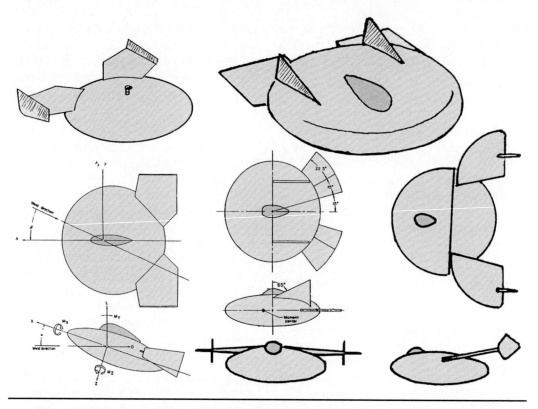

Figure A4.1: Reentry vehicle configurations considered under the Apollo moon program. *(Courtesy of NASA.)*

Giragosian and Hoffman while working at the Fairchild Stratos Corporation in 1962 - 1963. As shown in Figure A4.2, the basic concept involves the modification of the afterbody of a "proven" blunt-faced reentry vehicle (e.g., the Apollo CM) to achieve a lenticular cross-section that incorporates a movable lifting surface such as a horizontal stabilizer or flap.

Upon returning from space, the "landable disk" would enter the atmosphere at hypersonic speeds (Mach 25 - 35) in a high angle of attack attitude (perhaps 60 deg., like the Apollo CM), then nose over at supersonic speeds by unfolding control surfaces to initiate a forward flight mode.[36] Basically, Giragosian's goal was to generate a controllable disk configuration having favorable conventional landing characteristics.
(Note: Figure A4.3 shows a side view of the 15-m Lawrence spaceship equipped with a spherical segment heat shield for its lower hull, scaled-up from the Apollo CM geometry.)

These early studies proved that lenticular vehicles, following the hypersonic reentry maneuver, needed aft control surfaces to attain aerodynamic stability and satisfactory glide

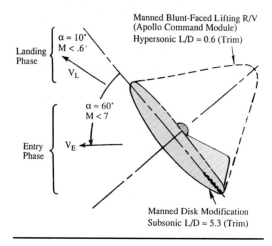

Figure A4.2: Manned disk reentry and landing phases. *(After P. Giragosian and W. Hoffman.)*

HISTORICAL ORIGINS OF LTI-15 TECHNOLOGY

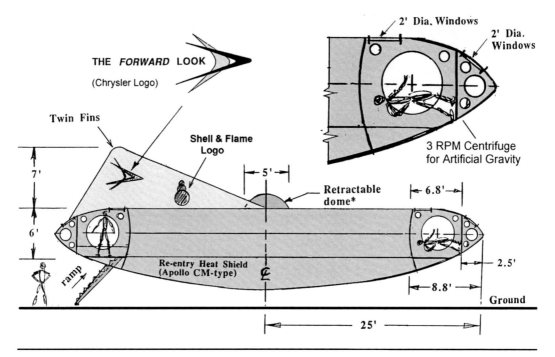

Figure A4.3: Cross-section of Lawrence spaceship with Apollo CM-type reentry heat shield (approximate dimensions in feet).

ratios for landing. Several Pyewacket tests investigated the stability characteristics of lenticular vehicles equipped with such control surfaces, deployed at low supersonic speeds by unfolding them from stored positions.

The basic problem with clean disks in forward flight at subsonic speeds is that the center of lift moves forward to a point ahead of the center of mass (which may or may not coincide with the axis of symmetry), causing the aircraft to become unstable. This instability materializes (Figure A4.4) as a strong tendency for the aircraft to nose up or down to very large angles of attack, and "go out of control." Obviously some kind of artificial stabilizing device is required, and the mainstream thinking at the time was to deploy large, physical rear stabilizers. Since the Lawrence ship in Figure A1.1 had twin vertical fins but no horizontal stabilizer, it would likely be unstable in lateral flight, barring the use of active flow control.

A4.3 Alternatives to Stabilizing Fins

Viable alternatives to physical stabilizing fins do exist today. The observed instability problems (appearing at certain Mach numbers and angles of attack) could be addressed with:

a) active flow control, such as the "Coanda" effect (air blowing across disk trailing edges);
b) rotation of all or part of the exterior aeroshell body to stabilize the disk by gyroscopic or magnus effects;
c) vectored thrust engines;
d) powered-lift devices;
e) electrostatic or electrodynamic torqueing;
f) Internal high speed gyros or momentum wheels;
g) various combinations of the above concepts.

A4.4 Frisbee Mode

Like the familiar Frisbee, flight stability can be greatly enhanced by simply rotating the entire body. A substantial body of theoretical and experimental research has investigated the subsonic aerodynamics, flight dynamics, stability and control of spinning disks. See, for example, References 27 to 30. [Note that microwave Lightcraft have specific design requirements which preclude the use of stabilizing fins, as discussed below.]

APPENDIX

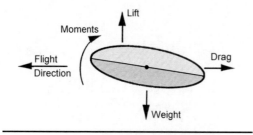

Figure A4.4: Aerodynamic instability of disc in flight.

A5.0 Plasma Shields against Solar Flares

Three alternatives to the obvious solution of a massive solid shield against solar proton storms have been explored in the literature:

a) purely magnetic shields,
b) purely electrostatic shields, and
c) hybrid magnetic / electrostatic shields.

The concept of the purely magnetic shield was first investigated by Levy[31] and has since been analyzed in significant detail during the 1960s and early 1970's.[32-37] Levy and Janes estimated that the purely electrostatic shield would require an unfeasibly large, 10-billion watt power supply to maintain an adequate positive potential of 200 million volts, largely because of the rate at which electrons would be scavenged from the interplanetary plasma. Levy and French[38] believed these approaches to be vastly inferior to the hybrid magnetic / electrostatic shield, since the limitations on both are sufficiently fundamental that technological developments are unlikely to change this conclusion.

A5.1 Best Option: Hybrid Plasma Shield

In 1963-64, Levy and Janes created a revolutionary, lightweight "plasma radiation shield" concept for protecting manned spacecraft against the energetic, solar-flare-produced protons that constitute the most serious radiation hazard in space.[39] In 1967, Levy and French[38,40] examined opportunities for integrating such radiation shields into realistic space vehicle designs. Detailed calculations in References 38 and 39 were performed for a toroidal spacecraft with a 13.3 m major diameter, and 3.33 m minor diameter (see Figure A5.1). Note the four superconducting cables that create a poloidal magnetic field around a toroid-shaped crew volume of roughly 275 cubic meters, and also the 300 keV electron gun used to eject electrons far from the spacecraft. (Incidentally, Levy and French recommended that an inflatable torus structure be seriously considered – historically significant for the LTI-15 Lightcraft.)

A5.2 Basic Principle behind the Plasma Shield

The basic principle behind the "hybrid" plasma radiation shield is that the spacecraft's exterior envelope becomes the shield, and repels protons and other positive particles by virtue of its being positively charged. Free space electrons are attracted by the positive charge, but cannot cross the magnetic field lines to discharge the hull. Instead, the electrons drift in an azimuthal direction around the spacecraft. Hence, a strong electric field is set up in the plasma radiation shield between the space vehicle hull (which is positively charged), and the free electron cloud (negative) that surrounds it. The charge on the vehicle is exactly equal to and opposite to the total charge in the electron cloud. (Note that the "hybrid" plasma shield concept was also investigated for radiation protection of space colonies – see Reference 41.)

Levy and French[38] asserted that the concept of a plasma radiation shield is physically sound, but important practical questions remain in two areas:

a) establishment and control of the high electrostatic voltages required; and,
b) integration of the concept into a realistic space vehicle design.

They also examined technical aspects of the required superconducting coil system, shielding voltage, vehicle electric power supply, and effects of the shield on communications, propulsion, attitude control, and life-support systems.

A5.3 Advances in Enabling Technology

In Reference 42, Landis suggested that recent advances in several technologies now make electromagnetic shielding a practical alternative for near-term future missions. These advances include: numerical computational solutions to particle transport in electromagnetic fields (e.g., Reference 43); high-strength composite

HISTORICAL ORIGINS OF LTI-15 TECHNOLOGY

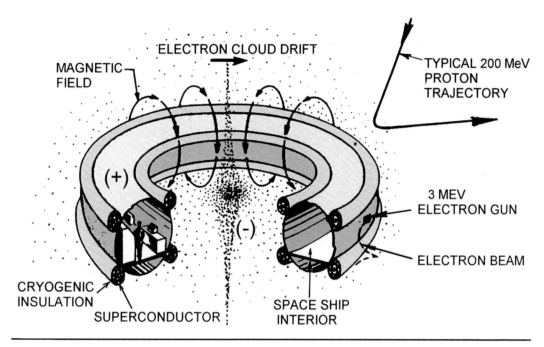

Figure A5.1: Space plasma radiation shield concept by Levy and Janes. *(Courtesy of AERL.)*

materials that can greatly reduce the mass of structures designed to keep magnets together;[44] high-temperature superconductors;[45-47] and a significant technology base for both constructing and operating large superconducting magnets.

A5.4 Plasma Shield / Spacecraft Integration

From the above discussion, it is evident that the overall plasma-shielded spacecraft design will be heavily influenced by the requirements imposed by the plasma radiation shield itself. In fact, the radiation shield is so critical to the mission that the entire design approach should probably begin with the shield. Levy and French[38] affirmed that the plasma radiation shield will strongly dictate the over-all configuration, power / propulsion plant, and the need for "leakage control."

Since the crew compartment of the Lawrence spacecraft is essentially toroidal, four superconducting magnetic coils can easily be accommodated (i.e., in the spirit of Figure A5.1), as well as the 300 keV electron gun. However, the twin vertical stabilizers would cut across magnetic field lines, and corona discharges would be triggered off the fin's sharp edges, disabling attempts to charge up the hull to a high potential. Furthermore, a carefully contoured hole must be opened through the center of the disk (i.e., "donut-hole") for the electron cloud in space, just before bringing the "shields up," as shown in Figure A5.2.

A5.5 Electric Power for Plasma Shield

A second shield-related issue is the electric power requirement for energizing the radiation shield and its superconducting coil system. Reference 38 indicates that a 5 to 10 kW electric power supply will be needed to run the cryogenic magnet cooling system, and that an additional 10 kW must be available for magnetic field energization (which they estimated could take up to 90 minutes) during every major solar flare. Reference 39 quotes a requirement for 30 kW. Also, this power supply must not vent exhaust gases of any kind into space whenever the radiation shield is operating, or the electrostatically charged spacecraft hull will be immediately discharged.

Obvious on-board power plant options are photovoltaics (exploiting solar or beamed-laser energy), self contained fuel cells, or microwave transmission to rectennas. The most logical

APPENDIX

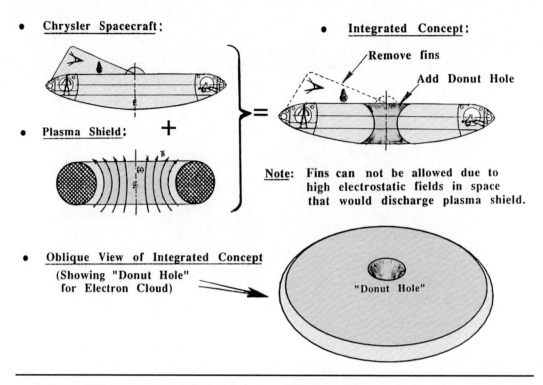

Figure A5.2: Integration of AERL space plasma shield into Lawrence spaceship concept.

option would be to cover one side of the 15 m spacecraft with efficient (~18% or higher) photovoltaic cells to provide ~35 kW of power.

A5.6 Energizing & Disabling Plasma Shields

A third issue relates to the process of energizing and disabling the plasma shield. The electron cloud itself presents a significant radiation hazard during the shut-down procedure. French and Levy[38] suggested that the shield can be turned on by a scheme called "inductive charge ejection." In this scheme, the electrons are ejected from the spacecraft at 300 keV, just as the magnetic field is being built up. The electrons are placed on magnetic flux surfaces near the spacecraft, and then carried away on these surfaces as the magnetic field is increased. By this mechanism, the power supply that energizes the magnet also energizes the electrostatic field.

If not turned off carefully, the electron cloud could conceivably become an X-ray radiation hazard. Properly executed, the shut-down procedure will emit a final flux dose roughly equivalent to a single medical X-ray. However, Reference 38 points out that even this could be shielded against by a barrier made of 1/8-inch thick lead plate, placed at a distance of several meters from the crew.

[For microwave Lightcraft applications, crewmembers must retreat to the safety of their individualized escape pods. There, a 1-cm thick layer of water or oxygenated perfluorocarbons (specific gravity = 1.6 to 1.8) that envelopes each crewmember will easily stop these low-energy electrons.]

A5.7 Attitude Control and Maneuvering in Space

A fourth shield-related issue impacts the choice of attitude control and maneuvering propulsion for the spacecraft. While the plasma shield is in operation, the attitude control / propulsion system must NOT have an exhaust stream, or the shield's charge will be quickly dissipated. If a spacecraft attitude adjustment is required in the mission, special torquing devices such as momentum wheels (gyro torquing), magnetic torquing (using planetary or interplanetary magnetic fields), or perhaps some variety of electrostatic torquing might be feasible.

Significant velocity vector changes such as mid-course corrections for a cislunar journey, present additional technical challenges. If high specific impulse, low-thrust electric rocket engines were to be applied over a long duration, the plasma shield would be disabled for the entire trip. Hence, higher thrust, lower I_{SP} rockets might be best for mid-course corrections, to minimize the "shields-down" time during solar flare activity.

A5.8 Human Factors and Life Support Issues

A fifth plasma shield issue relates to human factors and life support. Environmental life support systems must certainly be closed-cycle because ventilation-related out-gassing would disable the plasma shield system. Another human-factors issue that must be addressed is the penetration of strong magnetic fields (i.e., generated by the shield's on-board superconducting magnets) into the spacecraft interior. Through proper engineering design, however, the magnitude and gradient of these magnetic fields can be reduced to acceptable levels.

Note that Nuclear Magnetic Resonance Imaging (NMRI) machines commonly used to scan internal organs, normally expose patients to 1 or 2-Tesla fields for periods of 40 minutes or more. No deleterious effects have ever been observed. These powerful magnetic fields greatly exceed the levels expected within spacecraft crew quarters protected by a plasma shield.

A5.9 Internal Spacecraft Communications

The sixth issue related to space plasma shields is the effect of local, intense time-variant magnetic fields upon on-board electronic equipment. Fiber-optic cables will be mandatory for all sensor, computer data transmission, internal lighting, communications and control networks – i.e., the entire spacecraft "nervous system." All internal electronics and, in particular, the central processing units, will require special shielding and/or careful positioning. However, considering the small size of these electronic packages, the mass penalties associated with such shielding should be minimal.

A5.10 Compatible Structural Materials

The seventh plasma shield related issue impacts the choice of spacecraft structural materials. Since time-variant Tesla-level magnetic fields will be generated when the superconducting field coils are turned on (and off), or whenever the current is varied, powerful and potentially destructive eddy currents will be induced inside any nearby metallic or electrically conducting material. Hence, primary structural materials must certainly be non-metallic with excellent electrical insulator properties – an obvious application for advanced composites and exotic semiconductor structural films. The searing heat of reentry also places additional stringent requirements on the choice of external aeroshell hull materials. Refractory properties are particularly attractive, favoring high-temperature ceramic structural composites that can simultaneously serve as both the "hot structure" and reentry heat shield.

A6.0 Auxiliary Functions for Onboard Electromagnets

In a 1969 Popular Science article,[48] Wernher von Braun discussed the potentials of magnetic and plasma radiation shields for manned interplanetary spacecraft, and suggested several other auxiliary functions for these superconducting magnets in space:

- *"magnetic docking may offer advantages over mechanical coupling means"*
- *"forming a magnetic window in the hot plasma around a reentering spacecraft ... to avoid a communication blackout at that critical time"*
- *"magnetic aerobraking could ease g forces and heating during an interplanetary craft's high speed entry into a planet's atmosphere... for several minutes of the maneuver, a 10,000 gauss magnet could exert a substantial braking force against the atmosphere made electrically conductive by the shock wave of entry"*
- *"ion and plasma space propulsion engines under development, which use auxiliary magnetic fields, will have their performance improved by superconducting magnets."*

APPENDIX

A7.0 LTI-15 Propulsion Systems Integration

Briefly stated, the overriding engineering design objective was to integrate an ultra-energetic, microwave BEP engine into a 15-m single-stage, reusable spacecraft designed to transport a small crew in relative comfort and safety to the Moon and possibly Mars, in addition to standard suborbital and orbital missions. In sharp contrast with the chemical-fueled rocket engines of traditional launch vehicles, burdened with carrying their own propulsive energy source, this Lightcraft must exploit off-board, beamed-energy sources to energize its electromagnetic "motors."

The lenticular craft was to be based upon, or directly evolved from, the Lawrence "passenger spaceship" and make major concessions (as needed) to the BEP engine and human factors requirements – accommodating a crew complement of 3 to 5 space commandos. Human factors considerations dictated the need for artificial gravity and protection against the known hazards of space: e.g., high vacuum, micro-meteoroids, and lethal ionizing radiation from solar flares. The microwave propulsion system was to provide the spacecraft with ultra-energetic performance to accomplish its orbital mission direct from the Earth's surface, accelerating to at least 8 km/s (or more) without staging.

Figure A7.1 is a flowchart (or functional analysis diagram) of the engineering design process that generated the final vehicle configuration and conceptual "blueprints" for a 15-m microwave-boosted spacecraft. The design process exploited known physics and realistic extrapolations of existing SOA technology in materials, manufacturing processes, computer systems, microwave power-beaming systems, and the like.

A7.1 Choice of Frequency and Power Level

Ultra-energetic microwave-propelled launch vehicles cannot fly on low frequency (e.g., 2.45 GHz) power beams regardless of where the transmitter is located, for two principal reasons. First, the laws of diffraction dictate prohibitively large receiving antennas aboard the spacecraft, and very large transmitter apertures. Second, the low breakdown threshold of Earth's atmosphere (i.e., propagating through the Paschen minimum at ~30 km altitude) limits transmitted power densities (kW/cm^2) received at the spacecraft to infeasibly low levels.

The choice of higher microwave frequencies is, of course, limited to the Earth's atmospheric windows (e.g., 35 GHz, 94 GHz, 140 GHz, or 220 GHz), and the available state of the art rectenna technology. Note that atmospheric attenuation increases with increasing beam frequency (see Chapter 19). Also, the choice of thruster design can place additional demands on the frequency selection process. With pulsed detonation engines (PDE), for example, peak MSD over-pressures are a direct function of the microwave frequency which large measure, dictates the engine cycle efficiency.

Taken as a whole, the above restrictions demand that microwave launch vehicles exploit the millimeter wavelength range. For a near-term 15 m diameter aerospacecraft, 35 GHz was a logical choice because the requisite rectenna technology was close at hand. (As mentioned earlier, a rectifying antenna is a solid-state, microwave-to-electric power converter that provides DC electric power for the lightcraft's electromagnetic engines.)

At 35 GHz the atmospheric breakdown threshold (i.e., Paschen minimum at ~30 km altitude) is about 4 kW/cm^2. Given a 13.5 m diameter receiving antenna that covers 80% of the 15 m spacecraft upper surface area, a total beam power of 5.6 GW can be collected at 35 GHz. At higher frequencies, this power density limit increases to 30 kW/cm^2 for 94 GHz, and 66 kW/cm^2 for 140 GHz – i.e., by a factor of 7.5x to 16.5x, respectively (see Chapter 19). However, the task of developing efficient rectennas for these higher frequencies has historically presented greater technological challenges.

A7.2 Lightcraft Power Density and Disc Loading

So what kind of "creature" is a microwave-boosted launch vehicle or spacecraft? The

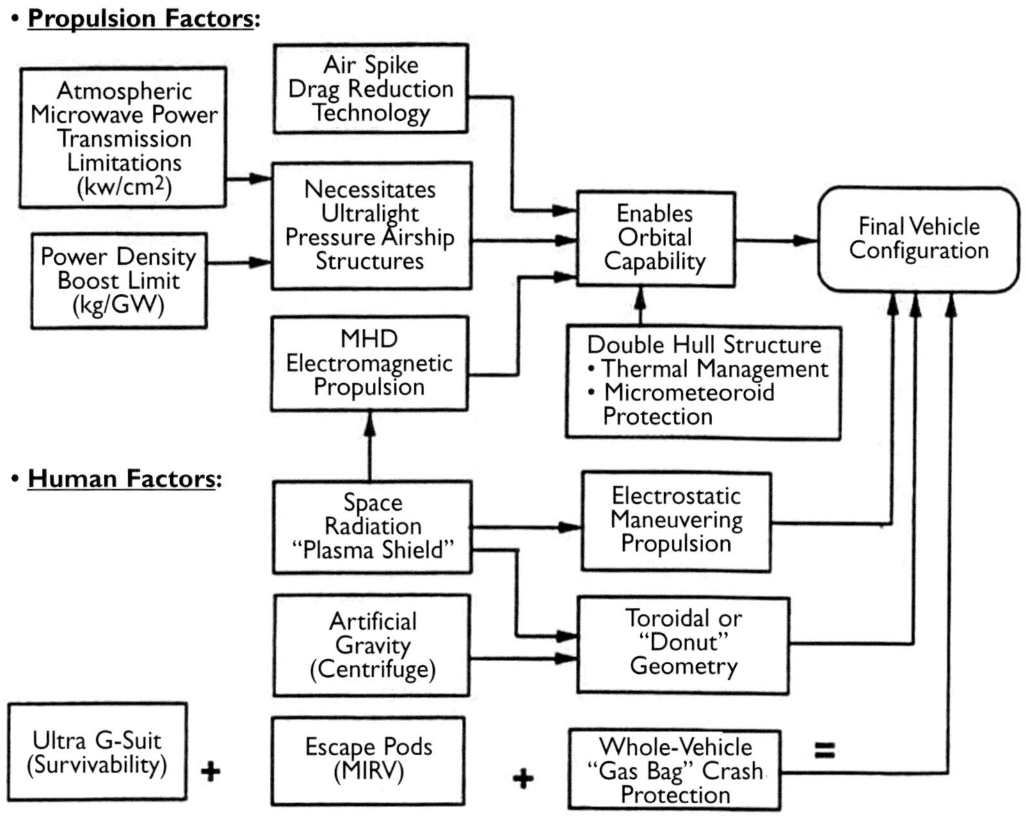

HISTORICAL ORIGINS OF LTI-15 TECHNOLOGY

- **Propulsion Factors:**

- **Human Factors:**

Figure A7.1: Functional analysis of propulsion system and human factors features that dictate LTI-15 configuration.

performance characteristics that define such BEP craft are in great measure driven by their power density and "disc-loading" (i.e., analogous to "wing-loading" for winged vehicles and lifting bodies). One revealing figure of merit addressed in the earliest laser propulsion studies by Kantrowitz[49-50] and others,[51-54] dating back to the early 1970's, was that 1 gigawatt of beam power can deliver one metric ton (1000 kg) of payload into low Earth orbit. Kantrowitz assumed that the entire solid propellant load would be ablated away during the laser boost, so that only the payload would be inserted into LEO – representing a power density of 1 MW/kg (or 600 HP/lb).

For hyper-energetic, microwave-boosted lightcraft, 1 MW/kg is optimistic by at least a factor of two[55-63] because the high-value payload (crewmembers) must be transported in their escape pods by a reusable spacecraft, and all this mass ends up in orbit. At best, an escape pod has the same mass as its occupant, and the lightcraft empty mass would generally equal or exceed (e.g., typically by ~2x) the total payload mass (i.e., crew compliment and escape pods). Assuming a conservative figure of 2 MW/kg (at launch) for the 15-m diameter craft, the maximum collectable beam power at 35 GHz is 5.6 GW, indicating a gross liftoff mass of ~2800 kg, of which half represents H_2O expendables for rectenna cooling. Note that 5 LCJ's (50 kg each) and their escape pods (50 kg each) represents a total payload of 500 kg, so the bare or "dry" spacecraft mass is roughly 1400 – 500 = 900 kg. The liftoff wing- or disc-loading is obtained simply by dividing the total launch mass by the cross-sectional area of a 15-m craft: i.e., 2800 kg/176.7 m^2 = 15.8 kg/m^2 (or 3.24 lb/ft^2).

It is useful to compare this figure with typical wing loadings of single-place ultralight aircraft, powered parachutes, and light helicopters at the turn of the millennium. Ultralights fell in the range of 2.6 to 5.9 lb/ft^2, or 4 lb/ft^2 on average. Ram-air canopies of single- and two-seat powered parachutes ranged from 1.16 to 1.39 lbs/ft^2. Rotor disc loadings on ultralight and light helicopters ran from 1.44 to 2.94 lb/ft^2, respectively. The "bottom line" here is that microwave Lightcraft have wing (or disc) loadings equivalent to early ultralight aircraft and airships, yet they must accelerate to hypersonic velocities and into space. So the Lightcraft design engineer is pressed to do *very much more, with very much less*. Mass is clearly THE enemy facing hyper-energetic Lightcraft performance.

A7.3 "Energy Rich and Mass Poor" Paradigm
Microwave-boosted Lightcraft have power loadings exceeding 2 GW/tonne (1200 hp/lb) of spacecraft weight, which is 100X larger than early ultralight aircraft (i.e., 12 to 24 lb/hp, with 27 to 80 hp engines). In contrast, the most heavily loaded engine component in the SSME rocket – the high-pressure fuel turbopump (HPFT) – processes 110 hp of shaft power per pound of mass at the 110% throttle setting (max rating). Take note: Lightcraft designers must deal with spacecraft power densities >10x beyond the most powerful rotating turbomachinery historically used in our best liquid chemical rockets. In this new "*energy rich, mass poor*" paradigm, every opportunity must be exploited for linking lightcraft structure to power handling roles, in shared multiple functions. To obtain such ultra-power densities, Lightcraft demand the creative exploitation of high-temperature plasmas and electromagnetic power processing technology.

A7.4 Advanced Tensile Structures
As discussed above, the low wing / disc loadings of microwave-propelled Lightcraft suggest exotic thin-film, high strength, balloon-type

Figure A7.2: Leik Myrabo with SRS inflatable space concentrator. *(Courtesy of SRS and NASA.)*

aeroshell structures that externally resemble pressure airships and early reentry lifting bodies. Inflatable space structures were variously proposed for high-altitude aerobraking,[25] emergency crew rescue with reentry capability, temporary space shelters, space stations, solar collectors, and space antenna functions. The first inflatable communications relay satellite to be launched into space in 1960 was *Echo 1*, a 30.5 meter diameter aluminum-coated balloon. Since then L'Garde, SRS (Figure A7.2), United Applied Technologies (Figure A7.3) and others have demonstrated a wide variety of inflatable devices both in space and on the ground.[64-73]

The logical extension of this proven "tensile" structures technology to microwave Lightcraft airframes required the exploitation of high-temperature ceramic matrix composites (CMC), such as C/SiC and SiC/SiC. Advanced, nano-fiber-reinforced CMC materials simultaneously satisfy both structural and thermal protection criteria in microwave Lightcraft. Furthermore, these exotic high-strength fibers present tremendous advantages on numerous fronts, such as in maintaining the integrity of superconducting magnets.

The Lightcraft's aeroshell tensile structure comprises sub-millimeter thick, nano-fiber reinforced "rip-stop" silicon carbide film (see Figure A7.4), manufactured with chemical vapor deposition (CVD) and other proprietary processes into semiconductor grade purity. Robotically formed into large on-axis, and off-axis parabolic shapes, such films can function as receiving rectennas, or as reflectors to concentrate microwave power for airspike support and/or pulsed detonation engine activation. Clearly, fine metallic coatings can provide the microwave reflecting function with an insignificant mass penalty beyond the basic tensile structure itself; this function is easily provided by special rectenna circuitry activated for the "reflect mode." Lightcraft tensile structures are made from semiconductor grade SiC so that solid state, massively-integrated electronics can be "grown" onto this substrate structure, where desired, to create rectifying

Figure A7.3: Rodney Bradford with inflatable polyimide toroid. *(Courtesy of United Applied Technologies.)*

Figure A7.4: First subscale mockup of LTI-20 double hull concept. *(Manufactured by ESLI.)*

antenna arrays that generate electric power for the Lightcraft's MHD slipstream engines, and other uses.

APPENDIX

a) Structurally Unsound

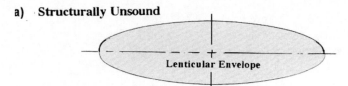

b) Stable (using toroid for pneumatic inflation)

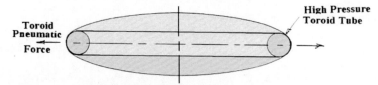

c) Stable (using SMES for magnetic inflation)

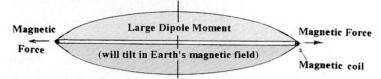

d) Spin-Stabilized with Dual-Coil SMES

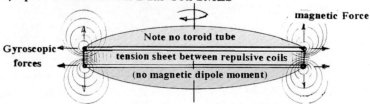

e) Toroid with Dual Coil SMES

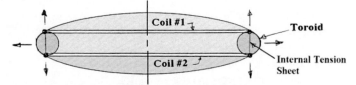

f) Add Photovoltaic Array and Retenna Arrays

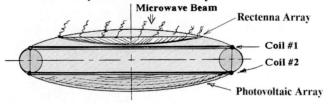

g) Add Pressurized Rectenna Plenums and Cooled Double Hull

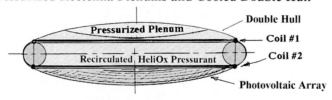

Figure A7.5: Stable and unstable lenticular geometries for microwave lightcraft tensile structures.

HISTORICAL ORIGINS OF LTI-15 TECHNOLOGY

Specially contoured rectennas can redirect the small amounts of "waste" microwave power that are normally reflected and dispersed from rectenna arrays, to enhance the Lightcraft's overall engine efficiency. This normally represents 2% to 5% of the incident power beam, and can be easily redirected and used for:

a) airspike support, or
b) air-plasma ionization (e.g., within the MHD slipstream accelerator engine).

With a little thought, every Lightcraft structural component can share multiple functions, and substitute beamed energy for mass at every opportunity.

Figure A7.5 clarifies the thought process for generating stable lenticular airframe geometries for microwave Lightcraft. Simple oblate-spheroid hulls (Figure A7.5a) will take on a spherical shape and "pucker up" at the perimeter, much like a common Mylar party balloon. This problem is easily resolved by attaching a high-pressure toroid to "anchor" the perimeter of oblate-spheroid segments (Figures A7.5b and A7.5c). Hence, the tension forces in the two spherical hull segments are transferred to the semi-rigid toroid that serves as a structural backbone for the assembly. See Reference 74 for a thorough introduction to the architectural design and engineering analysis of inflatable "tensile structures."

Self-repulsion forces of a superconducting magnet (Figure A7.5c) can also provide a perimeter reinforcing structure if the magnet is continuously energized, but geomagnetic torque will always tilt that magnet into alignment with the Earth's local magnetic field. Therefore, microwave Lightcraft require the twin-coil "bucking" magnet geometry (Figure A7.5d), wherein two magnets (carrying equal but opposing currents) repel each other and simultaneously cancel geomagnetic torque. The magnet housings must still be reinforced to support the large spherical hull segments when the electromagnets are discharged, unless the vehicle is continuously rotated to generate centrifugal stabilizing forces (i.e., clearly impractical). Since manned Lightcraft missions are incompatible with such high continuous rotation rates, the best solution is derived as in Figures A7.5e through A7.5g: i.e., the perimeter reinforcing structure is provided by the combination of:

a) high pressure toroids and
b) superconducting "bucking" magnets.

Figure A7.6 summarizes the tensile structures development process for microwave lightcraft.

[Note: A comprehensive technical discussion and three-dimensional, transient structural analyses of the LTI-20 is in References 75 through 80].

A7.5 Microwave Thruster Options

The matrix of viable "advanced accelerator" engine options compatible with beamed microwave (i.e., "off-board") power, and ultra-energetic launch vehicle applications has been explored in References 55 through 63. The three principal engine options are:

1) air-breathing;
2) rocket; and
3) compound- or combined-cycle systems that incorporate both rocket and air-breathing components.

Beamed energy propulsion inherently provides additional choices of:

a) microwave-thermal engines; and,
b) microwave-electric engines.

The former produces thrust as a result of beamed energy being directly absorbed into the working fluid (air or rocket propellant), whereas the latter type uses some kind of Microwave-to-Electric Power (MEP) converter to drive an electric thruster. Two viable candidates for high-power MEP converters are rocket-driven MHD open-cycle devices, and rectifying antennas with actively cooled, solid-state electronics.[56,61,62]

A7.5.1 Microwave-Thermal Engines

The pulsed detonation engine (PDE) mode for the LTI-15 was, in part, inspired by the German V1 "Buzz Bomb." Fueled with aviation gasoline, the V1's air-breathing pulsejet engine flew one-ton bombs to England during World War II. In this engine cycle, the flapping grid inlet (resembling a Venetian blind) admitted fresh air

APPENDIX

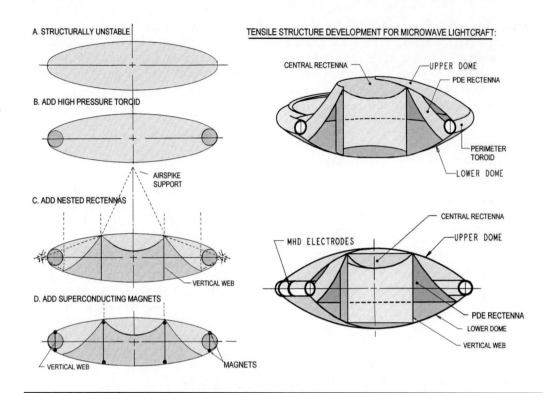

Figure A7.6: Tensile structure development for microwave lightcraft.

into the combustion chamber, and the subsequent spark-ignited explosion (a weak detonation) simultaneously closed the inlet and expelled the heated gases out a tuned exhaust pipe. The 1990's saw a rebirth of scientific research into both rocket and air-breathing versions of chemically-fueled PDEs, a body of research that has also influenced beamed-energy pulsejet investigations.

Note that beam-energized PDEs have been successfully demonstrated for both laser[81-91] and millimeter wavelength[92] radiation. Laboratory experiments and more than 150 flights of 12 - 15 cm laser-boosted Lightcraft with PDE engines were carried out at White Sands Missile range in 1996 - 2000.[81-91]

In the choice of BEP rocket vs. BEP air-breathing thrusters, it makes little sense to penalize Lightcraft acceleration performance with massive on-board propellant loads when *air-breathing* thermal and electromagnetic engines can provide superior specific impulse in transatmospheric flights. Air-breathing microwave PDEs can quickly accelerate a Lightcraft through Mach 1, but to exceed Mach 2 at low altitudes with the LTI-15 geometry and 35 GHz necessitates a directed-energy airspike to cut drag losses.[56]

With somewhat more streamlined forebodies, microwave PDE Lightcraft can reach Mach 6 at altitudes of 30 km.[58-60] But even with very slender conical forebodies, microwave scramjets will face difficulties exceeding Mach 10 - 12. In this hypersonic regime, the temperature of captured inlet air skyrockets as it is slowed and compressed prior to entering the BEP absorber / combustor; as a result, much of the input microwave energy is immediately lost to dissociation and ionization, instead of thrust production. Therefore, in the hypersonic regime, electromagnetic energy transfer (e.g., the MHD slipstream accelerator) provides the optimum solution.

A7.5.2 Microwave Electric Engines

Since an ultra-energetic microwave Lightcraft

must fly to orbit without staging (i.e., SSTO), the craft's overall configuration is heavily dominated by the needs of its air-breathing, hypersonic electric propulsion system. This electromagnetic engine concept owes its origins to the ground-breaking works of Stewart Way[93-96] and Richard Rosa.[4-6] [Note: The LTI-15's hypersonic slipstream accelerator design exploits features of both Rosa's and Way's MHD engines.]

In 1966, Way and his students built a 3-meter (10-ft) long, electromagnetically propelled submarine as a proof-of-concept test for their external MHD thruster geometry. Dubbed the EMS-1, the vehicle quietly accelerated the electrically-conducting sea water as it moved silently at 3.2 km/hr (2 mph) at a depth of one meter through the waters of California's Santa Barbara yacht basin. Electric power to run the external MHD slipstream accelerator was supplied by an onboard bank of lead-acid batteries. Way's engine was conceptually identical to the "caterpillar drive" popularized in Tom Clancy's novel *The Hunt for Red October,* which sported an internal MHD acceleration scheme.

In 1961 Rosa proposed his "MHD fanjet" for the transatmospheric flight, SSTO role. This engine concept exploited an air-breathing electric airturborocket cycle using an on-board nuclear electric power plant.[4-6] Rosa's MHD engine cannot be "started" at subsonic speeds (acting much like a stalled electric motor), because the air working fluid must first be moving at greater than Mach 1; otherwise, most of the power will be wasted in resistive heating of the atmosphere. Therefore, a very powerful PDE thruster mode is an absolutely essential ingredient for the LTI-15's multi-mode engine, to rapidly jump through Mach 1 before engaging the MHD mode.

The Lightcraft's MHD engine mode momentarily transforms the atmosphere into an ionized air-plasma "corridor" (for momentum exchange), using its Directed Energy Airspike to sweep inlet air to the periphery of this corridor. Propelled up this briefly-evacuated air channel, the Lightcraft effectively becomes the "armature" of an EML coil-gun, all the while experiencing greatly reduced aerodynamic drag levels.

Figures 5.1.3 and 5.5.1 show the basic MHD slipstream accelerator engine design in which powerful Lorentz forces exert a reaction force on the Lightcraft's superconducting magnets. Because these reaction forces are communicated at the speed of light, the MHD engine can efficiently deliver thrust at extreme velocities, up to escape velocity and beyond, while accelerating at the top of the atmosphere. Flight-weight superconducting magnets face stringent mass limits that exclude all but the most simple (and mass-efficient) geometries that simply encircle the vehicle perimeter. [Note that, equipped with such superconducting coil systems, Lightcraft are ideally compatible with 2-way electromagnetic launch / retrieval facilities. See Section A9 for elaboration.]

A7.5.3 Flow Control and Directed-Energy Airspikes

Off-board beamed-power and directed-energy airspikes (DEAS) streamline the blunt vehicle's external aerothermodynamics with energy, instead of the conventional approach of adding a physical (but massive) streamlined forebody. Reference 97 introduces the concept, physics, and performance of Directed Energy Airspikes. The first DEAS simulation was carried out on 24 April 1995 in Rensselaer's hypersonic shock tunnel, using an electrically-heated plasma torch to simulate the airspike's energy source.[98] Further combined experimental and theoretical / numerical research substantiated how such airspikes perform. DEAS physics is now very well established.[99-118]

In Lightcraft design it is of paramount importance to place the engine's hot air-plasmas and high-temperature propulsion conversion processes outside the vehicle hull, at a safe distance, where the full advantages of radiative cooling can be exploited with "external" thruster systems. Beamed energy can easily elevate air temperatures beyond 10,000° K, enabling efficient magnetohydrodynamic propulsion at almost any desired altitude. At these electrically-conducting plasma temperatures, MHD engines and electronically-vectorable magnetic nozzles can remotely manipulate the external slipstream air at a safe distance from

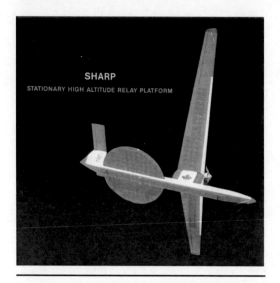

Figure A7.7: Stationary High Altitude Relay Platform – SHARP II. *(Courtesy of CRC.)*

the Lightcraft's aeroshell, thereby reducing thermal management loads to the minimum.

Simply stated, when engineering ultra-energetic Lightcraft engines, radiative cooling and hypersonic MHD propulsion must go "hand in hand." The combination enables revolutionary performance well beyond the limits of chemical rocketry.

Flight magneto-aerodynamics research emerged at the dawn of the Space Age, yielding a "mother lode" of creative ideas and technical reports through the 1960s and 1970s, all of significant historical relevance for microwave Lightcraft. First-of-a-kind experiments at the Avco Everett Research Laboratory (AERL), under the direction of Arthur Kantrowitz, demonstrated that powerful on-board magnets could indeed push reentry bow shock waves away from a spacecraft's reentry heat shields, and substantially reduce heat transfer rates.[119] Since then, the ability of on-board Lorentz forces to actively manipulate aerodynamic drag, modify lift / drag ratios, and create maneuvering forces for hypersonic reentry vehicles has been explored in great depth.

A7.6 High-Power Rectifying Antennas

One challenging conceptual problem facing microwave-boosted Lightcraft was the selection of microwave-to-electric power (MEP) conversion technology. This ultralight flightweight MEP converter must transform gigawatts of microwave power directly into electricity (on-board the spacecraft), yet weigh practically nothing. Only two viable candidates with sufficiently high power densities were identified, as mentioned earlier. The first concept to be explored was a laser-heated, rocket-driven, MHD generator using liquid hydrogen as the open-cycle working fluid.[56,120] The second was a microwave rectifying antenna array based on a high-strength, thin-film semiconductor substrate and solid-state electronics, actively cooled with water expelled as steam.[61-62]

Rectenna technology was ultimately "down-selected" because the 35 GHz technology was mature, and because its thin film arrays could be intimately integrated with the pressurized Lightcraft aeroshell structure with minimal substructure mass penalties. The result was a highly distributed propulsion system that would NOT be recognizable as a single entity in itself, wherein all system elements shared multiple functions. In sharp contrast, the alternative, high-temperature gas-dynamic conversion technology embodied in rocket-driven MHD generators, dictated clustered large masses that proved difficult to link into the thin-film Lightcraft structure.

Origins of the LTI-15 Lightcraft's 35 GHz rectenna technology can be traced back to the first 2.45 GHz rectifying antenna invented in 1963 by William Brown. In 1964, Brown demonstrated the first tethered flights of a rectenna-powered helicopter while employed at Raytheon Corporation in Massachusetts.[121] In 1986, under contract to NASA Lewis Research Center (now NASA Glenn), he developed a thin-film 2.45 GHz rectenna with an 85% efficiency, 1 kW/m^2 power density, and mass density of 160 g/m^2.[122] At this time, he also completed the preliminary design of a monolithic rectenna technology for use at 20 GHz and higher frequencies. In 1992, Brown produced the conceptual design for a LEO to GEO orbit transport vehicle, using rectenna-driven electric rocket thrusters, with 2.45 GHz power beamed from a ground-based transmitter.[123]

Figure A7.8: Sharp II aircraft showing thin film rectenna arrays. *(Courtesy of CRC.)*

Figure A7.9: Microwave powered electric airship demonstration. *(Courtesy of CRC.)*

In 1987 a Canadian research team composed of Joe Schlesak, Adrian Alden, James DeLaurier and others, under sponsorship of the Communications Research Center (CRC), designed, constructed, and flew a microwave-beam propelled aircraft called Sharp II[124-127] – see Figure A7.7. The lower surfaces of the 5-m wing and tail were covered with a lightweight "dual-polarization" version of Brown's thin-film 2.45 GHz rectenna, to supply power to an electric-motor-driven propeller (Figure A7.8). The CRC's 5.4-m diameter, 12-kW microwave transmitter supplied the beam. In flight, the Sharp II air-cooled rectennas had to passively reject waste heat into the moving boundary layer, which restricted the incident beam flux to ~1 kW/m². Somewhat later, CRC team also carried out microwave BEP experiments with a 3.35-m long microwave-powered *Pony Blimp B-11* airship (manufactured by Peck-Polymers in La Mesa, CA) using the same 2.45 GHz transmitter. The blimp's propulsion system consisted of a rectangular rectenna array suspended below the 2.3 m³ helium-filled envelope, and an electric-motor-driven prop (see Figure A7.9).

In stark contrast, ultra-energetic Lightcraft require rectennas designed for 35 GHz (and higher) with incident power densities of 4 kW/cm² – 40,000 times higher than the Sharp II's 2.45 GHz arrays. Hence these exotic Lightcraft rectennas require an active cooling system for the high density array elements, and advanced solid-state circuitry that can continue to function up to about 500°C. Elevated temperatures enable the efficient transfer of waste heat from solid-state diodes into the liquid coolant which, in the process, undergoes a phase change. Fortunately silicon-carbide based semiconductors demonstrated this 500°C ability[138] early on, and other viable competitors continue to emerge.

Research on high-power, high-frequency rectennas[129-132] has demonstrated rf-to-dc conversion efficiencies exceeding 70% at 35 GHz. For example, in 1991 Peter Koert demonstrated an rf-to-dc conversion efficiency of 80% with a prototype 35 GHz rectenna element, setting the baseline performance objective for future high-power rectenna arrays.

Alden[133-137] has produced a preliminary conceptual design for an ultra-high-power 35-GHz rectenna based on microchannel cooling using purified water as the open-cycle coolant, ejected as steam. Note that liquid-cooled microchannel heat exchangers (phase-change type) have experimentally demonstrated heat transfer rates of 1 kW/cm^2 and beyond. [The 35 GHz rectenna for the LTI-15 has a power density of 60 MW/kg, aerial density of 0.66 kg/m^2, and microwave-to-electric conversion efficiency of 75-85%.]

A8.0 Design Evolution of LTI-15 Concept

The general layout for the LTI-15 was inspired by the Lovell Lawrence spaceship, and incorporates most of its design features except for the nuclear propulsion system and twin stabilizing tail fins. Both vehicles are 15 m (50 ft) in diameter. In the process of adapting Lawrence's concept for ultra-energetic USSC Lightcraft missions, significant alterations were necessary, as clarified in this section.

A8.1 Basic Design Features

The principal design features for the LTI-15 are highlighted in the cross-sectional view of Figure A8.1, and further elaborated below:

Lenticular geometry, 15 m (50 ft) in diameter and 1.5 m (4.92 ft) thick at the rim.

Pressurized tensile structure (pressure airship), employing ultra-strength CMC thin films. Inflation pressure at sea level altitude is 1.5 bar, and the gauge pressure of 0.5 bar (7.35 psi) is maintained at all times.

Crew complement of 3 to 5 with comfortable accommodations (10 to 40 m^3 per person).

Hyper-energetic beamed energy propulsion system (5.6 GW at 35 GHz).

Vectored thrust and flow control; hover capability (VTOL); completely silent subsonic ion-propulsion system, good for 160 km/hr (100 mph) max speed at low altitudes.

Space plasma shield against solar proton storms.

30-kW photovoltaic array for autonomous electrical power in space.

Artificial gravity in space (rotates at 3 RPM; 1/5 Earth gravity).

Control room ("bridge" or "cockpit") has re-positionable chairs to accommodate either normal Earth or artificial gravity).

Individually pressurized compartments, like a Navy ship.

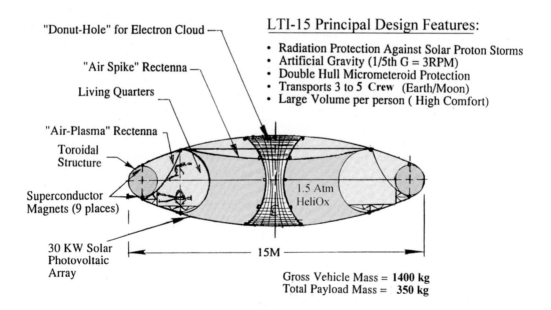

Figure A8.1: Principal design features of the LTI-15 lightcraft.

HISTORICAL ORIGINS OF LTI-15 TECHNOLOGY

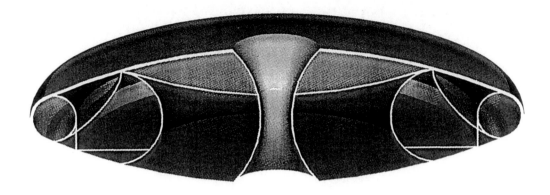

Figure A8.2: Oblique cutaway view of LTI-15 lightcraft, showing dual nested rectenna arrays, and interior partitions. *(Courtesy of RPI.)*

- Double outer hull with self-sealing coatings to halt small meteorites.
- Retractable, multi-function work surfaces (tables, benches) that fit smoothly into walls, ceilings, and floors.
- Closed-cycle HeliOx environmental support system.
- High-quality audio-visual communications (including virtual reality).
- Individualized escape pods with ultra-g protection, and reentry capability.
- Pods can function as landing gear to enable quick ingress / egress of crewmembers, since depressurization of Lightcraft compartments is not required.
- Two Maglev landers for rapid crew deployment from hovering Lightcraft. These plug into, and seal off, the central donut hole needed for the space-plasma shield.

A8.2 Final Integrated Configuration

As mentioned above, the final integrated design for the LTI-15 evolved from the functional analysis in Figure A7.1, by applying traditional engineering design processes. Note that human factors and propulsion factors take dominant roles in this process because the LTI-15 must transport very high value payloads (i.e., humans) at the highest safety levels consistent with USSC military missions in 2025.

The oblique cutaway view in Figure A8.2 reveals the upper and lower hull contours which were scaled up from the Apollo CM heat shield, resulting in a symmetric "double-Apollo" disc configuration. Note the placement of the high-pressure toroid (i.e., structural "backbone" at the rim), central parabolic rectenna, outer annular rectenna (lampshade-shaped), donut hole on centerline (for space plasma shield function), and nine superconducting magnets (~1 MA each). The 15 m toroid has a minor axis diameter of 1.5 m. Note that 50 % of the receiver cross-sectional area is occupied by the central rectenna, and the other half by the annular parabolic rectenna. Figure A8.3 presents side, top, and oblique exterior views of the LTI 15. Figure A8.4 shows an artificial gravity environment provided for the crew in space when the vehicle is rotating at 3 RPM about its axis of symmetry.

A8.3 MHD Engine / Vehicle Integration

Principal elements of the hypersonic propulsion system are:

a) two high-power 35 GHz water-cooled rectenna arrays,
b) microwave-induced airspike, and
c) MHD slipstream accelerator with its "flight weight" superconducting magnets and external electrodes.

Figure 5.1.3 shows the vehicle propelled by the MHD slipstream accelerator in hypersonic flight, highlighting functional relationships between the directed-energy airspike, MHD engine, and aerodynamic plug nozzle. The microwave power level needed for airspike support varies along the flight trajectory, and is

APPENDIX

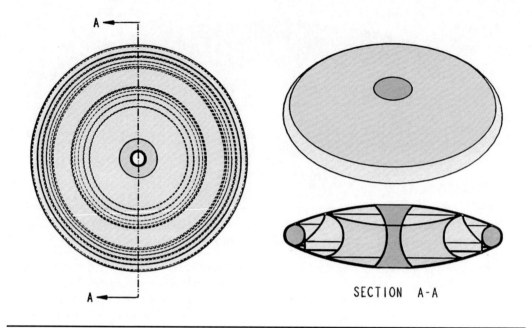

Figure A8.3: CAD 3-views of the LTI-15 lightcraft. *(Courtesy of RPI.)*

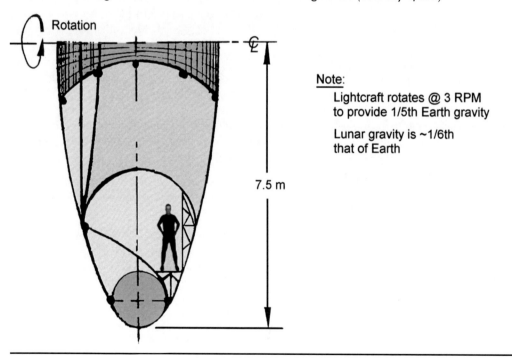

Note:
Lightcraft rotates @ 3 RPM to provide 1/5th Earth gravity

Lunar gravity is ~1/6th that of Earth

Figure A8.4: LTI-15 artificial gravity environment in space. *(Courtesy of RPI.)*

dictated by the Mach number and altitude on an "as-required" basis. By its very design, the MHD slipstream accelerator eliminates the sonic boom that is normally produced in supersonic flight through the atmosphere. Note the location (in Figure 5.1.3) just aft of the vehicle's

HISTORICAL ORIGINS OF LTI-15 TECHNOLOGY

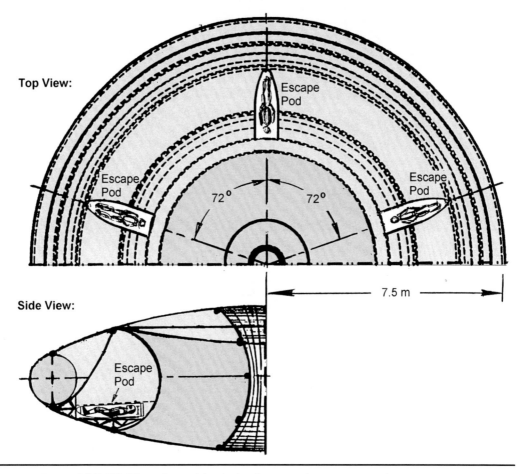

Figure A8.5: LTI-15 configured for high-g boost with crew secured in escape pods.

rim where the bow shock is annihilated (i.e., no sonic boom in flight).

A8.4 Conservation of Reflected Microwave Power
A small fraction of the microwave power incident upon the rectenna arrays is always reflected (roughly 5 to 7 %), and hence unavailable for conversion into DC power by the solid state electronics. The dual array is purposefully configured so that waste power reflected by the central rectenna helps to support the airpike; similarly, waste power reflected by the annular parabolic rectenna is focused out through the (microwave transparent) toroid to help ionize slipstream air entering the MHD accelerator. Microwave power requirements for airspike support and MHD engine ionization are monitored by the FMS computer, and automatically satisfied in flight.

A8.5 Phased-Array Microwave Transmitter Beam Patterns
Vanke, et al.[138] investigated novel microwave aperture distributions (phase, polarization, and intensity) for phased array transmitters of GBM and SBM stations that can project a central, on-axis "dip" in the beam intensity distribution at the receiver (in our case, the Lightcraft). Another work by Potter[139] analyzed transmitter aperture distributions that would project annular beams on target.

[Note: Hence, transmitting two different pulsed-power waveforms (intensity distributions that alternate in time) for the Lightcraft's central and annular rectenna arrays should be possible, if indeed the MHD propulsion system actually requires it. However, a more practical solution is to transmit a CW beam with uniform intensity

APPENDIX

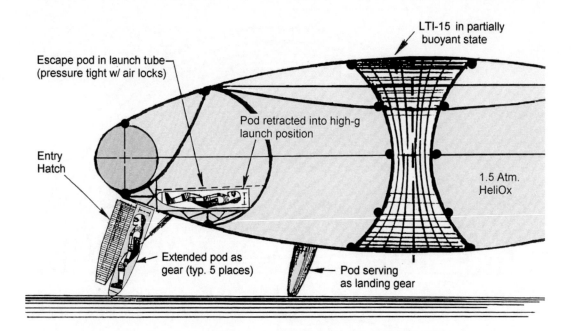

Figure A8.6: Dual use of escape pod for landing gear and crew boarding / egress.

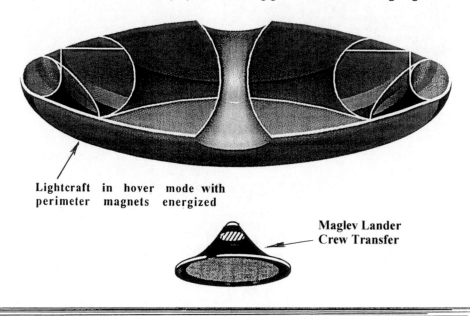

Figure A8.7: Maglev lander crew transfer to LTI-15 lightcraft in hover. *(Courtesy of RPI.)*

upon both rectennas, and employ on-board power processing and the rectenna's "transmit" functionality to deliver adequate microwave fluxes whenever and wherever the engine needs it. The latter option offers enhanced flexibility.]

HISTORICAL ORIGINS OF LTI-15 TECHNOLOGY

A8.6 Crew Accommodations and Pod Deployment

Figure A8.5 provides cutaway top and side views of the LTI-15, with the crew positioned and ready for a high-g boost. The five crewmembers are secured in their individual escape pods and pressurized g-suits, having just been retracted into the Lightcraft. Very short ingress / egress times are possible with this arrangement because the pods serve "dual use" as landing gear, as shown in Figure A8.6. This feature circumvents the lengthy delays associated with depressurizing large air-lock compartments in the Lightcraft hull, and the extension / retraction of landing ramps; such procedures are limited to loading and unloading of bulky cargo (and/or infirm personnel) when time is not an issue.

A8.7 Maglev Landers

Figure A8.7 portrays the LTI-15 in hover mode with its perimeter superconducting magnets energized for deployment of a maglev lander. Sized for transporting two or three crewmembers (Figure A8.8), each maglev lander may be lowered from hover altitudes up to 15 to 20 meters (depending upon the payload), to quickly deploy personnel or cargo. After contacting the ground, the lander may then (if desired) reverse its electromagnets and function as a magnetic landing gear for the hovering Lightcraft, supporting it by magnetic repulsion forces. The upper stabilizing coil on the Maglev lander assures proper alignment with the Lightcraft axis once it has lowered to within a few meters altitude; the assembly resembles a mushroom with its pedestal "foot." The LTI-15 maglev lander geometry is based on an early tension aeroshell design (see NASA TN-D-2994).

A8.8 Escape Pod Jettison System

On the rare occasion that an LTI-15 is faced with imminent destruction during a high-g boost, the FMS computers will automatically sense this and eject the escape pods. Standard USSC procedures require all crewmembers to mount their escape pods prior to any ultra-energetic maneuver. Figure A8.9 portrays the escape pod jettison system. Each escape pod is rapidly ejected from its launch tube, much like a torpedo is launched from a submarine, using compressed gas or a rocket gas generator. Once separated from the Lightcraft, the pod extends stabilizing fins and negotiates a controlled atmospheric reentry; then it deploys a drogue chute (inflatable ballute or other decelerator), and finally releases the main parachute for a safe, low-speed landing.

A9.0 Ultra-Fast Lunar Shuttle Mission

In the fall of 1994, a semester-long Horizon Mission Methodology (HMM) workshop was

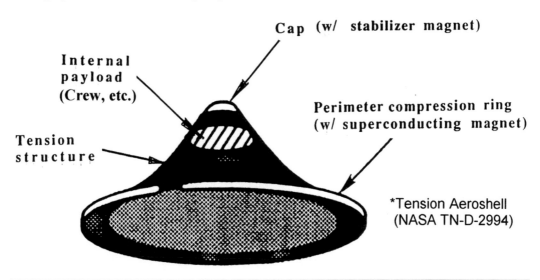

Figure A8.8: Design features of maglev lander for the LTI-15. *(Courtesy of RPI.)*

APPENDIX

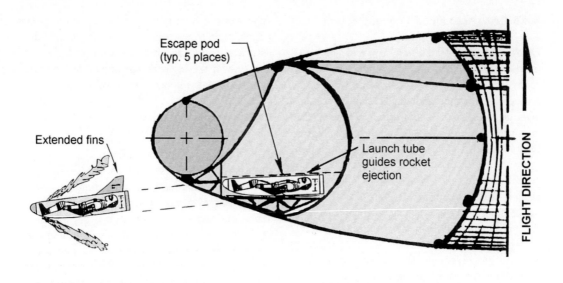

Figure A8.9: Escape pod rocket ejection system for LTI-15.

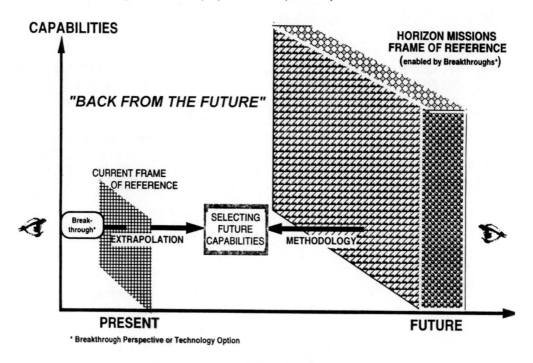

Figure A9.1: Anderson's Horizon Mission Methodology: "Back From the Future."
(J. Anderson; Courtesy of NASA.)

HISTORICAL ORIGINS OF LTI-15 TECHNOLOGY

convened at Rensselaer to create the vision for a mature *"Mid-21st Century Space Carrier Industry"* serving space-faring passengers in LEO / GEO, trans-global, and lunar markets. The mature lunar market necessitated a permanent presence on the Moon with all the attendant support services, of course, but the single "threshold" ingredient was deemed to be an *ultra-fast lunar shuttle* to ferry personnel with very high levels of safety.

A9.1 Workshop Approach

The workshop employed Anderson's Horizon Mission Methodology (HMM)[140-142] to investigate "high-leverage technology directions" that could enable ultra-fast shuttle missions, preferably with NO on-board fuel or propellant other than that required for midcourse corrections. The mission embodied performance requirements that could not be met by extrapolating chemical-fueled launch vehicle technologies. *"Highways of light"* and *"space links in a beamed-energy infrastructure"* were examples of thought-forms used to encourage leaps of imagination. The HMM methodology (Figure A9.1) involved four basic steps:

- Identify or define hypothetical "Horizon" space missions whose performance requirements exceed extrapolations of known technologies.
- Determine mission function and operational requirements, and performance requirements.
- Identify the implicit (and limiting) engineering assumptions associated with the performance requirements.
- Develop alternative engineering assumptions, and characterize new functions and technology capabilities that would be needed.

Three potent technologies (Figure A9.2) that could make this high speed lunar mission feasible were identified:

1) beamed power,
2) electromagnetic propulsion, and
3) extraterrestrial resource use.

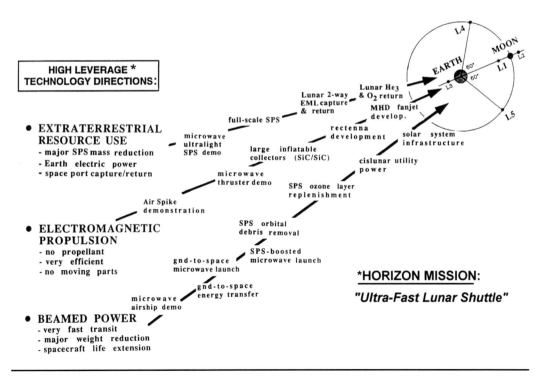

Figure A9.2: High leverage pathways to future technological capabilities for Ultra-fast Lunar Shuttle Mission.

APPENDIX

The existence and availability of crewed, beam-propelled Lightcraft, launched from Earth at escape velocity (or greater), was an underlying assumption in this HMM exercise. [The "*no on-board fuel or propellant*" requirement was intrinsically satisfied since Lightcraft use water as expendable coolant in transatmospheric flight to escape Earth's gravity, then coast to the moon.]

Participants proposed the break-through concept of placing a two-way (launch / capture) electromagnetic launcher (a passenger version of Gerard K. O'Neill's mass-driver[143] concept) at the L1 libration point (see Figure A9.3) as the <u>key ingredient</u>; electrical energy from a decelerated / captured Lightcraft shuttle could be temporarily stored at the L1 spaceport and used to accelerate the next departing shuttle back to Earth.

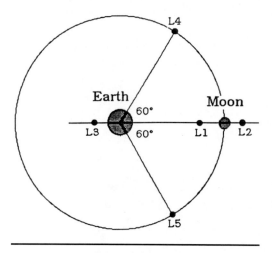

Figure A9.3: Libration points in the Earth-moon system.

A second EML space port was envisioned for use in low Earth orbit to capture a Lightcraft returning from the moon (at 11 km/s, or greater), and slowing it to circular orbit velocity; the stored electrical energy would then accelerate the next shuttle leaving for the moon at escape velocity (or more).

The integral superconducting magnets encircling a Lightcraft's perimeter should be intrinsically compatible with coil-gun type EML technology, simply scaled up to handle larger diameter payloads (e.g., 15 m bore).

Five libration points exist in the Earth-Moon system (see Figure A9.3) where EML space ports can reside without substantial expenditures of energy (or propellant). Libration point L1 is 56,300 km this side of the Moon and L2 is several lunar diameters beyond the Moon. L2 may be an excellent location for receiving payloads launched by a mass driver from the lunar surface.[144]

A9.2 Historical Foundations of EML Technology

Serious consideration of EML technology to launch payloads from the Moon and space stations dates from Arthur C. Clarke's proposals in 1950.[145] An excellent summary of the earliest ideas and work on *mass drivers* can be found in Gerard K. O'Neill's book, *The High Frontier*.[143] Nordley[146] has examined the station-keeping problems expected with two-way electromagnetic launchers positioned at L1.

In principle, EML facilities can be designed and engineered to either accelerate or decelerate payloads, even performing both functions equally well with the same device. And of course, trajectory correction coils placed at the EML's entrance and/or exit would automatically correct the payload's velocity vector as needed.

For lunar-surface destined payloads, two different concepts for EML "capture" mechanisms have been suggested by R. L. Forward[147] and G.D. Nordley.[146] Forward proposed that the incoming payloads first dock with a high-speed cradle (accelerated along the EML track to match the incoming vehicle velocity, before decelerating to a stop, much like magnetically levitated trains or catapults. Nordley's concept was analogous to aircraft carrier landings, and required a magnet-equipped spacecraft to "land" on its EML rails. The incoming velocities could be an order of magnitude higher than jet carrier landings, Nordley observed, but electronic reactions are >10x faster than human reactions, so the platform can easily be held stable. After landing on the rails, the spacecraft would essentially be a rapidly decelerating maglev rail coach.

HISTORICAL ORIGINS OF LTI-15 TECHNOLOGY

In the 1994 HMM exercise, a third "lunar capture" option was proposed for the L1 EML concept. Lightcraft coming from Earth would first be slowed at L1 and ejected at low velocity (less than lunar circular orbit speed) with a ballistic intercept trajectory, targeting a specific lunar EML-equipped landing site on the far side of the Moon. Near vertical descents into orientable (or fixed) maglev docking stations can be imagined, but further investigation of this concept is needed.

This two-way EML spaceport concept satisfied the HMM objective but demanded alternative engineering assumptions of what could be considered acceptable bounds for passenger G-tolerance vis-à-vis EM mass-driver lengths.

Figure A9.4 is a sketch of the two-way EML station based on push / pull electromagnetic coil-gun technology, and sized for a 15-m Lightcraft "bore" – in essence, a scaled-up version of the mass-driver proposed by Gerard O'Neill.

A9.3 Analysis of EML Facility Dynamics

Before beginning the HMM assessment, let us review the historic cislunar mission profile of, for example, Apollo 8 (21 Dec. 1968). Note that the Moon has a roughly circular orbit with a radius of 384,400 km and orbital speed of about 1 km/s. The circular orbit velocity at the Moon's surface is 1.68 km/s and the lunar escape speed is 2.4 km/s. As shown in Figure A9.5, if the translunar injection velocity is the speed of light, the trajectory is nearly a straight line with a time-of-flight of 1.28 seconds to the Moon. As the injection speed is lowered, the orbit goes from hyperbolic, to parabolic, to elliptical in shape with correspondingly higher flight times.

From an Earth parking orbit of 190.6 x 183.2 km, the Apollo 8 spacecraft accelerated from a circular orbit velocity of ~7.8 km/s to the escape velocity of 10.62 km/s, requiring a trans-lunar injection (TLI) delta-V of 2.82 km/s. After a couple mid-course corrections, the spacecraft arrived in the Moon's vicinity 66 hours later and

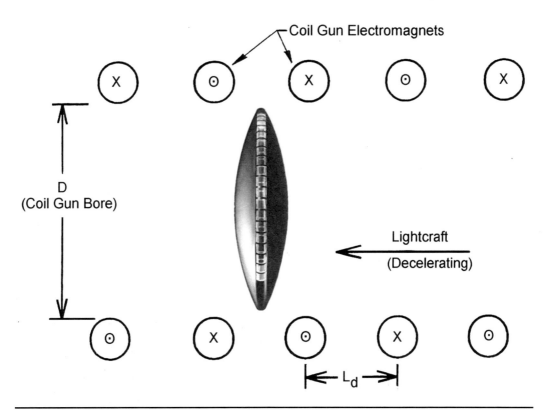

Figure A9.4: Schematic of 2-way EML (coil gun) spaceport at L1. *(Courtesy of RPI.)*

APPENDIX

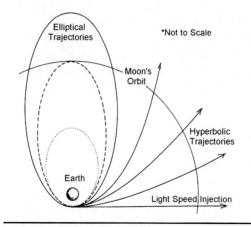

Figure A9.5: Effect of injection speed on trajectory shape for ultra-fast lunar shuttle.

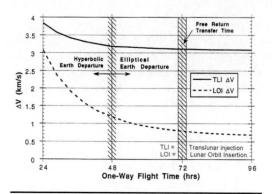

Figure A9.6: TLI and LOI delta-V variation with one-way flight time (300 km lunar orbit). *(Courtesy of S.K. Borowski and NASA.)*

fired the service module's engine for 4 minutes on the lunar orbit injection (LOI) burn. The spacecraft entered an elliptical orbit and two revolutions later, a second burn put the craft into a near circular, 110.4 x 112.3 km orbit for the final eight revolutions. Note that Apollo spacecraft arrived at the Moon with velocities of 875 to 915 m/s. After completing 16 hours in lunar orbit, *Apollo 8's* final trans-earth injection (TEI) burn returned the crew to Earth in a 57-hour flight.

Borowski et al.[148] carried out a first-order (idealized) assessment of how the combined TLI and LOI delta-V requirement varies with one-way flight time, assuming a 300 km lunar orbit. Note in Figure A9.6 that the ideal LTI and LOI total delta-V requirement for 24- and 36-hour trips are 6.9 and 5.1 km/s, respectively, whereas the typical Apollo one-way trip of 72 hours is approximately 4.1 km/s (i.e., TLI of 3.1 and LOI of 0.8 km/s or greater). For one-way transit times below 48 hours the combined TLI, and LOI delta-V (even more pronounced) increase significantly. Under such conditions the free-return-to-Earth option (used in the Apollo program) is no longer viable since the outbound trajectory requires a hyperbolic Earth departure.

Now imagine a convenient high-speed cislunar Lightcraft route that could service a permanent presence on the Moon, with 24-hour flights reminiscent of intercontinental jet routes at the turn of the century: e.g., New York to Sidney. If affordable, this could make the Moon accessible to large numbers of space adventurers, lunar explorers, scientists and engineers.

For a 24-hr flight, let us assume that the Lightcraft arrives at LI at 3.22 km/s (a little fast), and the two-way EML must slow the vehicle to 0.8 km/s; this requires a delta-V of 2.42 km/s. If decelerated at 30 g for 8.24 seconds, the EML must be 10 km in length. A 20-km long EML could accommodate arrival velocities to 4.23 km/s (delta-V = 3.43 km/s), by extending the 30-g deceleration to 11.7 seconds. [Note that 10 GW designs for GEO satellite solar power stations in the mid 1970s had overall dimensions ranging from 10 - 20 km.] With liquid immersion, centrifuge data indicates that five seconds is the maximum human tolerance at 30 g,[149] but future ultra-g suit developments should eventually permit sustained, 2 to 3 times longer exposures.

In contrast, a 36-hr flight has the Lightcraft arriving at 1.75 km/s, so slowing it to 0.8 km/s requires a delta-V of only 0.95 km/s. A deceleration of 15 g for 6.5 seconds accomplishes this in just over 3 km; 10 g for 9.7 seconds and 4.6 km.

[Space command's ultra-g suits make such decelerations not just survivable, but sufficiently comfortable as well, so one-way flight times to or from the moon of 24 to 36 hours are certainly feasible for 15-m lunar Lightcraft and their 3 to 5 person crews.]

HISTORICAL ORIGINS OF LTI-15 TECHNOLOGY

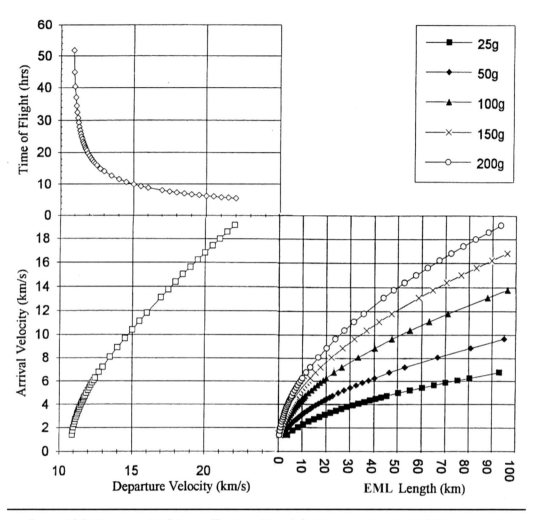

Figure A9.7: Nomograph of 1-way (Earth-to-Moon) flight parameters to L1 EML spaceport. *(Courtesy of RPI.)*

HMM participants suggested that for some travelers, 24 to 36 hours might still be too long to be cooped up in a small spaceship, so even higher injection speeds were entertained (see Figure A9.7). So, for example if a 15-hour flight time is desired, the injection speed climbs to 15 km/s, giving a lunar arrival velocity of 7 km/s. At 30 g's, a 24-second long deceleration would stop this vehicle in 83 km, but at 50 to 100 g, the EML length would shorten to 50- and 25 km, respectively. The greatest injection speed discussed in the workshop was 22 km/s, which gives a 5.5 hour flight time to the Moon. But of course, the required EML lengths may be completely out of reach unless ultra-g suits permit sustained decelerations of 200 g or more.

Finally, HMM participants suggested an alternative application of two-way EML facilities: to *accelerate an arriving Lightcraft*, instead of capturing it. Imagine ultra-quick flights arriving at L1 with 19 km/s, then getting an additional EML boost and depart at extreme velocities. Perhaps every two weeks when the planetary alignments are proper, robotic explorers and/or adventurous space tourists could be launched into the ecliptic plane of the solar system for fast excursions to the outer planets. Such propellantless *Grand Tours* of the solar system might even be aided by two-way EML "booster" stations built on the moons of Mars, Saturn, and Jupiter.

A9.4 Outcome of HMM Workshop

A major conclusion of the 1994 workshop was that ultra-fast lunar shuttles (i.e., based on beamed energy propulsion and Lightcraft) could indeed service a thriving *"Mid-21st Century Space Carrier Industry"* if and only if:

1) 10 to 20 km long mass-drivers can be built and maintained at L1 and LEO;
2) Lightcraft vehicle structures can withstand repeated 30- to 50-g loads (or more) from two-way EML spaceports; and
3) ultra-g-suit protection technology is developed for Lightcraft crews to extend human g-tolerance into the necessary regime.

A10.0 Microwave Power Transmission Limits

The physics of propagating high-power microwave radiation through the atmosphere constrains the operating regime within specific limits. Table A10.1 delineates the interaction regimes for increasingly higher power densities of microwave radiation.

Note that the air interaction is comprised of two components:

a) molecular absorption; and
b) an absorption and scattering process – principally caused by water vapor, aerosols, and solid particulates suspended in the air.

Of course, successful power transmission to the Lightcraft is the desired function of the radiation, so the intensity level striking the Lightcraft hull and receiving optics must remain BELOW that required for surface-induced electrical air breakdown (plasma formation) and/or the ignition of microwave-supported combustion (MSC) or detonation (MSD) waves. In the millimeter regime, air plasma temperatures within MSC and MSD waves can quickly reach 5,000 K and beyond, which will damage hull materials. One important function of the Lightcraft receptive optics is to act as concentrating reflectors, thereby bringing the microwave power beam to a focus, and triggering such plasmas at safe distances from the Lightcraft hull: e.g., within the airspike or perimeter MHD slipstream accelerators.

Four important factors in atmospheric transmittance are:

1) attenuation from absorption and scattering effects;
2) thermal blooming (non-linear);
3) microwave-induced electrical air breakdown (non-linear); and
4) turbulence-induced beam spreading.

Figure A10.1 shows the attenuation (2-way, dB/nm) of microwave radiation as a function of frequency for the standard atmosphere at sea level altitude, with 1% water vapor. Note the atmospheric "windows" at 35, 94, 140, and 220 GHz where transmission is especially desirable.

Any attempt to propagate a high-power microwave beam from the ground to space is heavily influenced by the integrated properties of the atmosphere within the beam path. For a given microwave beam path, the atmospheric transmittance (T_R) is:

$$T_R = \exp \left\{ -\int (k_a + \sigma_a + k_m + \sigma_m) \, dR \right\} \quad (A10.1)$$

where k_a and σ_a are the aerosol absorption and scattering coefficients; k_m and σ_m are the corresponding values for molecular constituents. The transmittance for a given wavelength is best for clear days with low water vapor and low aerosol contamination.

Table A10.1: High power microwave / air / matter interaction.[150] *(After J. Reilly; Courtesy of J. Benford)*

Beam Intensity (power/area)	Molecular Interaction	Particulate Interaction	Lightcraft / Target Interaction
Low Intensity	Attenuation	Attenuation	Absorption or Reflection
Higher	Blooming	Enhanced Scattering	Absorption/ Breakdown Threshold (MSC/MSD)
Still Higher	Blooming	Breakdown	Breakdown (MSD)
Very High	Breakdown	Breakdown	Breakdown (MSD)

HISTORICAL ORIGINS OF LTI-15 TECHNOLOGY

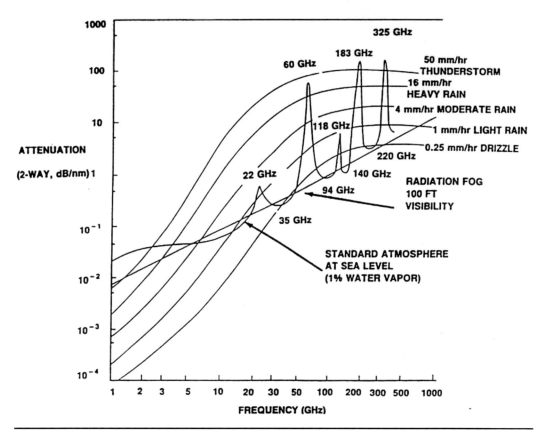

Figure A10.1: Two-way (i.e., double-pass) attenuation vs. frequency in dB per nautical mile. *(Courtesy of J. Benford.)*

A10.1 Vertical Transmission Efficiency

Table A10.2 from Reference 151 shows the vertical transmission efficiency from a 4-km GBM site (with 1 mm precipitable water vapor), where the ambient pressure is roughly 60% of sea level.

Note that at 220 GHz, for example, the vertical transmittance is 95% from the 4-km mountain site but only 45% from sea level, which is generally unacceptable for an uplink. In sharp contrast, a 35 GHz sea level GBM site can transmit 93% straight up through the atmosphere.

The height of a GBM site above sea level is the principal determinant of the overlying air mass and its molecular composition, which

Table A10.2: Microwave atmospheric transmission and breakdown.[151] *(Courtesy of J. Benford)*

Microwave Frequency (GHz)	Vertical Transmittance Through Atmosphere (Percent)		Power Density for Air Breakdown at 30-km Altitude (kW/cm^2)
	Sea level	4 km Peak	
35	93	99	4
94	79	98	30
140	71	99	66
220	45	95	163

APPENDIX

fundamentally affects the transmittance to space through both absorption and scattering processes. Furthermore, the prevailing weather patterns at the site heavily influence the aerosol content (vapors and dust), which can play a significant role in determining the scattering of power out of the beam.

The zenith angle (i.e., angle "Z" measured from the point vertically overhead) determines the total distance traversed through the atmosphere, which is proportional to secant Z to a very good approximation. Table A10.3 shows the effects of transmitting at various zenith angles with 220 GHz radiation from both sea level and 4-km mountain sites.

Note the transmittance falls to 93% at $Z = 45°$ and 86% at 70° at the 4-km site (i.e., quite acceptable); however, the sea-level siting performance is unacceptably poor at 220 GHz. Atmospheric transmittance to space is clearly a strong function of transmitter frequency, siting altitude, and zenith angle.

Other atmospheric factors impact the design of ground-based microwave transmitter optics. The projector aperture must be large enough to prevent the power density in the beam from exceeding the threshold where non-linear effects (e.g., electrical air breakdown) take place. The fundamental problem is presented in Figure A10.2, which specifies the beam intensity (ϕ) required to induce electrical air breakdown at various altitudes and microwave wavelengths. This relationship is given in Reference 152 as

$$\phi = 2.39\, p^2\, (1 + \omega_m^2/v^2),\, (watts/cm^2) \quad (A10.2)$$
$$= 2.39\, p^2\, (Torr) + 3.38\, f^2\, (GHz)$$

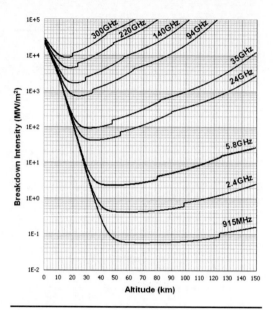

Figure A10.2: Breakdown intensity vs. altitude. *(Courtesy of K. Parkin.)*

where p is the pressure in Torr, ω_m is the angular frequency, $v = (5.3 \times 10^9)\, p$ is the molecular collision frequency of air at that pressure, and f is the microwave frequency in GHz. Table A10.2 shows the breakdown flux (or power density) at 30 km, which is a function of microwave frequency. The reduced atmospheric pressure at 30 km altitude (the so-called Paschen minimum) will result in air breakdown if the microwave flux exceeds a critical value of 4 kW/cm² at 35 GHz, for example. However, at 94 and 140 GHz the threshold climbs to 30- and 66 kW/cm², respectively, which is much less restrictive for these higher frequencies.

Table A10.3: GBM transmittance vs. zenith angle for 220 GHz.[151] *(Courtesy of J. Benford)*

Zenith Angle (Degrees)	Atmospheric Path Length (Relative to Vertical)	Transmittance Through Atmosphere (Percent)	
		Sea level	4 km Peak
0°	1.0	45	95
30°	1.15	40	94
45°	1.41	32	93
60°	2.0	20	90
70°	2.92	10	86

A10.2 Minimum Beam Spot Size vs. Range

In this section, the atmospheric "uplink" transmission ϕproblem is analyzed in sufficient detail to provide a good approximation the required GBM system parameters.

The relative and absolute importance of all the aforementioned transmittance factors is, in turn, strongly driven by the wavelength, which is the key system parameter.

The minimum beam spot size projected by a GBM upon the space relay mirror is driven by the beam spreading components of diffraction, atmospheric turbulence, thermal blooming, and jitter of the transmitting optics. The total beam spreading angle (σ_o) may be expressed as

$$\sigma_o^2 = \sigma_D^2 + \sigma_T^2 + \sigma_J^2 + \sigma_B^2 \quad \quad (A10.3)$$

where D, T, J, and B denote the contributions due to diffraction, turbulence, jitter, and thermal blooming.

Diffraction: The diffraction component is given by

$$\sigma_D = 0.45 \, \lambda \beta \, / \, D \quad \quad (A10.4)$$

where λ is the wavelength, β is the (narrow-angle) beam quality, and D is the diameter of the GBM transmitter aperture. Here, beam quality refers to the narrow-angle beam spread component, and not the wide-angle component associated with distant side-lobes (usually considered a power loss term).

Turbulence: Atmospheric turbulence and its beam-spreading effect on the microwave beam (σ_T) is generally dependent upon altitude, wind velocity, temperature, time of day, season of the year, zenith angle, and geographical features of the environment. For a given microwave wavelength, the magnitude of turbulence-induced beam-spreading effects is proportional to the scale of the turbulence and distribution of water vapor. The atmospheric coherence length also varies with the choice of microwave wavelength, and the associated disturbances can be corrected with adaptive optics of either the continuous membrane or multisegmented types; the coherence length determines the size, and thus the required number of the adaptive zones in the transmitter's sub-reflector surfaces. The effects of turbulence are generally small to negligible for microwaves, but very important at laser wavelengths.

Jitter: Jitter (σ_J) of the ~3000 constituent subapertures (9-m diameter each) of a large GBM transmitter array (say, 250- to 500 m across) is incurred as the entire beam is steered to follow the relay satellite passing overhead. With modern "Star Wars" targeting system technology (i.e., pointing and tracking mechanisms and software), σ_J beam spreading effects become negligible.

Thermal blooming: A primary consideration in designing a 5- to 20 gigawatt GBM installation is to deliberately select a transmitter diameter big enough to reduce thermal blooming effects (σ_B) to within limits correctable by adaptive optics on the secondary reflectors, given a specific known cross-wind velocity (derived from historical weather data at the GBM site) to clear the beam. Under such "clearing" conditions, with the power levels of interest, simple calculations reveal that electrical air breakdown will not occur within the beam, and no other non-linear atmospheric propagation effects are anticipated.

Hence, the power transmission problem reduces to the linear regime where the necessary relay satellite receiver aperture becomes a direct function of transmitter diameter, zenith angle, relay altitude and, of course, microwave frequency. Several additional assumptions must be made in the analysis:

a) the outgoing power beam has a specific beam quality "β" as in Eqn. A10.4; and
b) the relay receiver aperture is sized to intercept at least 95% of the incident radiation.

Assuming a gaussian distribution to represent the received beam intensity, the relay diameter ($d_o = 2 \, r_o$) must extend to the e^{-2} intensity level to capture that 95%. Figure A10.3 shows the typical atmospheric "uplink" geometry. Note that $(\sin \sigma_o) \cong \sigma_o$ when the target is at great range, so $\sigma_o \cong r_o/R = d_o/2R$.

Table A10.4 gives a list of the basic microwave system parameters used to evaluate Lightcraft (or target) engagements for beamed-energy propulsion and other applications.

APPENDIX

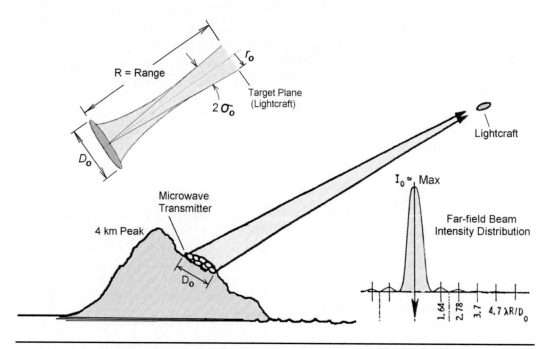

Figure A10.3: Microwave power transmission ("uplink") geometry.

Table A10.4: Basic microwave system parameters. *(Courtesy of J. Benford)*

Symbol	Parameter	Common Units (Metric)
B	Beam Brightness	Watts/ Sterradian
P	Microwave Power	Gigawatts (i.e., 10^9 watts)
σ_o	Total Beam Spread Angle (or "solid angle")	Radians
T_R	Power Losses in Transmission to Lightcraft	(dimensionless)
R	Range to Lightcraft (or Target)	Kilometers
Φ	Flux Intensity on Target	Watts/cm²
Δt	Target Irradiation Time	Seconds
F	Fluence Deposited on Target	Joules/cm²
d_o, r_o	Beam Diameter, Radius—at Target	Meters
A	Beam Spot Area on Target	cm²
E	Energy Deposited on Target	Joules
D	Transmitter Diameter (aperture)	Meters
β	Beam Quality	(dimensionless)
λ	Wavelength	(mm)

Hence the peak intensity (I_o) projected upon the relay mirror is

$$\sigma_o = PT_R / \{2\pi (\sigma R)^2\} \quad (A10.5)$$

where P is the total transmitted power at the GBM aperture, T_R is the transmissivity of the atmosphere, σ_o is the one-sigma beam spread angle, and R is the range between the GBM and relay satellite. Brightness (B), which is the best measure of system prowess / effectiveness for microwave power-beaming stations, can be expressed as

$$B = (P / 2\pi\sigma_o^2) T_R, \text{ (watts/ster.)} \quad (A10.6)$$

The power density, or flux, delivered onto the Lightcraft or target is

$$\phi = B / R^2, \text{ (watts/cm}^2\text{)} \quad (A10.7)$$

"Fluence" is the energy density delivered on target during an irradiation time of Δt:

$$F = \phi \Delta t, \text{ (joules/cm}^2\text{)} \quad (A10.8)$$

The total energy (E_T) transmitted to the target is simply the fluence integrated over the total target (or receptive optics) area:

$$E_T = FA, \text{ (joules)} \quad (A10.9)$$

where the spot area $A = \pi r_o^2$.

Figure A10.4 is a schematic of the transmission beam path / geometry for any microwave transmitter. Note the position of the focus, located at the center of a long beam "waist," which defines the minimum beam spot diameter addressed in the above analysis. Lightcraft normally fly in the "near-field" region that lies between the transmitter and the focus, and infrequently in the "far-field" region beyond the focus.

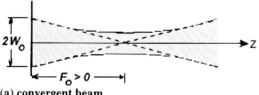
(a) convergent beam

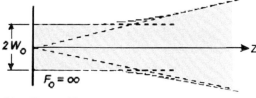
(b) collimated beam

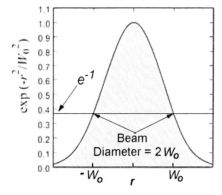
Amplitude profile of Gaussian-beam wave.

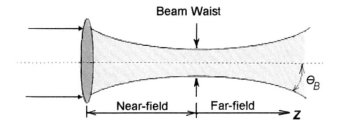

Figure A10.4: Near-field, beam waist, and far-field divergence angle for convergent beam.
(Courtesy of SPIE.)

REFERENCES (by Chapter)

CHAPTER 19: (POWER BEAMING INFRASTRUCTURE)

1. Gross, D.A. and Myrabo, L.N., "Metal Matrix Superconductor Composites for SMES-Driven, Ultra High Power BEP Applications: Part 1," in *4th International Symposium on Beamed Energy Propulsion*, edited by K. Komurasaki, AIP Conference Proceedings, New York: American Institute of Physics, 2006.
2. Gross, D.A. and Myrabo, L.N., "Metal Matrix Superconductor Composites for SMES-Driven, Ultra High Power BEP Applications: Part 2," in *4th International Symposium on Beamed Energy Propulsion*, edited by K. Komurasaki, AIP Conference Proceedings, New York: American Institute of Physics, 2006.
3. Young, D.P., Moldovan, M., and Adams, P.W., "Scaling behavior of the critical current density in $MgCNi_3$ microfibers," *Physical Review B* **70**, 064508 (2004).
4. Benford, J. and Myrabo, L.N., "Propulsion of Small Launch Vehicles Using High Power Millimeter Waves," *OE/LASE "94*, SPIE Proceedings Vol. 2154, International Society for Optical Engineering, Bellingham, WA, Jan. 1994, pp. 198-219.
5. Benford, J. and Dickinson, R., "Space Propulsion and Power Beaming Using Millimeter Systems," *OE/LASE "95*, SPIE Proceedings Vol. 2556, International Society for Optical Engineering, Bellingham, WA, Jan. 1994, pp. 179-192
6. Benford, J., by personal communication. (Forecast of gyrotron technology growth.)
7. Reilly, J.P., "Propagation of High Power Microwave Beams," W.J. Schafer Associates, Inc., Wakefield, MA, August 1979.
8. Weiting, T.J., et.al., "Plasma Mediated Coupling to Metals at 35 GHz/8.6 cm and 9.3 GHz/3.2 mm," *Proceedings of the 2nd National Conference on High Power Microwave Technology*, Harry Diamond Laboratories, Adelphi, MD, 1-3 March 1983.
9. Dickinson, R.M., "Cost Effectiveness of Spacecraft Pointing Antenna," JPL Technical Memorandum 33-390, 1968.
10. Hedgepeth, J.M., Miller, R.K., and Knapp, K., "Conceptual Design Studies for Large Free-Flying SolarReflector Spacecraft," NASA CR-3438, 1981.
11. Ehsani, M., Bilgic, O., and Patton, A.D., "Magnetically Inflatable SPS with Energy Storage Capability," SPS 91 - Power from Space, Societe des Ingenieurs et Scientifiques de France, Paris, France, 27-30 August 1991, pp. 248-252.
12. Flint, E.M., "Thin Film Disc-Shaped Large Space Structures (Sunsat or Solar Sail): A Proposed Construction / Assembly Method," *Proceedings of the 12th SSI Conference on Space Manufacturing*, held in Princeton, NY, 4-7 May 1995.
13. Scherfflins, J.H. and Lush, K.J., "ARIS Aperture Relay," LMSC-L030203, Contract No. DAAK90-76-C-DADE, 30 July 1977.
14. Decker, L., Aspinwall, D., Pohle, D., Dotson, R., and Bartosewcz, M., "Aperture Relay Experimental Definition," LMSC-L048510, Contract No. DASG60-78-C-0100, Lockheed Palo Alto Research Laboratory, CA, 28 August 1979.
15. Mattick, A.T., and Hertzberg, A. "Liquid Droplet Radiators for Heat Rejection in Space," *Journal of Energy*, Nov. 1981, Vol. 5, No. 6, pp. 387-393.
16. Mattick, A.T., and Hertzberg, A., "The Liquid Droplet Radiator: An Ultralightweight Heat Rejection System for Efficient Energy Conversion in Space," *Acta Astronautica*, 1982, Vol. 9, No. 3, pp. 165-172.
17. Mattick, A.T., by personal communication.
18. Dickerson, T., "Microwave Lightcraft Study: Mission Analysis for Propulsion Power Supply Station," PD31, NASA Marshall Space Flight Center, Huntsville, AL, 24 February 1999.

19. Dickerson, T., "Microwave Lightcraft Study: MWLC & Power Sat Ascent to Orbit & Power Beaming Relationships," PD31, NASA Marshall Space Flight Center, Huntsville, AL, 6 May 1999.
20. Frazier, S.R., "Trajectory Analysis of a Laser-Energized Transatmospheric Vehicle," M.S. Thesis, Rensselaer Polytechnic Institute, Troy, NY, August 1987.

CHAPTER 20: (HUMAN FACTORS AND G TOLERANCE)

1. "Technical Report by Aerospace Medical Panel," Advisory Group for Aerospace Research and Development (AGARD) Conference, Neuilly-Sur-Seine (France), 24-28 April 1989.
2. Simpura, S. F., "Change in Certain Indices of the Function of External Respiration During the Action of G-Forces," *Problems of Space Biology*, Vol. 16, pp. 66-79.
3. Hlastala, M. P., Chornuk, M.A., Self, D.A., et. al., "Pulmonary Blood Flow Redistribution by Increased Gravitational Force," 14 March 1996.
4. Glaister, D. H., "Centrifuge Training and Selection of Aircrew for High-G Tolerance," Proceedings of the RAS Symposium, Royal Aeronautical Society, London, England, 21 Oct. 1987.
5. Cammarota, J. P., Whinnery, J. E., "Enhancing Aircrew Centrifuge High-G Training Using On-line Videotape Documentation," *Aviation, Space, and Environmental Medicine*, Vol. 61, 1990, pp. 1153-1155.
6. Whinnery, J. E., Hamilton, R. J., "Aircrew Critique of High-G Centrifuge Training: Part 3: What Can We Change to Better Serve You?", Naval Air Development Center, Final Report, Oct. 1988 - May 1991.
7. Bischoff, D. E., Eyth, J. Jr., "F-14 Simulation in a Total G-Force Environment," *Society of Flight Test Engineers*, 1986; pp. 4.1-1 to 4.1-9.
8. Alberry, W. B., Van Patten, R. E., "Development of a Gravity Induced Loss-of-Consciousness (GLOC) Monitoring System," Proceedings of the IEEE National Aerospace and Electronics Conference, Vol. 2, 1990, pp. 831-837.
9. Gillingham, K. K., "High-G Stress and Orientational Stress - Physiological Effects of Aerial Maneuvering," *Aviation, Space, and Environmental Medicine*, Vol. 59, 1988, pp. A10-A20.
10. Wachs, T. J., Perry, C. J. G., "Psychogenic G-Force Intolerance Revisited," *Aviation, Space, and Environmental Medicine*, Vol. 49, 1978, pp. 76-77.
11. Webb, P., *Bioastronautics Data Book*, NASA SP-3006, 1964.
12. Ibid, (reporting work by Chambers in 1963).
13. Fraser, T.M., *Human Response to Sustained Acceleration*, NASA-SP-103, 1966, p. 55.
14. Bennett, P. and Elliott, D., *The Physiology and Medicine of Diving*, 4th ed, W. B. Saunders Co. Ltd., London, 1993.
15. Clark, L. C., Gollan, F., "Survival of Mammals Breathing Organic Liquids Equilibrated with Oxygen at Atmospheric Pressure," *Science*, Vol. 152, 1966, pp.1755-1756.
16. Gollan, F., Clark, L. C., "Prevention of Bends by Breathing an Organic Liquid," *Transactions, Association of American Physicians*, Vol. 29, 1967, pp. 102-109.
17. Kylstra, J.A., Nantz, R., Crowe, J., Wagner, W., Saltzman, H. A., "Hydraulic Compression of Mice to 166 Atmospheres," *Science*, Vol. 158, 1967, pp. 793-794.
18. Moskowitz, G.D., "A Mechanical Respirator for Control of Liquid Breathing," *Fed. Proc.*, Vol. 29, 1970, pp.1751-1752.
19. Shaffer, T. H., Moskowitz, G. D., "Demand-Controlled Liquid Ventilation of the Lungs," *J. Appl. Physiol.*, Vol. 36; 1974, pp. 208-215.
20. Shaffer, T. H., Moskowitz, G. D., "An Electrochemical Demand Regulated Liquid Breathing System," *IEEE Trans. Biomed. Eng.*, Vol. 22, 1975, pp. 24-28.
21. Modell, J. H., Hood, C. L, Kuck, E. J., Ruitz, B. C., "Oxygenation by Ventilation with Fluorocarbon Liquid (FX-80)," *Anesthesiology*, Vol. 34, 1971, pp. 312-320.
22. Shaffer, T. H., Forman, D., Wolfson, M. R. "The Physiological Effects of Breathing Fluorocarbon Liquids at Various Temperatures," *Undersea Biomed. Research*, Vol. 11, 1984, pp. 287-298.

REFERENCES

23. Deaton, J. E., Holmes. M., Warner, N., Hitchcock, E, "Development of Perceptual / Motor and Cognitive Performance Measures Under a High-G Environment: A Preliminary Study," Naval Air Development Center, Air and Crew Systems Technology Dept., Final rept. Dec 88-Apr 90.
24. Rupert, A. H., Guedry, F. E., Reschke, M. F., "The Use of a Tactile Interface to Convey Position and Motion Perceptions," AGARD conference publication, Sueilly sur Seine, France, 18-22 Oct. 1993.
25. Mawhinney, D. D., Kresky, T., "Non-invasive Physiological Monitoring with Microwaves," *Proceedings of the IEEE National Aerospace and Electronics Conference*, Vol. 3, 1986, pp. 746-51.
26. Myrabo, L.N. and Myrabo, K.A., "Human Factors Integration for Hyper-Energetic, Beam Boosted Aerospacecraft," AIAA 97-2795, 33rd AIAA / ASME / SAE / ASEE Joint Propulsion Conference & Exhibit, 6-9 July 1997.
27. Nordwall, B.D., "Mcdonnell Fluid-Filled Suit Can Help Pilots Withstand 10 G," *Aviation Week & Space Technology*, 19 Aug. 1991, p. 66.
28. Webb, P., "The Space Activity Suit: An Elastic Leotard for Extravehicular Activity," *Aerospace Medicine*, April 1968, pp. 376-383.
29. Webb, P. and Annis, J.F., "The Principle of the Space Activity Suit," NASA CR-973, National Aeronautics and Space Administration, Washington, D.C., 1967.
30. Annis, J.F. and P. Webb, "Development of a Space Activity Suit," NASA CR-1892, National Aeronautics and Space Administration, Washington, D.C., Nov. 1971.
31. Ordway, F.I., *Visions of Spaceflight – Images from the Ordway Collection*, publ. by Four Walls Eight Windows, NY, 2001, p. 147.
32. Tuttle, J., *Eject! The complete History of U.S. Aircraft Escape Systems*, MBI Publishing Co., St. Paul, MN, 2002.
33. Frisbee, R.H., "Propulsion Systems," *Space Education*, Vol. 1, No. 10, October 1985, pp. 448-459.
34. Frisbee, R.H., "Spacecraft Propulsion Systems – What They Are and How They Work," *Foundation Astronautics Notebook – 6*, World Space Foundation, Jan-July 1983.
35. French, J.R. and Cruz, M.I., "Aerobraking and Aerocapture for Planetary Missions," *Astronautics and Aeronautics*, Washington, D.C., Feb. 1980, pp. 48-55, 71.

APPENDIX: (ORIGINS OF LTI-15 TECHNOLOGY)

1. Winter, F.H., *Rockets into Space*, Harvard University Press, Cambridge, MA, 1990, p.40.
2. Alway, P., *Retro Rockets – Experimental Rockets 1926-1941*, Saturn Press, Ann Arbor, MI, 1996, p. 50.
3. Blakeslee, A., "Space Ship," Associated Press, NY, 29 Dec. 1956.
4. Bussard, R.W. and DeLauer, R.D., *Fundamentals of Nuclear Flight*, McGraw-Hill, NY, 1965.
5. Rosa, R.J., "Propulsion System Using a Cavity Reactor and Magnetohydrodynamic Generator," *ARS Journal*, July 1961, 884-885.
6. Rosa, R.J., "The Application of Magnetohydrodynamic Generators in Nuclear Rocket Propulsion", AERL Research Report 111, AFBSD-TR-61-58, Contract No. AF 04(647)-278, Avco-Everett Research Laboratory, Everett, MA, Aug. 1961.
7. Rosa, R.J., "Magnetohydrodynamic Generators and Nuclear Propulsion," *ARS Journal*, Aug. 1962, pp. 1221-1230.
8. Wild, J.M., "Nuclear Propulsion for Aircraft," *Astronautics & Aeronautics*, March 1968, pp. 24-30.
9. Rom, F.E., "Airbreathing Nuclear Propulsion - A New Look," NASA TM X-67837; see also, *Sixth Symposium on Advanced Propulsion Concepts*, sponsored by Air Force Office of Scientific Research (AFOSR), Niagara, NY, 4-5 May 1971.
10. Miller, R. *The Dream Machines – An Illustrated History of the Spaceship in Art, Science, and Literature*, Krieger Publ. Co., FL, 1993.
11. Hall, T.W., "The Architecture of Artificial-Gravity Environments for Long Duration Space Habitation", Ph.D. Thesis, University of Michigan, Ann Arbor, MI, 1994.
12. Ware, G.M., "Static Stability and Control Characteristics at Low-Subsonic Speeds of

a Lenticular Reentry Configuration," NASA TM X-431, 1960.

13. Mugler, J.F., Jr., and Olstad, W.B., "Static Longitudinal Aerodynamic Characteristics at Transonic Speeds of a Lenticular-Shaped Reentry Vehicle," NASA TM X-423, Dec. 1960.

14. C.M. Jackson Jr. and R.V. Harris Jr., "Static Longitudinal Stability and Control Characteristics at a Mach Number of 1.99 of a Lenticular-Shaped Reentry Vehicle," NASA TN D-514, 1960.

15. Demele F.A. and Brownson, J.J., "Subsonic Longitudinal Aerodynamic Characteristics of Disks With Elliptic Cross-sections and Thickness-Diameter Ratios From 0.225 to 0.425," NASA TN D-788, 1961.

16. Demele, F.A. and Lazzeroni, F.A., "Effects of Control Surfaces on the Aerodynamic Characteristics of a Disk Reentry Shape at Large Angles of Attack and a Mach Number of 3.5," NASA TM X-576, 1961.

17. Demele, F.A. and Brownson, J.J., "Subsonic Aerodynamic Characteristics of Disk Reentry Configurations with Elliptic Cross Sections and Thickness-Diameter Ratios of 0.225 and 0.325," NASA TM X-566, 1961.

18. Lazeroni, F.A., "Aerodynamic Characteristics of Two Disk Reentry Configurations at a Mach Number of 2.2," NASA TM X-567, 1961.

19. Demele, F.A. and Brownson, J.J., "Subsonic Aerodynamic Characteristics of Disk Reentry Configurations with Elliptic Cross Sections and Thickness-Diameter Ratios of 0.255 and 0.325," NASA TM X-566, 1961.

20. Olstad W.B. and Wornom, D.E., "Static Longitudinal Stability and Control Characteristics at Mach Numbers of 2.86 and 6.02 and Angles of Attack up to 95 Deg. of a Lenticular-Shaped Reentry Vehicle," NASA-TM X-621, 1961.

21. Lazzeroni, F.A., "Experimental Investigation of a Disk-Shaped Reentry Configuration at Transonic and Low Supersonic Speeds," NASA TM X-652, May 1962.

22. Ware, G.M., "Investigation of the Low-Subsonic Aerodynamic Characteristics of a Model of a Modified Lenticular Reentry Configuration." NASA TM X-758, Dec. 1962.

23. McShera J.T. and Lowery, J.L., "Static Stability and Longitudinal Control Characteristics of a Lenticular-Shaped Reentry Vehicle at Mach Numbers of 3.5 and 4.65," NASA TM X-763, 1963.

24. Keys, J.W., "Aerodynamic Characteristics of Lenticular and Elliptic Shaped Configurations at a Mach Number of 6," NASA TN D-2606, Feb. 1965.

25. Robinson, J.C. and Jordan, A.W., "Exploratory Experimental Aerodynamic Investigation of Tension Shell Shapes at Mach 7", NASA TN D-2994, NASA, Wash., D.C., September 1965.

26. Giragosian, P.A. and Hoffman, W.D., "Landable Disk Reentry Vehicles," *Proceedings of the Symposium on Dynamics of Manned Lifting Planetary Entry*, Ed. by S.M. Scal, A.C. Harrison, and M. Roger, John Wiley & Sons, 1963, pp. 729-749.

27. Katz, P., "The Free Flight of a Rotating Disc," *Israel Journal of Technology*, Vol. 6, No. 1-2, 1965, pp. 150-155; see also, *Proc. X, Israel Annual Conf. on Aviation and Astronautics*, Feb. 1965.

28. Potts, J.R. and Crowther, W.J., "Visualisation of the Flow over a Disc-Wing," Paper No. 247, 9th International Symposium on Flow Visualization, 2000.

29. Potts, J.R. and Crowther, W.J., "Flight Control of a Spin Stabilized Axi-symmetric Disc-Wing," AIAA-2001-0253.

30. Potts, J.R. and Crowther, W.J., "Frisbee TM Aerodynamics," AIAA-2002-3150, 20th AIAA Applied Aerodynamics Conference, 24-26 June 2002.

31. Levy, R.H., "Radiation Shielding of Space Vehicles by Means of Superconducting Coils," *ARS Journal*, Vol. 31, No. 11, 1961, pp. 1568-1570.

32. Bernert, R.E. and Stekley, Z.J.J., "Magnetic Shielding Using Superconducting Coils," *2nd Symposium on Protection Against Radiation in Space*, (A. Reetz, Jr., Editor), NASA SP-71, Gatlinburg, TN, Oct. 1964, pp. 199-209.

33. Prescott, A.D., Urban, E.W. and Sheldon, R.D., "Application of the Liouville Theorem to Magnetic Shielding Problems," 2nd

REFERENCES

Symposium on Protection Against Radiation in Space, (A. Reetz, Jr., Editor), NASA SP-71, Gatlinburg, TN, Oct. 1964, pp. 189-198.

34. "Feasibility of a Magnetic Orbital Shielding System," Lockheed Missiles and Space Co., Technical Report, Contract AF 04(685)-252, May 1964.

35. Helgesen, J.O. and Spagnolo, F.A., "The Motion of a Charged Particle in a Magnetic Field Due to a Finite Solenoid With Application to Solar Radiation Protection," AIAA 1966-512, 4th Aerospace Sciences Meeting, Los Angeles, CA, 1966.

36. Brown, G.V., "Magnetic Radiation Shielding," Chapter 40, *High Magnetic Fields*, (Kolm, Lax, Bitter, and Mills, Editors), MIT Press, Cambridge, MA, 1962, pp. 370-378.

37. E.C. Hannah, "Meteoroid and Cosmic-Ray Protection," in *Space Manufacturing Facilities, Proceedings of the Princeton / AIAA / NASA Conference*, (J. Grey, editor), May 7-9 1975, AIAA, 1977, pp. 151-157.

38. Levy, R.H. and French, F.W., "Plasma Radiation Shield: Concept and Applications to Space Vehicles," *J. Spacecraft*, Vol. 5, No. 5, May 1968, pp. 570-577.

39. Levy, R.H. and Janes, G.S., "Plasma Radiation Shielding," *AIAA Journal*, Vol. 2., No. 10, Oct. 1964, pp. 1835-1838; also presented at 2nd Symposium on Protection Against Radiation in Space, Gatlinburg, TN, Oct. 1964, NASA SP-71, pp. 211-215.

40. French, F.W., "Solar Flare Radiation Protection Requirements for Passive and Active Shields," *J. Spacecraft*, Vol. 7, No. 7, July 1970, pp. 794-800.

41. Hannah, E.C., "Radiation Protection for Space Colonies," *J. British Interplanetary Soc.*, Vol. 30, 1977, pp. 310-314.

42. Landis, G.A., "Magnetic Radiation Shielding: An Idea Whose Time Has Returned?", *Space Manufacturing 8: The High Frontier*, Proceedings of the 10th SSI-Princeton Conference, AIAA, 1991, pp. 383-386.

43. Andrews, D.G. and Zubrin, R.M., "Use of Magnetic Sails for Advanced Exploration Missions," *Vision-21: Space Travel for the Next Millennium*, (G. Landis - ed.), NASA CP-10059, April 1990, pp. 202-210.

44. Leung, E.M.W., Hilal, M.A., Parmer, J.F., and Peck, S.D., "Lightweight Magnet for Space Applications," *IEEE: Transactions on Magnetics*, Vol. MAGG-23, No. 2, March 1987, pp. 1331-1335.

45. Connolly, D.J., Heinen, V.O., Aron, P.R., Lazar, J., and Romanofsky, R.R., "Aerospace Applications of High Temperature Superconductivity," 41st Congress of the Intnl. Astronautical Federation, Paper IAF-90-054, Dresden, GDR, Oct. 6-12, 1990.

46. Hull, J.R., "High Temperature Superconductors for Space Power Transmission Lines," ASME Winter Annual Meeting Conf., San Francisco, CA, Dec. 10-15, 1989.

47. Cocks, F.H., "A Deployable High Temperature Superconducting Coil (DHTSCI)," *J. Brit. Interplanetary Soc.*, Vol. 44, 1991, pp. 98-102.

48. von Braun, W., "Will Mighty Magnets Protect Voyagers to Planets?", *Popular Science Magazine*, January 1969, pp. 98-100, 198.

49. Kantrowitz, A., "Propulsion to Orbit by Ground-Based Laser," *Astronautics & Aeronautics*, Vol. 10, No. 5, May 1972, p. 74-76; see also: Kantrowitz, A., "The Relevance of Space," *Astronautics & Aeronautics*, Vol. 9, No. 3, March 1971, pp. 34-35.

50. Douglas-Hamilton, D.H., Kantrowitz, A.R., and Reilly, D.A., "Laser Assisted Propulsion Research," *Radiation Energy Conversion in Space*, Vol. 61, Progress in Astronautics and Aeronautics, K.W. Billman, Ed., M. Summerfield, Series Editor-in-Chief, American Institute of Aeronautics and Astronautics, NY, 1978, pp. 271-278.

51. *Proceedings of SDIO / DARPA Workshop on Laser Propulsion*, 7-18 July 1986, J.T., Kare, ed., LLNL CONF-860778, Vol. 1-3, Lawrence Livermore National Laboratory, 1987.

52. *Proceedings, 1987 SDIO Workshop on Laser Propulsion*, J.T., Kare, ed., LLNL CONF-9710452, Lawrence Livermore National Laboratory, CA, 1990.

53. *Proceedings of the AFOSR Workshop on Laser Propulsion*, M. Birkin, ed., AFOSR-TR-88-1430, Feb. 1988.

54. Myrabo, L.N., et.al., "Lightcraft Technology Demonstrator - Transatmospheric Laser

Propulsion," Final Technical Report, prepared under Contract No. 2073803 for Lawrence Livermore National Laboratory and the SDIO Laser Propulsion Program, June 30, 1989.

55. Myrabo, L.N., "Lightcraft Propulsion Concepts for Space Exploration," briefing given at the Fourth Advanced Space Propulsion Research Workshop, held at the Jet Propulsion Laboratory, Pasadena, CA, April 5-7, 1993.

56. Myrabo, L.N., Limmer, J.D., and Rahn, M., "Transatmospheric Microwave-Boosted Lightcraft with Inflatable Tensile Structures," *Space Manufacturing 9: The High Frontier - Accession, Development and Utilization,* Proceedings of the Eleventh SSI-Princeton Conference, Princeton, NJ, published by AIAA, Sept. 1993, pp. 46-70.

57. Myrabo, L.N., "Microwave-Boosted Spacecraft," briefing given at the Fifth Advanced Space Propulsion Research Workshop, at the Jet Propulsion Laboratory, Pasadena, CA, May 18-20, 1993.

58. Benford, J. and Myrabo, L.N., "Propulsion of Small Launch Vehicles Using High Power Millimeter Waves," *OE/LASE Proceedings,* SPIE Vol. 2154-23, SPIE, Bellingham, WA, Jan. 1994, pp. 198-219.

59. Myrabo, L.N. and Benford, J., "Propulsion of Microspacecraft to Orbit Using High Power Millimeter Waves," *Proceedings of the 29th Microwave Power Symposium,* ISSN: 1070-0129, International Microwave Power Institute, Monassas, VA, 25-27 July 1994, pp. 116-120.

60. Myrabo, L.N., "Microwave Launches of Small Payloads," *SSI Update – The High Frontier Newsletter,* Vol. 20, Issue 6, Nov./Dec. 1994. pp. 1-4.

61. Myrabo, L.N., "Hyper-Energetic Manned Aerospacecraft Propelled by Intense Pulsed Microwave Power Beam," *OE/LASE Proceedings,* SPIE Vol. 2557-26, SPIE, Bellingham, WA, 1995.

62. Myrabo, L.N., "Technological Evolution of Manned Microwave-Boosted Spacecraft with Hyper-Energetic Transatmospheric Performance," *Space Manufacturing 10: Pathways to the High Frontier,* Proceedings of the 12th SSI-Princeton Conference (May 4-7, 1995), Space Studies Institute, Princeton, NJ, 1995, pp. 71-87.

63. Myrabo, L.N., Seo, J., Head, D., Marsh, J., and Cassenti, B.N., "Thermal Management System for an Ultralight Microwave Propelled Transatmospheric Vehicle," AIAA 94-2924, 30th AIAA / ASME / SAE / ASEE Joint Propulsion Conference, Indianapolis, IN, 27-29 June 1994.

64. "Solar Rocket System Concept Analysis," F.G. Etheridge – study manager, Final Report (Feb. 1979-Sept. 1979), AFRPL TR-79-79, by Rockwell International for Air Force Rocket Propulsion Laboratory, Edwards AFB, CA, Dec. 1979.

65. Thomas, M. and Veil, G., "Highly Accurate Inflatable Reflectors," Final Report (Sept 1983- Feb. 1984), AFRPL TR-84-021, by L'Garde, Inc. for Air Force Rocket Propulsion Laboratory, Edwards AFB, CA, May 1984.

66. Shoji, J.M., "Solar Rocket Component Study," Final Report (Oct. 1980-June 1984), AFRPL TR-84-057, by Rockwell International, Rocketdyne Div., for Air Force Rocket Propulsion Laboratory, Edwards AFB, CA, Feb. 1985.

67. Pettit, R.G., "Geometric and Optical Properties of Inflatable Point-Focusing Reflectors," M.S. Thesis, Brigham Young University, UT, Aug. 1985.

68. Bradford, R., "Research and Development of Large, Highly Accurate Inflatable Reflectors," Final Report (Sept. 1984-Sept. 1986), AFRPL TR-86-074, by SRS Technologies for Air Force Rocket Propulsion Laboratory, Edwards AFB, CA, Feb. 1987.

69. Veal, G.R., "Highly Accurate Inflatable Reflectors, Phase II," Final Report (Oct. 1984-Oct. 1986), AFRPL TR-86-089, by L'Garde, Inc. for Air Force Rocket Propulsion Laboratory, Edwards AFB, CA, March 1987.

70. Gierow, P.A. and Myers, R.S., "Thin Film Creep-Forming for Solar Propulsion Applications," Final Report (June 1986

REFERENCES

Dec. 1986), AFAL TR-87-009, by SRS Technologies for Air Force Astronautics Laboratory, August 1987.

71. Williams, G.T., "Inflatable Solar Concentrator Flight Test Experiment," Final Report (June 1986-Jan. 1987), AFAL TR-87-035, by L'Garde, Inc. for Air Force Astronautics Laboratory, Edwards AFB, CA, Aug. 1987.

72. Renk, K., Jacques, Y., Felts, C., and Chovit, A., "Holographic Solar Energy Concentrators for Solar Thermal Rocket Engines," Final Report (June 1987-March 1988), AFAL TR-88-025, by NTS Engineering for Air Force Astronautics Laboratory, Edwards AFB, CA May 1988.

73. Thompsen, C., Dalbey, S., Friese, G., Zaccardelli, E., and McDaniel, B., "Concentrator Flight Test Experiment," Final Report (Sept. 1987-May 1990), by L'Garde, Inc. for Phillips Laboratory, Edwards AFB, CA, May 1991.

74. Otto, F., editor, *Tensile Structures*, The MIT Press, Cambridge, MA, 1973.

75. Myrabo, L.N., and Cassenti, B.N., "Structural Analysis of a Transatmospheric Beam-Boosted, Super-Pressure Aerospacecraft," AIAA #96-2893, AIAA, Reston, VA, 1996.

76. Myrabo, L.N. and Cassenti, B.N., "Three-Dimensional Structural Analysis of a Transatmospheric, Beam-Boosted Super-Pressure Aerospacecraft," AIAA # 973211, AIAA Joint Propulsion Conference, AIAA, Reston, VA, July 7-9, 1997.

77. Cassenti, B.N., and Myrabo, L.N., "Three Dimensional Transient Structural Response of a Microwave Boosted Lightcraft," AIAA 98-3742, 34th AIAA Joint Propulsion Conf. & Exhibit, Cleveland, OH, 13-15 July 1998.

78. Myrabo, L.N. and Cassenti, B.N., "Transient Structural Analysis of a 20-m Diameter, Hyper-Energetic Lightcraft: Part I - Axisymmetric Model," *Beamed Energy Propulsion III*, 3rd International Symposium on Beamed Energy Propulsion, AIP Conference Proceedings Vol. 766, American Institute of Physics, NY, 2005, pp. 320-333.

79. Myrabo, L.N. and Cassenti, B.N.," Transient Structural Analysis of a 20-m Diameter, Hyper-Energetic Lightcraft: Part 2 –Three Dimensional Model," *Beamed Energy Propulsion III*, 3rd International Symposium on Beamed Energy Propulsion, AIP Conference Proceedings Vol. 766, American Institute of Physics, NY, 2005, pp. 333-346.

80. Poole, E., "Structural Assessment of a Microwave Lightcraft Conceptual Design," ED22-02-093, Strength Analysis Group, NASA Marshall Space Flight Center, Huntsville, AL, 5 Sept. 2002.

81. Myrabo, L.N., Messitt, D.G. and Mead, F.B., Jr. "Ground and Flight Tests of a Laser Propelled Vehicle," AIAA 98-1001, 36th AIAA Aerospace Sciences Meeting & Exhibit, Reno, NV, 12-15 Jan. 1998.

82. Mead, F.B., Jr., Myrabo, L.N., and Messitt, D.G., "Flight Experiments and Evolutionary Development of a Laser Propelled, Transatmospheric Vehicle," Space technology & Applications International Forum (STAIF-98), Albuquerque, NM, 2529 Jan. 1998.

83. Mead, F.B., Jr., Myrabo, L.N., and Messitt, D.G., "Flight and Ground Tests of a Laser-Boosted Vehicle," AIAA 98-3735, 34th AIAA Joint Propulsion Conf. & Exhibit, Cleveland, OH, 13-15 July 1998.

84. Myrabo, L.N., Mead, F.B., Jr. and Messitt, D.G.,"Flight and Ground Experiments with a Laser-Boosted Lightcraft," International Conference on Lasers "98, Tucson, AZ, 711 Dec. 1998.

85. Wang, T.-S., Cheng, Y.-S., Liu, J., Myrabo, L.N., and Mead, F.B., Jr., "Performance Modeling of an Experimental Laser Propelled Lightcraft," AIAA 2000-2347, 31st AIAA Plasmadynamics and Lasers Conference, 19-22 June 2000, Denver, CO.

86. Messitt, D.G., Myrabo, L.N., and Mead, F.G., Jr., "Laser Initiated Blast Wave for Launch Vehicle Propulsion," AIAA 2000-3848, 36th AIAA Joint Propulsion Conference, 16-19 July 2000, Huntsville, AL.

87. Myrabo, L.N., "World Record flights of a Beam Riding Rocket Lightcraft: Demonstration of "Disruptive" Propulsion

Technology," AIAA Paper 2001-3798, 37th AIAA Joint Propulsion Conference, 8-11 July 2001, Salt Lake City, Utah.
88. Wang, T.-S., Cheng, Y.-S., Liu, J., Myrabo, L.N., and Mead, F.B., Jr., "Advanced Performance Modeling of Experimental Laser Lightcrafts," *Journal of Propulsion and Power*, Nov.-Dec. 2002, pp. 1129-1138.
89. Myrabo, L.N., "Brief History of the Lightcraft Technology Demonstrator (LTD) Project," *Beamed Energy Propulsion I*, 1st International Symposium on Beamed Energy Propulsion, AIP Conference Proceedings Vol. 664, American Institute of Physics, NY, 2003, pp. 49-60.
90. Libeau, M.A., Myrabo, L.N., and Filippelli, M., "Combined Experimental & Theoretical Flight Dynamics Investigation of a Laser-Propelled Vehicle," *Beamed Energy Propulsion I*, 1st International Symposium on Beamed Energy Propulsion, AIP Conference Proceedings Vol. 664, American Institute of Physics, NY, 2003, pp. 125-137.
91. Libeau, M. and Myrabo, L.N., "Off-Axis and Angular Impulse Measurements on a Lightcraft Engine," *Beamed Energy Propulsion III*, 3rd International Symposium on Beamed Energy Propulsion, AIP Conference Proceedings Vol. 766, American Institute of Physics, NY, 2005, pp. 166-177.
92. Nakagawa, T., Mihara, Y., Matsui, M., Komurasaki, K, Takahashi, K, Sakamoto, K, and Imai, T., "A Microwave Beaming Thruster Powered by a 1 MW Microwave," AIAA 2003-4430, 39th AIAA Joint Propulsion Conference, 20-23 July 2003.
93. Way, S., "Propulsion of Submarines by Lorentz Forces in the Surrounding Sea," ASME Paper No. 64-WA/ENER-7, ASME Winter Annual Meeting, Nov. 29 - Dec. 4, 1964, American Society of Mechanical Engineers, NY.
94. Way, S. and Devlin, C., "Prospects for the Electromagnetic Submarine," AIAA Paper No. 76-432, AIAA 3rd Joint Propulsion Specialist Conference, July 17-21, 1967.
95. Way, S., "Electromagnetic Propulsion for Cargo Submarines," *Journal of Hydronautics*, Vol. 2, No. 2, April 1968, pp. 49-57.
96. Way, S., "Research Submarines with Minimal Ocean Disturbance," SAE Paper 690028, presented at SAE Conf., 13-17 Jan. 1969.
97. L.N. Myrabo, and Yu.P. Raizer, "Laser-Induced Air Spike for Advanced Transatmospheric Vehicles," AIAA Paper No. 94-2451, 25th AIAA Plasmadynamaics and Lasers Conference, Colorado Springs, CO, 20-23 June 1994.
98. Myrabo, L.N., Head, D.R., Seo, J., Marsh, J.J. and Cassenti, B.N., "Estimation of Gasdynamic and Heat Transfer Conditions within Laser-Induced Air Spikes," AIAA Paper 96-0317, AIAA, Reston, VA, 1996.
99. Marsh, J.J., Myrabo, L.N., Messitt, D.G., Nagamatsu, H.T., and Raizer, Yu.P., "Experimental Investigation of the Hypersonic "Air Spike" Inlet at Mach 10," AIAA # 96-0721, AIAA, Reston, VA, 1996.
100. Diaz, E., Toro, P.G.P., Myrabo, L.N., Nagamatsu, H.T., and Messitt, D.G., "Experimental Pressure Survey of the Hypersonic Air Spike Inlet at Mach 10," AIAA # 96-3143, AIAA, Reston, VA, 1996.
101. Toro, P.G.P., Myrabo, L.N., and Nagamatsu, H.T., "Experimental Investigation of Hypersonic "Directed-Engery Air Spike" at Mach 10-20," AIAA # 97-0795, AIAA, Reston, VA, Jan. 6-10, 1997.
102. Toro, P.G.P., Myrabo, L.N., and Nagamatsu, H.T., "Directed-Energy Air Spike Inlet at Mach 10 with 15-25 kW Arc Power," COBEM "97 Conference, Brazil, Dec. 1997.
103. Toro, P.G.P., Myrabo, L.N.,and Nagamatsu, H.T. Pressure Investigation of the Hypersonic Directed-Energy-Air-Spike Inlet at Mach 10 with Arc Power up to 75 kW, AIAA # 98-0991, 36th Aerospace Sciences Meeting and Exhibit, AIAA, Reston, VA, Jan 12-15, 1998.
104. Toro, P.G.P., "Experimental Pressure and Heat Transfer Investigation Over a "Directed-Energy Air Spike" Inlet at Flow Mach Numbers of 10 to 20, Stagnation Temperature of 1000 K, and Arc Power up

REFERENCES

to 127 kW," Ph.D. Thesis, Rensselaer Polytechnic Institute, Troy, NY, 11 June 1998.

105. Myrabo, L.N., Raizer, Yu.P., and Shneider, M.N., "The calculation and similarity theory of the experiment simulating the air-spike effect in hypersonic aerodynamics," *High Temperature*, v.36, No. 2, 1998, pp. 287-292.

106. Toro, P.G.P., Nagamatsu, H.T., Minucci, M.A.S., and Myrabo, L.N., "Experimental Pressure Investigation of a "Directed-Energy Air Spike" Inlet at Mach 10," AIAA 99-2843, 35th AIAA / ASME / SAE / ASEE Joint Propulsion Conference, 20-23 June 1999.

107. Toro, P.G.P., Nagamatsu, H.T., Myrabo, L.N., and Minucci, M.A.S., "Experimental Heat Transfer Investigation of a "Directed-Energy Air Spike" Inlet at Mach 10," AIAA 99-2844, 35th AIAA / ASME / SAE / ASEE Joint Propulsion Conference, 20-23 June 1999.

108. Toro, P.G.P., Minucci, M.A.S., Myrabo, L.N., and Nagamatsu, H.T., "Experimental Heat Transfer and Pressure Investigations of a "Double Apollo Disc" at Mach 10," Proceedings of COBEM 99, Nov. 1999, Brazil.

109. Minucci, M.A.S., Bracken, R.M., Nagamatsu, H.T., Myrabo, L.N., and Shanahan, K.J., "Experimental Investigation of an Electric Arc Simulated Air Spike in Hypersonic Flow," AIAA 00-0715, 38th Aerospace Sciences Meeting & Exhibit, 10-13 Jan. 2000, Reno, NV.

110. Toro, P.G.P., Minucci, M.A.S., Korzenowski, H., Nagamatsu, H.T., and Myrabo, L.N., "Numerical and Experimental Investigation of "Double Apollo Disc" at Mach 10," AIAA 2000-3490, 36th AIAA Joint Propulsion Conference, 16-19 July 2000, Huntsville, AL.

111. Minucci, M.A.S., Toro, P.G.P., Chanes, J.B., Jr., Ramos, A.G., Pereira, A.L., Nagamatsu, H.T., and Myrabo, L.N., "Experimental Investigation of a Laser-Supported Directed-Energy "Air Spike in Hypersonic Flow Preliminary Results," 7th International Workshop on Shock Tube Technology, Sept. 18-20, 2000, Long Island - NY.

112. Minucci, M.A.S., Toro, P.G.P., Chanes, J.B., Jr., Pereira, A.L., Nagamatsu, H.T., and Myrabo, L.N., "Investigation of a Laser Supported Directed-Energy "Air Spike" at Mach 10 Flow," AIAA 2001-0641, 39th AIAA Aerospace Sciences Meeting and Exhibit, 8-11 January 2001, Reno, NV.

113. Toro, P.G.P., Minucci, M.A.S., Nagamatsu, H.T., and Myrabo, L.N., "Experimental Pressure Investigation of Blunt Body at Mach 10," AIAA-2001-0644, 39th AIAA Aerospace Sciences Meeting and Exhibit, 8-11 January 2001, Reno, NV; see also: Toro, P.G.P., Minucci, M.A.S., Korzenowski, H., Nagamatsu, H.T., and Myrabo, L.N., "Numerical and Experimental Investigation of "Double Apollo Disc" from Mach 10 to 20," AIAA 2001-0567, 39th AIAA Aerospace Sciences Meeting and Exhibit, 8-11 January 2001, Reno, NV.

114. Bracken, R., Myrabo, L.N., Nagamatsu, H.T., and Meloney, E., "Experimental Investigations of an Electric Arc "Air Spike" With and Without Blunt Body in Hypersonic flow," AIAA-2001-0796, 9th AIAA Aerospace Sciences Meeting and Exhibit, 8-11 January 2001, Reno, NV.

115. Myrabo, L.N., Raizer, Yu.P., Shneider, M.N. and Bracken, R., "Reduction of Drag and Energy Consumption during Energy Release Preceding a Blunt Body in Supersonic Flow," *High Temperature*, Vol. 42, No. 6, 2004, pp. 901-910.

116. Hartley, C.S., Portwood, T.W., Filippelli, M.V., Myrabo, L.N., Nagamatsu, H.T., Shneider, M.N., and Razier, Yu.P., "Experimental and Computational Investigation of Drag Reduction by Electric-Arc Airspikes at Mach 10," *Beamed Energy Propulsion III*, 3rd International Symposium on Beamed Energy Propulsion, AIP Conference Proceedings Vol. 766, American Institute of Physics, NY, 2005, pp. 499-513.

117. Myrabo, L.N., Raizer, Yu.P., and Shneider, M.N., "Drag and Total Power Reduction for Artificial Heat Input in Front of Hypersonic Blunt Bodies," *Beamed Energy Propulsion III*, 3rd International Symposium on Beamed Energy Propulsion, AIP Conference Proceedings Vol. 766,

American Institute of Physics, NY, 2005, pp. 485-498.
118. C. Misiewicz, C., Myrabo, L.N., Shneider, M.N., and Raizer, Yu.P., "Combined Experimental and Numerical Investigation of Electric-Arc Airspikes For Blunt Body at Mach 3," *Beamed Energy Propulsion III*, 3rd International Symposium on Beamed Energy Propulsion, AIP Conference Proceedings Vol. 766, American Institute of Physics, NY, 2005, pp. 528-541.
119. Kantrowitz, A. "Introducing Magnetohydrodynamics," *Astronautics*, Oct. 1958, pp. 18-20 and 74-77,
120. Moder, J.P., Myrabo, L.N. and Kaminski, D.A., "Analysis and Design of an Ultrahigh Temperature Hydrogen-Fueled MHD Generator," *Journal of Propulsion and Power*, Vol. 9, No. 5, Sept.-Oct. 1993, pp. 739-748.
121. Brown, W.C., "The History of Power Transmission by Radio Waves," *IEEE Transactions on Microwave Theory and Techniques*, Vol. MTT-32, No. 9, Sept. 1984, pp 1230-1242.
122. Brown, W.C., "Rectenna Technology Program: Ultra Light 2.45 GHz Rectenna and 20 GHz Rectenna," PT-6902, NASA CR-179558, Final Report under Contract NAS3-22764, by Raytheon Company, for NASA Lewis Research Center, 11 March 1987.
123. Brown, W.C., "A Transportronic Solution to the Problem of Interorbital Tansportation," PT-7452, NASA CR-191152, Final Report under Contract NAS3-25066, by Raytheon Company, for NASA Lewis Research Center, July 1992.
124. Rogers, M. and Springen, K., "Can a Plane Fly Forever? An aviation first: powered flight without fuel," *Newsweek*, 28 Sept. 1987, pp. 42-43.
125. Fisher, A., "Secret of Beam-Power Plane," *Popular Science*, Jan. 1988, pp. 62-65, 106-107.
126. Alden, A., "Dual Polarization Rectenna," *Canadian Research*, Feb. 1988, p. 26; see also, Martin, J., "On a Wing and a Microwave," pp. 24-25 (same issue); and, Schlesak, J., "The Airplane Design, p. 27 (same issue).
127. DeLaurier, J., Gagnon, B., Wong, J., Williams, R., Hayball, C., and Advani, S., "Progress Report on Research for Stationary High Altitude Relay Platform (SHARP)," *Unmanned Systems*, Vol. 5, No. 3, Winter 1987, pp. 26-46.
128. Peillisch, R., "Silicon carbide takes the heat." *Aerospace America*, Oct. 1994, pp. 28-37.
129. P. Koert, P. and Cha, J.T., "35 GHz Rectenna Development," *Proceedings of the 1st Annual Wireless Power Transmission Conference*, held in San Antonio Texas on 23-25 Feb. 1993, avail. from Center for Space Power, Texas A&M University, College Station, Texas, pp. 457-466.
130. Koert, P., "Millimeter Wave Technology for Space Power Beaming." IEEE *Transactions on Microwave Theory and Techniques*, Vol. 40, No. 6, June 1992, pp. 251-258.
131. Yoo, T.W. and Chang, K., "Theoretical and Experimental Developments of 10 and 35 GHz," *IEEE Transactions on Microwave Theory and Techniques*, Vol. 40, No. 6, June 1992.
132. Genuario, R. and Koert, P., "High Power UHF Rectenna for Energy Recovery in the HCRF System," *Microwave and Particle Beams II*, SPIE Proceedings Vol. 1407, 1991, pp. 553-565.
133. Alden, A., "A 35 GHz Extremely High Power Rectenna for the Microwave Lightcraft," CRC Contract Report No. CRC-VPRS-00-03, Communications Research Centre, Ottawa, Canada, March 2001.
134. Alden, A., "A 35 GHz Extremely High Power Rectenna for the Microwave Lightcraft," *Beamed Energy Propulsion I*, 1st International Symposium on Beamed Energy Propulsion, AIP Conference Proceedings Vol. 664, American Institute of Physics, NY, 2003, pp. 292-300.
135. Cummings, T., Janssen, J, Karnesky, J., Laks, D., Santillo, M., Strause, B., Myrabo, L.N., Alden, A., Bouliane, P., and Zhang, M., "6 GHz Microwave Power-Beaming Demonstration with 6-kV Rectenna and Ion-Breeze Thruster," *Beamed Energy Propulsion II*, 2nd International Beamed Energy Propulsion Symposium, AIP Conference Proceedings Vol. 702,

REFERENCES

American Institute of Physics, NY, 2004, pp. 430-442.

136. Alden, A., Bouliane, P., and Zhang, M., "Some Recent Developments in Wireless Power Transmission to Micro Air Vehicles," *Beamed Energy Propulsion III*, 3rd International Symposium on Beamed Energy Propulsion, AIP Conference Proceedings Vol. 766, American Institute of Physics, NY, 2005, pp. 303-307.

137. Alden, A., Bouliane, P., and Zhang, M., "The Use of Semiconductor Devices in Wireless Power Transmission," *Proceedings of the European Solid State Device Research Conference*, Sept. 2005.

138. Vanke, V.A., Zaporozhets, A.A., and Rachnikov, A.V., "Antenna Synthesis for the SPS Microwave Transmission System," *Proceedings of the 2nd International Symposium: SPS 91*, held in Paris on 27-29 Aug. 1991, pp. 528-534.

139. Potter, S., "Specialized Phased-Array Antenna Patterns for Wireless Power and Information Transmission," *Space Manufacturing 10: Pathways to the High Frontier*, Proceedings of the Twelfth SSI-Princeton Conference, Space Studies Institute, Princeton, NJ, 1995, pp. 371-379.

140. Anderson, J.L., "Horizon Mission Methodology: A Tool for the Study of Technolology Innnovation and New Paradigms," AIAA 93-1134, AIAA / AHS / ASEE Aerospace Design Conference, 1619 Feb. 1993.

141. Anderson, J.L., "Back from the Future," *Aerospace America*, AIAA, Wash., D.C., Nov. 1993.

142. Anderson, J.L., Rather, J.D.G., and Powell, J.R., "Beamed Energy for Fast Space Transport," AIAA 96-2785, 32nd AIAA / ASME / SAE / ASEE Joint Propulsion Conference, 1-3 July 1996.

143. O'Neill, G.K., *The High Frontier – Human Colonies in Space*, 3rd Edition, Apogee Books (in conjunction with the Space Studies Institute and the Space Frontier Foundation), Ontario, Canada, 2000.

144. *Pioneering the Space Frontier – An Exciting Vision of Our Next Fifty Years in Space*, The Report of the National Commission on Space, Bantam Books, NY, 1986.

145. Clarke, A.C., "Electromagnetic Launching as a Major Contribution to Space Flight," *Journal of the British Interplanetary Society*, Vol. 9, No. 6, 1950, pp. 261-267.

146. Nordley, G.D., "Stationkeeping with Two-Way Electromagnetic Launchers," *J. Propulsion*, Vol. 10, No. 6: Technical Notes, pp. 912-913.

147. Forward, R.L., Conley, R., Stanek, C., and Ramsey, W., "The Cable Catapult: Putting It There and Keeping It There," AIAA 92-3077, July 1992.

148. Borowski, S.K., Corban, R.R., Culver, D.W., Bulman, M.J., and McIlwain, M.C., "A Revolutionary Lunar Space Transportation System Architecture Using Extraterrestrial LOX-Augmented NTR Propulsion," AIAA 94-3343, 30th AIAA / SAE / ASME / ASEE Joint Propulsion Conference, 27-29 June 1994.

149. Fraser, T.M., "Human Response to Sustained Acceleration," NASA-SP-103, 1966.

150. Reilly, J.P., "Propagation of High Power Microwave Beams," W.J. Schafer Associates, Inc., Wakefield, MA, August 1979.

151. Benford, J. and Myrabo, L.N., "Propulsion of Small Launch Vehicles Using High Power Millimeter Waves," *OE/LASE "94*, SPIE Proceedings Vol. 2154, International Society for Optical Engineering, Bellingham, WA, Jan. 1994, pp. 198-219.

152. Weiting, T.J., et.al., "Plasma Mediated Coupling to Metals at 35 GHz/8.6 cm and 9.3 GHz/3.2 mm," *Proceedings of the 2nd National Conference on High Power Microwave Technology*, Harry Diamond Laboratories, Adelphi, MD, 1-3 March 1983.

ACRONYMS

AC	alternating current
ACEP	advanced crew escape pod
ADS	auto-destruct system
AEC	Atomic Energy Commission
AERL	Avco Everett Research Laboratory
AFB	Air Force Base
ANP	Aircraft Nuclear Propulsion Program
AP	Associated Press
ATA	alternative technical approaches
BCS	beam control system
BCEP	bi-conic escape pod
BEP	beamed energy propulsion
BIS	British Interplanetary Society
BTC	breakthrough concepts
BWG	beam waveguide
CCN	cloud condensation nuclei
CM	Apollo command module (reentry capsule)
CMC	ceramic matrix composites
CONUS	continental United States
CPR	cardiopulmonary resuscitation
CRV	crew return vehicle
CTO	communications / tactical officer
CVD	chemical vapor deposition
DC	direct current
DE	directed energy
DEAS	directed energy airspike
DET	directed energy transmission
DFS	dynamic flight simulator
E-Beam	electron beam
ECCS	environmental control and cooling system
ECS	environmental control system
EVA	extravehicular activity
FM	Frisbee mode
G	gravitational constant (9.80665 m/s^2)
GBM	ground-based microwave (power station)
GEO	geostationary orbit; geosynchronous orbit
GLC	g-induced loss of consciousness
GLOC	g-induced loss of consciousness
GLONASS	global navigation satellite system
GPS	global positioning system
HeliOx	helium and oxygen mixture (breathable)
Hg	Mercury
HIT	Human Interface Technology Laboratory, Washington Technology Center
HMM	Horizon Mission Methodology
HPFTP	high pressure fuel turbopump in SSME rocket engine
HPNS	high pressure nervous syndrome
IFF	identification friend or foe
ISS	International Space Station
L1	security level 1 (all hands to battle stations)
L2	security level 2 (normal operation)
LASL	Los Alamos Scientific Laboratory (now, Los Alamos National Laboratory)
LED	light emitting diodes
LEO	low Earth orbit
LIDAR	laser imaging, detection and ranging
LINAC	linear accelerator
LNA	low noise amplifier
LMA	laryngeal mask airway
LMO	low Mars Orbit
LOI	lunar orbit insertion
LOM	low observables mode
LTI	Lightcraft Technologies International
LV	liquid ventilation
MAGLEV	magnetic levitation
MBS	magnetic bearing system
MEP	microwave-to-electric power converter
MHD	magnetohydrodynamics
MIRV	multiple independently (targeted) reentry vehicle
MLO	mission leading officer
MM_Sc	metal matrix superconductor
MMSC	metal matrix superconductor
MMU	manned maneuvering unit
MRI	magnetic resonance imaging
MRS	microwave relay satellite
MSC	magnetic station-keeping control
MSO	Mars synchronous orbit

ACRONYMS

NADC	Naval Air Development Center	SBM	space-based microwave (power station)
NASA	National Aeronautics and Space Administration	SBR	space based relay
		SC	superconductor
NEPA	Nuclear Energy for Propulsion of Aircraft program	SCO	second in command officer
		SEAL	sea, air, and land
NEO	near Earth object	SiC	silicon carbide
NSPO	NASA Space Nuclear Propulsion Office	SiC/SiC	silicon carbide composite material
		SIS	structural integrity system
OMS	orbital maneuvering system engines	SMES	superconducting magnetic energy storage
PBF	pulmonary blood flow		
PD	pulsed detonation	SPRS	space plasma radiation shield
PDE	pulsed detonation engine	SSI	Space Studies Institute (Princeton, NJ)
PDS	planetary defense system	SSME	space shuttle main engine
PFC	perfluorocarbon	STS	shuttle transportation system
PFOB	perfluoro-octyl-bromide	TEI	trans-earth injection
PLV	partial liquid ventilation	TLI	trans-lunar injection
PLWS	pulsed laser weapon system	TMS	thermal management system
PMS	plenum management system	US, USA	United States of America
PSW	personal stun weapon	USAF	United States Air Force
PV	photovoltaic	USSC	United States Space Command
Q	dynamic pressure	VCN	virtual computer network
Rectenna	rectifying antenna	VR	virtual reality
RO	reverse osmosis	VRD	virtual retinal display
RPM	revolutions per minute	WAL	Westinghouse Astronuclear Laboratory
SAR	search and rescue		
SARSAT	search and rescue satellite	WPS	water purification system
SAU	space activity unitard (uniform)		

INDEX

1/6th g artificial gravity, 17, 97, 105, 106, 215
2025 Space Command Mission, 34, 35, 127, 129
2-way EML: facilities, 240, 243
    booster stations, 243
Abduction, 100, 121, 122, 128, 139-40
    ground transport vehicles, 139
    small military targets, 128
Acceleration, 16, 27, 30, 35, 39, 49, 52, 63, 66, 68, 72-73, 100, 127, 128-29, 132-34, 147, 154, 188, 195-99
    transverse supine, 196-97
    transverse prone, 196-97
    negative, 196
Acquisition of Lightcraft fleet, 15
Action-at-a-distance, 50, 52
Active denial, 100-01
    microwave non-lethal weapon, 139
    microwave weapon system, 17
Active-structure systems, intelligent, 29
Advanced Crew Escape Pod (ACEP), project at RPI, 9, 87, 154
Aerial crane, 121
Aerial de-boarding program, 95
Aerobraking maneuver, 209
Aerobraking, 10, 119, 150-52, 225
    atmospheric re-entry, 59
    controlled deceleration, 150
    magnetic, 16, 151-52, 221
Aerocapture, 150
    biconic, 209
    Grumman, 209
Aerodynamic drag, 9, 52, 63, 67, 76, 168, 171, 177, 185, 229, 230
Aerodynamic flap, 116-17, 155, 209
    retractable, 116
Aerodynamic lift, 64, 74, 78, 83, 130, 132, 133, 142
Aerojet General, 214
Aerospace superiority, 139
    role, 17
    roles, 15
Aerospike engine, 51
Aerothermodynamics, 8, 37, 52, 53, 143, 229
    active control, 8, 59

    massive scale, 8, 143
    hypersonic, 8, 37
AFO 141, 200, 263
Air Force Office of Scientific Research (AFOSR), 10, 252, 254
Air mattress, 110
Air plasma generation, 58-59, 69, 141
Airbreathing microwave PDE, 228
Aircraft carrier, 170, 174, 175, 214, 240
    Nimitz-class, 170, 214
Aircraft Nuclear Propulsion (ANP), 213-14
Airlock (air lock), 29, 88, 94, 95, 97, 118, 119, 123, 125, 155, 157, 158, 159, 170, 237
    doors, 95, 158, 159
Airship, 5, 9, 16, 38, 39, 74, 88, 144, 223, 224, 225, 231, 232
    hypersonic, 5, 7, 9
Airspike, 8, 10, 24, 38, 43, 49, 50, 51, 52, 53, 55, 58, 59-60, 61, 63, 67, 68, 89, 100, 119, 121, 128-29, 132-33, 142, 147, 149-50, 152, 159, 161, 166, 229, 233
    concept, 8, 51, 59, 229
    experimental data, 8
    physics, 8, 68, 161, 229
    maintenance, 40, 59
    microwave-induced, 49, 233, 235
    production, 59-60, 65, 67, 159
    system, 43, 53
    auxiliary functions, 89, 121
Alden, Adrian, 10, 54, 161, 231-32
All weather, 15, 69, 100, 131
    capability, 15
Alliance Pharmaceutical Corporation, 200
Anderson, John, 8, 10, 238, 239
Annihilation, 49
    bow shock wave, 52
Antenna, 16, 30, 49, 54, 71, 88, 146, 150, 154, 166, 167, 173, 191, 222, 225, 227, 230
Apollo 8, 241-42
Apollo CM, 197, 216
    heat shield, 215, 216, 217, 233
Apollo Command Module (CM), 150, 181, 215
Apollo spacecraft, 215, 242
Architecture, 105, 173, 215
    computer systems, 46

INDEX

Lightcraft, 10, 115, 162
Ares V, 176, 177, 182, 191
Army Ordnance Corps, 213
Artificial cloud, 141
   curtain, 85
Artificial gravity, 87, 173, 177, 213, 222, 223
   in space, 87, 97, 103, 107, 212, 215, 232, 233, 234
   comfort zone, 105-06, 215
   generation, 17, 105-06, 184
Artificial intelligence, 16, 107, 203
Artificial lifeform, 205
   exoskelleton, 205
Assets, 89, 132, 139
   GEO, 173
   LEO, 173
Assignments, 41
   crew, 35-36
Associated Press, 211, 212, 213
Astronaut, 165, 206
   corps, 8
   rescue, 15, 100
   Apollo, 197
Atlantis Warrior, 205
Atmosphere skipping, 150
   hypersonic, 64, 65, 127, 133, 151
   boost/glide maneuvers, 64
Atmospheric models, 165
   sensors, skipping
   transmission losses, 81, 164
   transmittance, 166, 168, 244-47
   turbulence, 39, 81, 244, 247
   windows, 67, 222, 244
Atomic Energy Commission (AEC), 213, 214
Atomic reactor, 212
Attitude control, 220
   system, 87, 218
   thrusters, 116, 117, 207
Audible range limit, 33, 72
   human, 16
Auto-destruct system, 102, 127
Automatic flight system, 43
Automatic self-destruct system (ADS), 102
Automobile abduction, 100, 121, 122, 139-40
Autopilot mode, 37-38, 43
   remote-controlled, 16
Auxiliary Lightcraft systems, 32, 87-89, 121-26, 135, 156, 221
AVCO Corporation, 215

AVCO Everett Research Laboratory, 152, 214, 230
Axis of rotation, 87
Ball lightning, 141
   air plasma, 141
   decoy, 141
Balloon, 8, 107, 224-25
   hypersonic, 7, 8, 39
Beacon system, 53
   Lightcraft, 164-65, 168, 182, 188
Beam control system (BCS), 53
Beam propagation, 170, 247, 249
   near-field region, 249
   beam waist, 249
   far-field region, 249
Beam waveguide (BWG), 167
   parabolic antenna, 167
Beam weapons, 17, 65, 159
Beamed Energy Propulsion (BEP), 7-8, 38, 188, 222-23, 228
   technology, 5-6, 51, 231
Becky, Ivan, 10
Bell X-1 rocket plane, 212
Benford, James, 10, 168, 244, 245, 246, 248
Beryllium, 163
   matrix, 56
Biconic aerobraking, 208
   concept, 207
   Grumman, 10, 207, 209
Binocle laser relay satellite, 80
Bioelectronics technology, 206
BioPulmonics Incorporated, 200
Bi-propellant tanks, 88, 116-17
Black box, 38, 127-28, 139
Black program, 21
Bladder, 135-36
   liquid filled, 205
   g-suit, 206
Blink out, 68, 84
Board of Directors, SSI, 8
Boarding options, 93, 97, 115, 118, 146-47
   Lightcraft, 93, 114-15, 158, 236
Boarding ramp, 42, 115
Boost/glide, 64, 263
   trajectories, 64
   suborbital, 162
Borowski, Stan, 242
Brain-computer interaction, 204
Brayton topping cycle, 177

high temperature, 177, 178
Breakdown intensity, 71, 72, 246
   threshold, 71
Breathing, 40, 107, 112, 133, 195, 198
   positive-pressure, 112, 198
Briant, R.C., 214
Bridge officers, 34, 40
   operations, 36, 40, 41
British Interplanetary Society, 215
Brown, William, 10, 230, 231
Brownout, 72
Brush-fire, wars, 15
Bucking magnets, 51, 171, 227
   superconducting, 52, 57, 175, 117
Buoyancy, 74, 105, 132, 136, 143
   partial, 74, 144
Cables, 32-33, 39, 157-58, 174, 185, 188, 218
   fiber-optic, 43, 221
Calcium depletion problem, 177, 180, 215
Carbon-fiber filaments, 56
Carbon-fiber $MgCNi_3$ superconductors, 56, 163-64, 170
Cardiopulmonary resuscitation (CPR), 109, 195
Cardiopulmonary, system, 102
Case Study, microwave Lightcraft, 7, 8, 9
Cassenti, Brice, 10
Catapult, 122, 136, 175, 240
Catastrophic failure, 18, 30, 43, 66, 128, 139, 153, 154, 173
   structure, 31, 65, 115, 119, 153, 154
Catenary SiC curtains, 26
Cathodes, thermionic, 182
Center of mass, 87, 137, 217
Ceramic matrix composite (CMC) sandwich, 26, 32, 50, 58, 221, 225
Chain of command, general, 40
Charge clouds, ion, 77-78, 80, 144
Charge ejection, inductive, 77, 79, 82, 220
Chemical vapor deposition (CVD), 16, 28, 54, 225
Chrysler Missile Operations, 211, 213
Chrysler, 212
   Imperial parade cars, 213
   Thunderbolts, 213
   The Forward Look, 213
   cranked arrow, 213
Circular flight, loiter mode, 15, 36, 42, 74, 127, 130, 131, 140, 265
Cislunar mission, 241-42

Cislunar space, 16, 92, 211, 221
Clancy, Tom, 229
Clarke, Arthur C., 240
Classified material, DOD, 5, 47, 99, 210, 215
Cloud, 77-79, 84, 85, 127, 131
   "drilling" with laser, 85
   condensation nuclei (CCN), 137-38, 144-45, 165-66
   evaporation procedure, 145
   mining, 137-38
   seeding, 138
CMC laser armor, 175, 180, 182
Coanda effect, 217
Coatings, 81, 114, 163, 225
   self-sealing, 212, 233
Coil protection circuitry, 56-57
Cole, John, 10
Colliers Magazine, 207
Collision avoidance, 44, 89, 125, 141, 153, 154
Comfort zone, artificial gravity, 105-06, 215
Command and control systems, 7, 206
Command systems, 34-45
Communications Research Center (CRC), 230-31
Communications, 90-92
   crew-to-crew, 90
   intra-ship, 90
   ship-to-ground, 90
   ship-to-ship, 90
   personal, 91
   long-range transceiver, 92
Communications/Tactical Officer (CTO), 36, 38, 100, 109
Computational fluid dynamics (CFD), 8, 10
Computer network, systems, 29-31, 40-41, 46-48
Concentrating reflector, 16, 244
Consoles, (see also, terminals, control panels), 37, 48
   flight-deck, 35, 37, 41, 43, 46, 90
Constellations, of Lightcraft, 139-40, 171, 172
Construction chronology, 10, 18-25
   LTI-20, 19
   SPS-01, 19-21
Contamination sensor, 103-04
Continental United States (CONUS), 15, 16, 61
Control panel, 35, 41, 42
   FMS, 41
   reconfigurable, 46

INDEX

Convair/Astronautics Division of General Dynamics, 215
Cooling system, 50, 69, 71, 103, 104
    cryogenic magnets, 126, 219
    open cycle, 53, 58, 136, 138, 168, 169
    rectenna array, 53-54, 58, 117, 231
Cork-screw maneuver, 130
Corona discharge, 80, 86, 142, 156, 219
Coulomb repulsion, 77
Count_Sheep, 48
Coupling coefficient, 72-73, 74, 83
Covert operations, 15, 16, 67, 74, 100
    hover, 127, 131
    military missions, 16, 70, 100, 127
    survivability, 17
    surveillance, 15, 16, 127
Crew, 15, 17, 34, 35-39
    accommodations, 106, 232, 237
    ejection pods, 153, 155
    ejection procedures, 66
    quarters, 29, 37, 41, 73, 88, 109, 110, 221
Crew return vehicle (CRV), 174, 181
Crew support, 109-120
    functions, 7
    systems, 109-120
    medical, 109
    quarters, 110
    food rehydration, 110
    life support options, 112
Cruise propulsive modes, 64, 84, 127, 128
    subsonic, 15, 36, 128
    supersonic, 127, 128
    hypersonic, 127, 128
Cryogenic fluid transfer, 87
    refrigeration, 60, 175
Cryogens, 87
Culligan® industrial systems (zeolite), 63
    water purification, 62, 88
Cumulonimbus clouds, 144
Cumulus clouds, 77, 78, 83, 127, 131, 137, 145
Daily repeating orbit, 19, 21, 171, 185
Data links, optical, 46
Dawn of the space age, 211, 212, 213, 230
De Seversky ionocraft, 74
DEAS "wedge", 8, 9, 52, 55, 59-60, 63, 67-68, 142, 149, 166, 168, 229
Declassification, of LTI-20, 5, 19
Decompression time, 105, 112, 198-99
Decoy, air plasma, 141

Deep ocean drilling, operations, 201
Deep reconnaissance, 15
Deformations, 30, 195
    human body, 195, 197, 206
    density differences, 195
DeLaurier, James, 231
Delta IV, 171
Delta-V requirement, 117, 118, 242
    TLI, 241-242
    LOI, 242
Design evolution, of LTI-15, 232-37
Design Lineage, 18
Design specifications, 15-18
Detectors, collision avoidance, 154
Development milestones, 20-21, 24-25
Dickerson, Tom, 10
Dickinson, Richard, 10
Diffraction, 89, 165, 222, 247
Dipole field, Lightcraft, 145
Direct retinal projection, 111, 114, 203-04
Directed energy airspike (DEAS), 9, 52, 55, 59, 60, 142, 149, 166, 168, 229
    theoretical foundations, 8
Directed energy sources, 5, 169
Directed energy weapons, 38
Directed-Energy Transmission (DET), records, 169
Disabled Lightcraft, 121, 127, 135
Disc aerodynamics, 84, 142
Disintegrating Lightcraft, 153, 154, 212
Disorientation, pathologies, 195
Display panels, 42, 47-48
    screens, 46, 48
    thin-film, 46
    personal access displays, 48, 132
Divers, professional and technical, 198
Diving environment, 198
Docking bay hatch, 157
Dog tags, electronic, 92
Donald Duck, phenomenon, 199
Donut hole, central, 33, 88, 95, 122, 123, 126, 137, 153, 155, 157, 219, 233
Dorsal hatch, 94, 95
Double hull, 27, 62, 73, 103, 153, 223, 225
    actively cooled, 27
    thermal protection, 16, 62
    thin film, 58
Double-Apollo disc, 233
Dow 705 silicone oil, 174, 177, 178, 179, 180, 181

INDEX

Downed Lightcraft, 89, 121, 127, 138-39
Drogue parachute, 46, 237
Dropout, 72
Duties, crew, 35-39
Dynamic Flight Simulator, 196
Earth departure, 141, 145, 242
   hyperbolic, 242
Echo 1, 225
Eco-friendly, 265
EGPS, 46
Ehrike, Kraft, 171
Electric breakdown limit, 39, 83
Electric grid, terrestrial, 5, 168, 179, 183-84
Electric RCS rockets, 89, 116-17, 118, 188, 190
Electrical air breakdown, 59
   laser-induced, 247
   microwave-induced, 56, 244, 246
Electrocardiography, implant, 92
Electrodynamic airship, 38
   engines, 182, 185, 188
   torqueing, 217
Electroencephalography implant, 92
Electroluminescent interior surfaces, 16
Electromagnetic aircraft launch system (EMALS), 175
Electromagnetic engines, 7, 59, 222, 228-29
   forces, 50
   Lorentz, 53
Electromagnetic launcher (EML), 222, 240-44
   two-way, 240-245
   launch/capture, 240
Electromagnetic abduction, 121, 128, 139-40
Electromagnets, 221
   on-board, 221
   auxiliary functions, 221
Electron accelerators (see also, E-guns), 77, 79, 85
Electron avalanche, cascade, 59
Electron guns (E-guns), 49, 51, 58-59, 77, 79, 85
EMALS catapult, 175
Emergency de-orbit from LEO, 89
Emergency environmental control system, 103, 106-07
Emergency operations, 153-161
   introduction, 153
   crew retrieval methods, 155-158
   fire suppression, 158-59
   maglev lander use, 159
   ejection of biconic escape pods, 154-55

Emergency systems, 7
Emergency, repair, 33
EML rails, 240
EML space port, 2-way, 240, 242, 244
Emmart, Carter, 207, 208
Endoatmospheric ion drive (EID), 38-39
   diagnostic program, 39
   ion propulsion, 10
Endurance, 196-97
   voluntary, 196-97
   g-force, 196
Energy rich & mass poor, paradigm, 224
Energy-beaming infrastructure, 6, 7, 183
Entertainment utilities, 41-42
Environmental systems, 103-08
   control, 103-05
   emergency, 106-07
Ergonomics, 109
Escape pod, 34-35, 40-41, 154-55, 235, 237, 238
   architecture, 114-15, 236
   biconic, 10, 206-09
   computers, 46
   custom-fit, 115, 116, 205
   individual, 17, 63, 72, 115, 205
   geometry, 118-19, 136
EVA recovery, 121
   magnetic crane, 121
Evasive maneuvers, 84-85
Evil Eye, 202
Exoatmospheric flight, 19, 24
   vehicle, 15
Exoskeleton, 205
Expendable fluids, 30, 87, 96, 122, 133
Extinction (attenuation), 165-66, 168
   beamed-energy, 165
Extraction maneuver, 96, 97-98
Extraterrestrial threats, 5, 7
Extravehicular activity (EVA), in Space, 124-25
Eyeball retention function, 203
Eyeballs in, 195-95, 206
Eyeballs out, 195-96
Eye-tracking, 202
   device, 202
   video-based, 202
FailSafe NeuralNet, 38, 43
Fairchild Stratos Corporation, 216
Fairings, pod nose, 117
Falling-leaf, maneuver, 82-83, 127, 129-30
Faraday cage, 72

INDEX

Feedback sensors, 47
   body-position, 47
Fiber-optic, 38, 43
   cables, 43, 221
   sensor network, 31
Fighting mirror, 17, 100, 101
Figure controls, 32
Film cooling, 49, 58, 62
Filtration processes, 28, 117, 137
   electrical, 107
   mechanical, 107
Fire control: solutions, 100
Fireballs, 141
Firepower: tactical, 17, 38
   laser weapons, 100, 175
   microwave weapons, 17
Fishbowl helmet, 201
Fleet maneuvers, 139-41
Flight control, basic, 42-43
Flight deck, bridge, 34, 36, 46
Flight dynamics: aerobraking, 150-52
   hypersonic regime, 149-50
   introduction, 142-43
   subsonic regime, 143-45
   supersonic regime, 146-49
   re-entry, 150, 152
Flight Engineer/ (pilot crew-member), 36, 38-39, 42, 107, 109, 124
Flight management functions: autonomous guidance, 37, 44
   re-entry, 46
Flight Management System (FMS), computer, 16, 34, 40
Flight maneuvers: circular-flight loiter, 130
   flip, 82, 85, 129, 130, 131
   ion-propulsion, 15, 38, 74, 76, 130, 132, 134, 142, 144, 145, 157
   low-observables mode, 127, 130-31, 134
   pendulum maneuver, 127, 129-30
   pitch and roll, 70, 129, 146, 150
   spin stabilization, 129, 177
Flight Medical Officer (FMO), 36, 39, 47, 109
Flight modes: axial, 50-51, 57, 60, 63-64, 127, 129, 133-34, 142, 147
   high-performance, 36, 37, 38, 40-41, 46, 63, 65, 67, 72, 133
   lateral, 49, 51, 59, 61, 63-65, 67-68, 70, 74, 76, 81, 84, 86, 97, 127, 128-29, 142-43, 147
   vertical, 15, 49, 63-65, 127

Flight operations: introduction, 127
   flight maneuvers, 129-35
   flight modes, 128-29
   mission types, 127-28
   takeoff, landing, 135-36
Flight path: circular-arc, 64-65
   planning, 37
Flight principles, 15
Flight stability and control, 143, 150, 217
Flight tests, 18-19, 84, 166, 193
   tethered, 18
Flight training, 15
Float-out: ceremony, 5
   declassification, 19
   unveiling, 5
Flow control, 212, 229, 232
   active, 142-43, 217
   external, 146
Fly-by-light, 38, 43
Flying bed springs, 74
Flying wing, 212
FMS: autopilot control, 37-38
   console, 35, 41
   craft integrity, 73
   diagnostic programs, 39
   maintenance, 40
   monitoring, 37, 40, 43, 44, 45, 72, 73, 138
   remote viewing system, 35
Food hydration, 87
Foot pad, inflatable bladders, 135
Forward, Robert L., 240
Fossil fuel supplies, 7-8
Frassanito & Associates, 10, 183
Free electron laser (FEL), 80, 84, 168-69, 173, 180-81
Freemann, Fred, 207
French and Levy, 218-20
Frisbee mode (FM), 49, 59, 61, 74, 84, 86, 97, 103, 142, 146, 217
Full suit, 202
Functional residual capacity (FRC), 200
Furman, Tom, 203
Future of flight, 8, 263
G forces: hyper-g, 195
   low-g, 195
Gallium indium phosphide, 33
   PV array material, 80
Gas-bag, impact protection system, 16
Gear extension actuators, 135

Gemini, 150
General Electric, 214, 215
Geomagnetic torqueing, 143, 144, 145, 182, 227
German V1 "Buzz Bomb," 227
G-force environment, total, 196
G-induced loss of consciousness (GLOC or GLC), 195, 196
Giragosian and Hoffman, 216
Glaser, Peter, 10, 182
Global autonomous guidance, 37, 44
   hot spots, 139
Global Positioning System (GPS), 37, 44, 89, 165
GLOC, detection, 196
GLONASS2, 46
Government and industry collaboration, 6
GPS-INS navigation, 37, 44
Grand Tour, solar system, 243
Gross, Dan, 10, 163, 176
Grumman, 10, 207, 209
G-suits, 205, 207, 223, 237, 244
   liquid-immersion, 197, 205, 207
G-tolerance, 35, 195-209, 244
Guidance and navigation, 44-45
Guidance, global autonomous, 37, 44
Gyration radii, 106
Gyroscopic stabilization, 97, 123, 127, 129, 143
Gyrostabilization, 143-44
Gyrotrons, 164, 166-69, 180, 182, 184
Habitat modules, 175
   inflatables, 182
Handbook, pre-flight briefing, 7
Hand-pad, 46
Hardpoints, 88
Hatch, 42, 88, 94, 97, 114, 116-18, 123, 125, 137, 156-57
Hazardous waste, 107
Headhunter tracking system, 202
Heat transfer, 28, 49, 58-61, 63, 67, 129, 147, 149, 151, 230, 232
Heavy-lift launch vehicles (HLLV), 176
Heimlich maneuver, 109
Heinlein, Robert A., 215
Helicopter, 82, 128, 144, 230
   light, 224
   ultralight, 77
HeliOx: coolant cycle, 28
   coolant, 27-28, 33, 61-62, 69, 72, 103
   helium-oxygen mixture, 105

HeliOx plenum management system (PMS), 103-04
HeliOx thermal management system (TMS), 16, 60-61, 64, 66, 69
Hertzberg, Abe, 10
High Energy Microwave Systems Test Facility (HEMSTF), 18, 19, 166-67
High Pressure Nervous Syndrome (HPNS), 198
High-g training curriculum, 196
High-pressure fuel turbopump (HPFT), 224
Highways of Light, 6, 19, 162, 183, 239
Hollomon Air Force Base, 196
Home Run Pictures, 10
Hoop stresses, SMES, 176
Horizon Mission Methodology (HMM), 8, 10, 237-44
Horizon Mission, 7-8, 10, 237-39
Hostile airspace, 15, 17, 132-33
Hull: catastrophic collapse, 65, 153
   contamination, 85-86
   electrodes, 30, 49-51, 56-58
   integrity, 31, 139, 159
   layers, 27-28
   lenticular, 16, 72, 73, 77, 103, 130, 141, 142-43, 153, 226-27
Hull sensor array: acoustic, 31
   electric potential, 30, 80
   fiber optic web, 30-31
   magnetic flux, 30, 80
   structure loading, 31
   temperature, 29-30
   radiation, 30
   vibration, 31
Human central nervous system, 201
Human comfort zone, 105-06
   artificial gravity environment, 215
Human factors and g-tolerance: biconic escape pod, 206-09
   bioelectronics technology, 206
   communication, 202-05
   introduction, 195-96
   super-human survivability, 205-06
   space activity suit, 206
Human factors: g-tolerance, 196-98
   immune response, 91
   respiration, 198-202
   structural anatomy, 195
Human Interface Technology Laboratory, 203
Human/machine interface, 46-47, 202

INDEX

Human-computer interface (HCI), 46
Hunt for Red October, The, 229
Hydrogen electrolytic production unit, 87-88
Hygroscopic particles, 138
Hyper-energetic: mission, 39
   spacecraft, 5, 15, 53, 73, 127, 129, 223-24, 232
Hyper-g range, 195
Hyperjump, 61
   axial, 63, 147
   departure, 145
   evade detection, 127, 132-33
   lateral, 63
   maneuver, 16, 103
Hypersonic: airship, 5
   atmospheric skipping, 64, 127, 133
   cruise, 127
Hypersonic magnetohydrodynamic (MHD) flight, 43, 62, 143, 149, 233
Hypersonic shock tunnel, at RPI, tests, 8, 24, 229
I-beam trusses, ultra-light, 26
Identification Friend or Foe (IFF), 89
Imagineering, 9, 10
Implants, 47, 113, 204, 206
   micro-electronic, 39
   surgical, 90, 91-92, 112
Indium-Gallium-Arsenide Nitride (InGaAsN), 79-81, 84
   photovoltaic cells, 79
Inertial Navigation System (INS), 37, 44
   sensors, 45
Inflatable structures, 225
   ballute decelerator, 237
   dome, 26, 166
   tensile, 52, 107, 225-27, 232
   torus, 218
Infrastructure: commercial, 5
   deployment, 19-21
   energy-beaming, 6, 7, 183
   power-beaming, 162-94
Insect swarms, 85
Institute for Problems in Mechanics, 8
Integrated microelectronics, SiC, 16, 24, 26, 27-28, 32, 54, 58, 67, 69, 79-80, 182, 184, 225
Intelligence amplification technology, 203
Intelligence gathering, 15, 16
Intelligent systems, 29, 34, 40, 153
Intensive Images, Inc., 10

Interdiction, of terrorists, 100
Interface
   crew-machine, 47
   man-machine, 46-48, 202
   wearable, 202
International Space Station (ISS), 169, 174-75
International terrorism, war against, 5, 7
Interplanetary missions, 7, 92, 150
Interplanetary spacecraft, 221
Intervention, real-time human, 40, 206
Introduction to Space Technology, course at RPI, 9-10
Inverse Bremmstrahlung absorption, 58
Ion breeze, propulsion and control system, 18
Ion cloud, 75, 77, 79-82, 83-83, 144
Ion drive, 74, 77-79, 80, 83-86
Ion plasma cone, 75-76
   fog-like tail, 75
Ion propulsion mode, 30, 33, 72-79, 82-83, 128, 136, 142-45, 157
Ion thrusters, 17, 36, 83, 144-45, 171
Ion-propulsion, 15, 38, 74-86, 130-31, 132, 134, 142, 144-45, 157, 232
ISCAN, 202
Jettison system, 237
Jitter, 39, 247
Joystick interface, 43, 47
Kantrowitz, Arthur, 10, 151-52, 223, 230
Kinetic energy munitions, 17, 100, 173
KIWI program, 214
Kiwi-A, nuclear rocket reactor, 214
Kneeling/unkneeling, 39
Knowles, Brook, 10
Koert, Peter, 231
L'Garde, 225
L1 libration points (L1, L2, L5), 240
Laboratory modules, 175
Landable disk, 215-16
Landing gear, legs, telescoping, 135-36
Landing ramp, 29, 39, 136, 155, 157-58, 213, 237
Landing zone, hostile, 159
Laryngeal Mask Airway (LMA), 201
Laser armor, lightweight, 180, 182
Laser DataLink, 37, 38
Laser imaging, detection, and ranging (LIDAR), 44, 89, 100, 153
Laser mirrors, 171-72
   "fighting," 100-01
   free-flying, 171

INDEX

lightweight, 171, 180, 182
Laser radar (LIDAR), 44, 89, 100, 153
Laser weapons system, 100, 175
Lateral flight mode, 61, 63-65, 70, 81, 123
Lattice matching, 163
Launch tube, 29, 117-19, 125, 154-56, 237
    bays, 154
Launch-assist, vertical, 89
Lawrence passenger spaceship, 211-217, 219-20, 222, 232
Lawrence, Lovell, Jr., 211-14, 232
Lehner, Jack, 10, 20-25
Lenticular domes, 33
Lenticular double-hull, 73, 103, 153
Lenticular hull envelope, 27, 29, 103, 105, 141, 142
Lenticular wing, 132, 212
LEO relay, 18, 53, 100, 101, 168
Lethal beam intensities, 100
Levitation range, 95
Life support, 103-05, 107, 136, 221
    system, 112
    HeliOx, 103-05
Life-sign monitoring, 39
Lift triangle, 74-75, 132
    dynamic lift, 74
    powered static, 75
    static lift, 74-75
LightCraft Jockey (LCJ), 35, 211, 93-94, 105-06, 223
Lightcraft revolution, 5-6, 7, 15
Lightcraft Technologies International, 5, 15
Lightcraft, definition, 15
Lightning strike, storms, 144-45
LightPort, 7
Limited strike and defensive roles, 127
Limits, thermo-structural, 28, 57
Linear accelerator (LINAC), 125
    relativistic electrons, 38, 58, 74, 77, 144
Linear electric motor, unwrapped, 57
Linear electromagnetic actuators, 175, 180
Liquid droplet radiator (LDR), 10, 173-75, 177-82
    heat rejection, 178
Liquid immersion, 17, 197, 205, 207, 242
Liquid ventilation system, 201
    demand-regulated, 200
    partial (PLV), 17, 39, 47, 90, 107, 112, 113, 133, 198, 200, 207

total (TLV), 91, 112, 114, 133, 198, 200, 205, 207
Liquid waste recycling, 107-08
Liquid-filled lung, 112, 199-202
Liquid-helium coolant, 32, 33
Liquivent, 200
Lithium battery, 93, 101
Load Master / Scanner (LMS), 36, 39, 109
Loading/unloading, 39, 237
LOM flight, 130
Long-range communications, 92, 123
Lorentz force, 50-53, 57, 64, 149, 229-30
Los Alamos Scientific Laboratory (LASL), 214
Loss of consciousness, 102, 195
Lounge area, 109
Low-noise amplifier (LNA), 167
Low-observables mode (LOM), flight, 130-31, 134
Low-power mode, 61, 85
LTI-X10, 18-19
LTI-X15 Lightcraft, 18-19
Lunar base, USSC, 42
Lunar capture option, 241
Lunar gravity equivalent, 17
Lunar orbit injection (LOI), 242
Lunar shuttle mission, ultra-fast, 9, 211, 237, 239, 242, 244
M1 (a flat-faced cone and a half-cone), 215
Macro-pulse, RF-type FEL, 80
Maglev belt, flying belt, 32-33, 39, 89, 93-95, 96, 101-02, 114, 121, 125, 153, 155-57, 205
Maglev lander, 16-17, 88-89, 95-96, 100, 101, 112, 118, 121-22, 126, 134, 136-37, 153, 157, 237
    central "donut hole," 88, 95, 123, 233
    configurations, 123-24
    de-spun, 129, 142, 143, 147
    emergency pickup, 128
    mated mode, 97, 143
    operations, 122, 159
    portal, 155
Maglev: introduction, 93
    limitations, 96
    ship boarding, 93-98
    system operation, 93-96
Maglev-belt, 93-96, 114, 155-57, 205
Magnetic aerobraking, 16, 151, 221
Magnetic bearing system (MBS), 97, 123
Magnetic docking, bays, 122-23, 157

INDEX

Magnetic Energy Storage (SMES), battery, 10, 17, 32, 56-57, 59, 65, 71-72, 121, 159-60, 162-64, 166, 168, 170, 174-76, 184-88, 191
Magnetic grapple, 89, 121
Magnetic landing gear, 237
Magnetic levitation (maglev): landers, 16-17, 88-89, 95-96, 100, 101, 112, 118, 121-22, 126, 134, 136-37, 153, 157, 237
  magnetic docking, 121-22, 221
Magnetic lock, 88, 97
Magnetic mirrors, 70
Magnetic moment, 145, 180
Magnetic nozzle, 69-71, 147, 229
Magnetic resonance imaging (MRI), 109, 221
Magnetic station-keeping control (MSC), 92
Magnetic torquing, 85, 143-45, 180, 182, 184, 220
Magnetic ultrasonic cleaning, 33
Magnetic window, 221
Magneto-aerodynamics, research, 230
Magnetohydrodynamic (MHD) slipstream accelerator, 49-66, 104, 129, 133-34, 149-50, 190, 228-29, 233-34
Magnus, effect, 145, 217
Main bridge, 34-35
Maintenance, 28, 33, 40, 117, 159, 161, 177
Maneuvering, low-G, 36, 109
Maneuvers: high performance, 63, 65, 72
  hyper-jump, 61, 63, 162, 164
  orbital boost, 37, 58, 127, 133-34, 171
  orbital flight, 63-64, 134, 152
  re-entry, 28, 152, 216-17
  suborbital boost, 63, 162, 164
  ultra-energetic, 62, 206
Mankins, John, 10
Man-machine interfaces, 46-48, 202
Manned maneuvering unit (MMU), 125
Manned Mars mission, 177, 212, 222
Manual control, 43, 73
Mars lander: concept, 207
  manned, 208
Martin Company, 215
Mass driver, 240-41, 244
Mass spectrometer, 87
Master oscillator, 168
Mattick, Tom, 10
Maximum performance maneuver, 147, 152
Media Fusion, 10, 60, 141, 167, 194
Medical evacuation (MEDEVAC), 97

Mercury capsule, 150, 215
Mercury (Hg), 17, 105
Metal matrix superconductor composites (MMSC), 32-33, 56
Meteoroid protection, 28, 125, 212, 222-23
MHD accelerator electrode, 49-51, 53, 55-58, 233
MHD aerobraking, 59, 151
MHD engine, accelerator-class, 16, 24, 38-39, 49-53, 56-60, 63-64, 129, 132-34, 149-50, 152
MHD generator, 49
  rocket-driven, 230
MHD propulsion mode, 40-41, 65, 122, 134, 168, 183
MHD slipstream accelerator, 24, 38-39, 49-51 57-58, 60, 63-64, 72, 104, 129, 133-34, 149-52, 190
  external, 49, 52, 229
MHD-fanjet, 214, 229
Microbial treatment, 107, 137
Microchannel heat exchanger, 54, 58, 62, 136, 168
Microfiber superconductors, 163, 166, 170, 184
Micro-gravity, 110
Micrometeoroid, armor, 125
  bumper, 154, 182
  protection, 125
  protection scheme, 28
Micro-organisms, 63, 88
Micro-pulses, 102
Microswitch, tongue-activated, 205
MicroWave LightCraft (MWLC), 8, 9-10, 15, 21, 24, 54-55, 162, 164, 217, 220, 224-28, 230
  demonstrator, 21
Microwave rectennas, 10, 21, 27, 32-33, 49-69, 71-72, 85, 88-89, 92, 103, 115, 133-36, 141, 159-60
Microwave relay satellite, 19, 168, 171
Microwave transmission limits, 166, 244-49
Microwave windows, perimeter, 58
Microwave-electric engine, 227
Microwave-supported combustion (MSC), 244
Microwave-supported detonation (MSD), 222, 244
Microwave-thermal Engine, 227
Microwave-to-Electric Power (MEP), 222, 227, 230
  converter, 227, 230

flightweight, 230
Mid-21st Century Space Carrier Industry, 239, 244
Mie scattering, 71
Military training center, 169
Mirror: fighting, 17
   high-power laser, 100
   orbital relay, 19, 80, 101, 106, 170-72, 247-48
Mission Leading Officer (MLO), 36-38, 40, 100, 102, 107, 109, 132
Mission, objectives, 15, 37, 43-44, 90
   operations, 43-44
   specialist, 36-39, 109
   time-critical, 136
   worthiness, 29
Mode instability, 77
Molecular absorption and scattering, 244
Momentum coupling coefficient (see also, coupling coefficient), 72-74, 83
Momentum exchange, 75, 229
   with atmosphere, 52
Monitoring, 29, 37, 39-41, 43-45, 72-73, 138, 204
   non-invasive, 202
Monocle, laser relay station, 19, 80, 101, 171-72
Moon Maker, 215
Moon, permanent presence, 239, 242
MRS-01, microwave relay satellite, 19, 168, 171, 173, 183
MRS-02, microwave relay satellite, 171
Mucous membranes, 105
Multiple independent re-entry vehicle (MIRV), 28-29, 223
MWLC (MicroWave LightCraft), development plans, 21-22, 24
   materials technology, 23
   technology systems, 20, 22, 25
   propulsion system, 54, 228
   vehicle structure 23-24, 224-28
Myrabo, Kenneth, 10
Nano-fibers, 26
   structure reinforcement, 225
Nanosatellite, 263, 265
Narcotic effects, 198-99
NASA Ames Research Center, 215, 264
NASA Headquarters, Washington, D.C., 8
NASA Langley Research Center, 215
NASA Lewis Research Center, (now NASA Glenn), 230

Nasal tubes, 112
   liquid breathing tubes, 112
Navigation systems, 37, 44, 46
Naval Air Development Center (NADC), 196
Navigation officer, 36-37, 109
NERVA program, 214
Neural interface, 206
Neutron exposure, 173
NewAbilities Systems, 204, 206
Nimitz class, nuclear aircraft carrier, 170, 214
Nitrogen narcosis, 105, 198
Nitrox 21, 198
Nitrox 32, 198
Noordung, Herman, 215
Nordley, Gerald, 240
NRX A-6 reactor, 216
Nuclear aircraft, manned, 214
Nuclear Energy for Propulsion of Aircraft (NEPA), 213
Nuclear Magnetic Resonance Imaging (NMRI), 109, 221
Nuclear propulsion, spacecraft, 213-14
Nuclear thermal rocket, 177
Nuclear-powered, 170, 175
Nuclear-ramjet, direct cycle, 214
Nuclear-turbojet, direct cycle, 214
O'Neil, Gerard, 240-41
Oak Ridge National Laboratory, 213-14
Observation deck, 17, 34, 104-05, 143
On-station loiter periods, 42
Orbital boost, 19, 37, 58, 63, 72, 127, 133-35, 162, 164, 171
Orbital debris, 19, 44, 153, 173, 184
Orbital maneuvering system (OMS), 153
Orbital mechanics, 37, 172, 184-86
Orbital power station, 73, 100, 173
Oxygenated perfluorocarbons, 17, 220
Paddle, conducting, 49, 51
Parachute, 46, 124, 224, 237
   ram-air, 209
   mattress-type, 155
Partial liquid ventilation (PLV), 17, 39, 47, 90, 107, 112-13, 198, 200-02, 205, 207
Particulate absorption and scattering, 52, 244, 246
Particulate contaminants, 28, 63, 86, 87
Paschen minimum, 222, 246
Passenger spaceship, 211, 215, 222
Passive relay station, 19, 171

INDEX

PDE mode, 67-73
PDE thrusters, 16, 19, 36, 38-39, 65, 69-73, 132, 138, 144, 146-49
PDS-01, 18-19, 164, 166-71
PDS-02, 18-19, 53, 80, 135, 164, 166, 168-71, 175, 177, 211, 214
Peck Polymers, 231
Pedestal foot, 237
   landing gear, 136
   mushroom, 237
Pendulum motion (see also, "falling leaf"), 127
Pendulum oscillation, descent, 130
Penetration distance, 77
Perflubron, 200
Perfluorocarbon: liquid, 17, 40, 112, 199-200, 220
   vapors, 105, 107
Perfluorooctylbromide (PFOB), 200
Performance specifications, 72-73, 81-84
Personal communicator, 2-way, 37, 41, 91-92
Personal stun weapon (PSW), 100-02
PFC: breathing, 199-200
   evacuation, 202
   computer controlled, 201, 202, 205
PFC liquids, 17, 40, 112, 199-200, 220
   oxygen-rich, 206
Phased array, 8, 65, 68, 89, 92, 159, 161, 168, 180, 190-91, 235
   microwave radar, 89
Phoebus 2B reactor, 214
Photon emission, ultraviolet, 77
Photovoltaic (PV) array, 16, 33, 78, 80-81, 84-85, 103, 126, 130-31, 135, 171, 232
   thin-film, 16, 184
Physiological response to g-forces, 106, 195
Pierce, H. Franklin, 211
Piezoelectric, actuators, 18, 30, 32
   films, 30
Pilot, helmsperson, 34, 36-38, 43, 85, 97, 109, 132-33, 196-97, 212
Pinned-fin, heat exchanger, 62
Planetary defense: Extraterrestrial threats, 5, 7
   internal threats, 7
Plasma filaments, high conductivity, 102
Plasma paddle, 52-53, 58-59
Plasma radiation shield: space, 125, 219
   hybrid, 218
   pure electrostatic, 218
   pure magnetic, 218, 221

Plasma, electrically conductive, 51, 147
Plenum management system, HeliOx, 103-05, 159
Plug nozzle, 233
Pod boarding and de-boarding, 115, 188-19
Pod: deployment, 118, 136, 237
   launch tube, 29, 117-19, 125, 154-56, 237
   propulsion system, 115-17
Pointing: coarse, 180
   fine, 71, 182
Pony Blimp B-11, 231
Poor-man's HeliOx, 198
Posterior-to-anterior, acceleration, 195, 203
Power beaming infrastructure, 162-94
   basing considerations, 164
   ground-based stations, 166-70
   microwave power transmission, 163-64
   nuclear-powered, 173-82
   orbital power relays, 170-72
   SMES, 162-64
   solar-powered, 182-94
   space-based stations, 172-73
Power reception, microwave, 71-72
Power transmission: "cross-link," 53, 162, 171
   "down-link," 19, 164
   limits, 222, 244
   time-shared, 61
   "uplink," 53, 162, 164, 165, 171, 172, 245, 247, 248
Power-beaming experiments, 183, 228, 231
Power-beaming Infrastructure, 5, 7, 19-21, 162, 165, 166, 193
   commercial, 5, 182-83
   "tractor," 53, 164-166, 171, 193
   "pusher," 164-66
   space, 5, 165, 172-73
Power-beaming station, 19, 53-54, 166-69, 248
   rechargeable, 5, 53, 182
Powered parachutes, 224
Powered re-entry, 152
   maximum performance descent, 152
Predicted flight plan, 37
Pressure airship, 9, 225, 232
   structural shell, 16, 265
   structure, 7, 223
Pressure vessel: lenticular, 26, 77, 103, 105
   toroidal, 16, 26, 30, 32, 56, 103, 153, 154
Pressurization, 26, 27-29
   bulkheads, 39, 106, 116, 119

INDEX

door, 29, 39, 106, 123, 156, 207
Pressurized tensile structure, 232
Pressurized water reactor (PWR), 62, 170, 177, 214-15
Princeton University, 8
Problems of Space Flying, 215
Procedures, emergency, 65-66, 71, 73, 81, 85-86
Project Apollo spacecraft, 150, 176, 177, 181, 197, 215-17, 233, 242
Project Moon Base, 215
Prone position, 195
   "eyeballs in," 195-96, 206
   "eyeballs out," 195-96
Propulsion energy converters, 162
Propulsion systems integration, 222
Propulsion, ion, 74-86
   application, 74-77
   components, 79-80
   emergency procedures, 85-86
   evasive maneuvers, 84-85
   physical processes, 77-79
   performance specifications, 81-84
   system, 80-81
   theory, 74-77
Propulsion, MHD, 49-66
   conversion, 52-58
   physics, 50-52
   system, 58-63
Propulsion, PDE, 67-73
   emergency procedures, 73
   performance specifications, 72-73
   pulsed detonation engine, 69-72
   theory and application, 67-69
Propulsive efficiency, 49, 59, 63, 71, 149
Protective gas bag, 153-54
Psychological aversion, PLV/TLV use, 210
Pulmonary blood flow (PBF), 195
Pulsed laser weapons system (PLWS), 100, 101
PulseTimeDomain communication, 38
Pusher beam, 164, 166
PV array, 28, 80-81, 88
   thin film, 28
Pyewacket Program, 215
   tests, 217
Radar signature, 85, 130, 141
Radar: laser, 100
   short-range, 89
   system, 89
Radiation shield, 50, 125, 218-19

shadow-type, 173, 178
Radiation shielding, 121
   nuclear, 213-14
Raizer, Yuri, Prof., 8, 10
Ram-air parachutte, 155, 209
   canopy, 224
Ramp, 29, 42, 115, 157-58
   entrance, 39, 157
   extension/retraction, 39, 237
Rapid decompression, 105, 112
Rapid extraction procedure, 97
Rappelling harness, 93
Rapture of the deep, 198
Rather, John, 10
Raytheon Corporation, 230
Reaction control system (RCS): in-space, 89, 125, 188, 190
   rockets, 89, 116-18
Reaction Motors, Inc, 212
Real-time flight path, 37
Real-time spin balancing, 87
Reception, beamed power, 58
Reconfigurable seating, 106
Reconnaissance/ surveillance/ rescue missions, 10, 44, 127, 129-30, 144
Recovery of covert military personnel, 100
Rectenna: arrays, 32, 54-55
   central, 49, 55, 58, 59, 89, 123, 149, 159, 233, 235
   off-axis annular, 89
   "lampshade," 55, 56, 58, 233
   nested, 54, 58, 141, 159, 233
   next generation, 193
   truss structure, 135
Rectifying antenna, (see also, rectenna): actively-cooled, 16
   microchannel heat exchanger, 54, 58, 62, 136, 168
   nested antennas, 54, 58, 141, 159, 233
   reflecting backplane, 54
   SiC wafer carrier, 54
   solid-state electronics, 54, 62, 230
Reentry: atmospheric, 37, 150, 237
   hypersonic, 119, 216, 230
   lenticular disk, 215
   lifting bodies, 223, 225
   maneuvers, 28
   options, 215
   physics, 37

INDEX

vehicle, 29, 150, 152, 155, 215-16, 230
Rehydration: food, 87, 110
   unit, 110
Reinhardt, Senter, 9-10
Relaxation and socialization, crew, 109
Relay satellite: laser, 80
   LEO, 19, 100, 101, 168
   microwave, 19, 168, 171
   mirror, 19, 80, 100, 101, 166, 172, 247-48
   passive, 171
Remote manipular cranes, 180
Rensselaer Polytechnic Institute (RPI), 7, 59
Rescue, of downed Lightcraft, procedure, 127, 138-39
Rescue: downed aeronaut, 15
   downed astronaut, 15
   downed personnel, 128
Retinal identification, 47
Retrieval: downed Lightcraft, 127
   sensitive black boxes, 127-28
Retro rocket, 88, 89, 116-18, 207
Retro-braking, 117, 206
Reverse osmosis (RO), 63, 88, 137-38
   filter, 88, 137, 138
Rim electrodes, 53, 55, 57, 58, 148
Rim magnets: bucking, 51-52, 57, 171, 227
   superconducting, 58, 59, 67, 85, 93, 145-46, 147
Rip-stop, 225
Robotic equipment, 173
Rocket gas generator, 119, 154-55, 237
Rocket thrusters: electric, 188, 230
   RCS, 116
Rocket-driven: MHD generator, 230
   open-cycle converter, 227
Roles, crew, 35-36
Rosa MHD fanjet, 229
Rosa, Richard J., 10, 214, 229
Rotating lift vector, 82
Russian Academy of Sciences, Moscow, 8
Safety Officer, 39
Safety protocols, 165
Santa Barbara yacht basin, 229
SARSAT, 46, 92
Satellite attack or defense, 100
Satellite power stations, GEO, LEO, 7
Saturn S-1, 212
Saturn S-1B, 212
Saturn V, 176-77, 197

Sauls, Bob, 10, 183
Scans: finger print, 46
   of crew, 48, 109, 221
   retinal, 46, 47, 111, 114, 203
Schlesak, Joe, 231
Schottky diode, 54
Scud, 173
SEAL insertion and recovery, 100
Search-and-rescue (SAR), 46, 100
Second-in-Command Officer (SCO), 36-38, 40, 100, 102, 109
Security levels, L1 (battle stations; L2 normal operation), 36, 37
Seed electrons, relativistic, 58
Self-defense, 17, 132
Self-magnetic pinch pressures, 56
Sensor array, 30-31, 46, 94
Sensor grid: inner hull, 28
   outer hull, 28
Sensor suites, 44, 206
Sensor-net, microwave, 202
Shadow-shield, 179
   reactor, 214
Sharp II, 230-31
Shell and flame, 213
Shesta, John, 211
Shields down, 221
Shields up, 219
Shneyder, Mikhail, 8
Showstopper, 8
Shuttle orbiter, 176-7
Silent running, mode, 17, 84, 127, 131
Silicon-carbide (SiC), 16, 24, 26-28, 32, 54, 58, 67, 69, 79, 80, 182, 184, 225, 231
Silver iodide dispersal, 138
Skip-glide trajectory, 150
Sleep-floating, 110
Sleeping bag, 110
Smart materials, 16, 32
Smart tubes, computer controlled, 201-02
SMES battery, 170, 180, 184
   rechargeable, 175-76
SMES unit, 17, 33, 43, 56, 62, 65, 71, 80, 85, 101-02, 126, 127, 131-32, 159-60, 164, 166, 176, 180, 181, 184, 191
Snow-plow MHD accelerator, 53, 58
   Lorentz force thruster, 52
Soaring/thermaling mode (riding thermals), 127
Solar power stations, 8, 21, 173, 182, 183, 242

INDEX

Solar proton storms, 17, 30, 142, 182, 218, 232
Solar wind plasma, 125
Sommerscales, Euan, 10
Sonic boom: annihilation, 49, 53, 129, 234-35
Space access: highly reliable, 6
   low cost, 7
   safe, 7
Space activity suit (SAS), 63, 206
Space activity unitard (SAU), see also, liquid-filled ultra-g suit, 110, 113-14, 203, 204, 206
Space age, dawn, 211-13, 230
Space Nuclear Propulsion Office, NASA, 214
Space plasma radiation shield (SPRS), 125-26, 219
Space plasma shield, operations, system, 125-26
Space shuttle main engine (SSME), 224
Space solar power, 5
Space structure, inflatable, 115
Space Studies Institute (SSI), 8
Space superiority, 7, 15, 17
Space tourism, 183
Space-based microwave (SBM), 164, 172, 235
   power-beaming stations, 162, 248
Spacecraft Design Studio, 9
Special ops, 35, 39, 42
Speech synthesis, 46
Spin-off, 6
Spin-stabilzation, 146
Spiral-wrapping, 164
SPS-01, 182-94
SRS, 224-25
Stabilizing coil, 237
Station-keeping, 15, 44, 92, 130, 170, 174, 182, 185, 240
Stealth: mode, 79, 130
   procedures, 17, 141
   propulsion system, 15
Stealthy entrance / exit, 17
Stone skipping over water, 64
Storage tanks, water, 62, 87, 96, 123, 137
Storm cellar, 174, 181-82
Strategic Defense Initiatives Office (SDIO), 10
Structural backbone, 16, 30, 56, 227
Structural deformation, 195
Structural integrity system (SIS), 30-31
Structure: active, 29-32
   aeroshell, 225, 230
   balloon-type, 224
   main, 26-27
   open-cell, 27
   pressure-airship, 7, 223, 224-25, 232
   secondary, 27
Stun gun, 139
Stun weapons, 100, 101-02
Submarine: electromagnetically propelled, 229
   Stewart Way, 229
Subsonic cruise, 15, 36, 128, 214
Subsonic ion propulsion (IP) flight, 127, 142, 232
Subvocalization sensor, 47
Superconducting magnet, 17, 32-33, 50, 55-56, 93-97, 121-22
   bucking, 51-52, 57, 171, 175, 227
   flight-weight, 32, 52, 164, 229, 233
   perimeter, 32, 56, 137, 145, 237
Superconducting magnetic energy storage (SMES), 17, 32, 56, 121, 162-164
   battery, 170, 175-76, 180, 184
   system, 56, 65, 69, 84, 121, 162
Superhuman capabilities, 17, 205
Supersonic cruise, 127
Supersonic pulse detonation engine (PDE) flight, 127
Surface contamination, hull, 86
Surface-induced electrical air breakdown, 244
Surveillance, 15, 16, 44, 127, 129, 130, 152, 165
Survivability, super-human, 205
Suspended particulates, 137, 166
Systems development milestones, 21-25
Tactical: activities, 100
   doctrine, 100
   systems, 100-02
   weaponry, 100
Techno-babble, 7
Technology directions, high leverage, 239
Technology Readiness Level (TRL), 21
Tensile structures: advanced, 224-25
   aeroshell, 225-228
Tensing, 205
Tension aeroshell, 237
Terminal interfaces: distributed, 46
   use, 41-42
Terrestrial electric grid, 168, 170, 183-84
   applications, 6
Terrorist activity, 15
The Abyss, 199, 201
The bends, 105, 112
The Case for Mars, 207, 208

INDEX

The Forward Look, 213
The High Frontier, 240
Theory of Propulsion, course at RPI, 9
Thermal blooming, 39, 81, 244, 247
Thermal management system (TMS), 16, 60-64, 66, 69
   high-power, 61, 62, 63
   low-power, 61, 62
Thermal protection system (TPS), 155
Thermal runaway, magnets, 30
Thermal updrafts, riding, 131
Thin-film structures, 17
Threat negation, 17
Thrust vectoring, 67, 69, 132, 142-43, 150
   active, 65
   engines, 69, 129
Thunderbolts (Chrysler Imperial parade cars), 213
Tie-down, 88, 136
TongueTouch Keypad (TTK), 204, 206
Top secret, mission, 16
Toroidal pressure vessel, 26, 30, 153-54
   perimeter, 16, 32, 56, 103
Tory 2-C, nuclear-ramjet reactor, 214
Total liquid ventilation (TLV), 112, 133, 200, 202, 205
Tracking beacon, on-board, 164-65
Tractor beam, 164, 166
Tractor/repulsor field, magnetic, 89
Trajectory, 150-51
   boost/glide, 64
   quasi-ballistic arcs, 64
Transatmospheric: boost, 16, 17, 28, 73, 93, 136, 138
   flight, 26, 49, 50, 52-53, 103, 127, 128, 149, 211, 228, 229, 240
   trajectory, 35, 134
   vehicles, 15, 150
TransAtmospheric Vehicle Design (TAVD), course at RPI, 8, 9-10
Trans-earth injection (TEI), burn, 242
Translunar injection (TLI), velocity, 241-42
Transmitter: Ground-based microwave (GBM), 71, 164, 166, 168, 171, 183, 230, 246
   phased array, 68, 180, 235
   solid-state, 184, 188
Transport, emergency supplies, 128
Triangular truss "keel," 173, 177
Trimix, 198

Tripod landing gear, 124, 135-36
   auxiliary, 135
Turbo-generators, 170, 182
Two-stream instability, 77
Ultra-capacitor, 93
Ultra-energetic, vision, 7
Ultra-g suit, 46-48, 110-11, 203
   liquid filled, 133
   personal protection, 112-14, 119, 155, 243
Ultra-high-g maneuvers, 115
Ultralight: aircraft, 144, 224
   airship, 224
   helicopter, 77, 224
   powered parachute, 224
Ultrasonic cleaning, 33, 86
Ultra-stealth, 38
   capabilities, 5
Umbilical cords, 88
Unidentified spacecraft, 165
Unitard, 90, 110, 113-14, 202-06
United Applied Technologies, 225
United States Air Force (USAF), 196, 213
Universal translator, 38
(USRA) University Space Research Association, 10
Uranium, enriched, 170
USAF School of Aerospace Medicine, 196
USS Enterprise (CVN-65), 174-75
USS George H.W. Bush (CVN77), 170
USSC special-ops, 35
Utilities and auxiliary systems, 87-89
   networks, 87
V1 Buzz Bomb, 227
Valentine, Lee, 8
Variable geometry inlet, 149
Vectored thrust, 212, 217, 232
Vehicle structures, 7, 244
Velcro, fasteners, 114
Ventilator system, TLV, PLV, 200-02
Vertical flight mode, 49, 63-65, 70, 74-76
Vertical takeoff and landing, 15, 127, 142, 212, 232
Vertical transmission efficiency, 245
Vertigo, 105-06
Video games, 42
Viewing window, to exterior, 34
Virtual cathode instability, 77
Virtual Computer Network, 46-47
Virtual jet engine, 51

INDEX

Virtual reality (VR), 40, 42, 203
   display, 46, 111
   drivers, 46
   environment, 47, 110-12
   eye-movement tracking, 47
   goggles, 46, 47, 111, 114, 155
   high quality, 110
   iris scans, 48
   proprioceptor sensors, 48
Virtual Retinal Display (VRD), 203
Virtual vision, 203
Virtual-reality tours, system, 42
Visual-audio communications (VideoCom), 91
Voice recognition, 46
Voice unscrambler, 199
von Braun, Magnus, 212
von Braun, Werhner, 10, 207, 209, 212, 215, 221
VR immersion, 203
VR link, 112
VR systems, 110-12, 115-16, 203
   interface, 110, 112, 203
Walk on the walls, 212
War against terrorism, 7
Warren, Matthew, 10
Washington Technology Center, 203
Waste heat, 16, 29-30, 54, 58, 61-62, 66, 72, 80, 103, 133, 136, 140, 168, 170, 172-73, 175, 179, 231
Waste management control system, 103
Waste management, system, 107-08
Waste removal, 87

Water collection from cumulus clouds, 137-38
Water purification system (WPS), 61-62, 88
Water storage system, 32, 123, 137
Water,
   coolant, 17, 32, 54, 62, 87, 88, 95-96, 103, 107, 121, 122, 123, 134, 136, 170
   de-ionized, 16, 88, 168
   expendable coolant, 87, 122, 240
   filtered, 62, 87, 137, 138
   steam, 16, 49, 53, 57-58, 61-63, 69, 85, 103, 133-34, 168-70
Water, ultra-purified, 62, 168, 169
Water-filled maglev lander, 136-37
Way, Stewart, 229
Way's MHD engine, 229
Weakly ionized air, 75
Weapons, personal stun, 101-02
Wearable cockpit, in a helmet, 203
Webb, Paul, 206
Weight and balance, 39
Westinghouse Astronuclear Laboratory, 214
White Sands Missile Range (WSMR), 18-19, 166, 168, 171, 228
Winch cable, 157
Windows, microwave transparent, 56
Wireless power transmission, 214
Womb environment, 200
WorldNet Virtual (WNV), 5
X-15, rocket plane, 212
Zero-g, 87, 110, 177, 180, 215

ABOUT THE AUTHORS

Leik N. Myrabo:

Dr. Leik N. Myrabo teaches doctoral candidates in Engineering Physics at the Rensselear Polytechic Institute in New York. Before joining the faculty at RPI in 1983, he spent a total of seven years pursuing "Star Wars" research with Physical Sciences, Inc., W.J. Schafer Associates, and the BDM Corporation. He has authored and co-authored more than 200 journal, symposium, and conference articles, and one book – The Future of Flight. Over the past 25 years, he has given hundreds of invited presentations on lightcraft technology to a wide variety of audiences. His lightcraft research has been covered in 18 television documentaries and over 70 print media articles.

From Sept. 1996 to Sept. 1999, Leik brought laser propulsion from the raw concept stage to flight reality by flying his laser lightcraft prototypes on a 10 kW high-power infrared laser at White Sands Missile Range (WSMR) in New Mexico, funded with $1 million of combined USAF and NASA support. Since his first WSMR test in July 1996, he has conducted 24 separate test campaigns at the High Energy Laser Systems Test Facility in New Mexico. On 2 October 2000, sponsored under a grant to his company, Lightcraft Technologies, Inc., he established a new world altitude record of 71 meters for laser-boosted vehicles in free flight. On 2 Dec. 2002, he was awarded U.S. Patent #6488233 – "Laser Propelled Vehicle" – covering that successful lightcraft design.

In December 1999, he demonstrated the first-ever vacuum photonic thrust measurements for a laser-accelerated lightsail using 5-cm diameter prototypes, constructed from exotic carbon micro-truss fabrics by ESLI. These experiments employed the 100 kW LHMEL II infrared laser at Wright Patterson Air Force Base, within an evacuated 7x9-ft tank. In Dec. 2000, he returned to the LHMEL II facility to carry out vertical wire-guided flights of even more advanced ESLI laser sail materials.

Leik's research is focused on innovative aeronautical, aerospace, and space flight propulsion concepts for the 21st Century and beyond. This advanced energetics research takes a long-term, high-risk approach to identifying areas of potential technological breakthrough. The primary research emphasis is on the application of beamed energy (e.g., laser, microwave, millimeter wave sources) and field propulsion engines for future craft designed for a variety of hypersonic flight missions. These revolutionary beam-powered vehicles have their propulsive energy source on the ground or in space, and carry minimal on-board propellant or expendable coolant. Among other promising engine concepts (i.e., compatible with beamed electromagnetic power), he has investigated pulsed detonation engines, various rocket-based combined cycle engines, the directed-energy AirSpike, magnetohydrodynamic slipstream accelerators, rotary pulsejet, scramjet, and a unique air-breathing "Ion-Breeze" thruster.

ABOUT THE AUTHORS

John S. Lewis:

John S. Lewis is Professor Emeritus of Planetary Sciences and Co-Director of the Space Engineering Research Center at the University of Arizona. He was previously a Professor of Planetary Sciences at MIT and Visiting Professor at the California Institute of Technology, and he and his wife Peg were Visiting Professors at Tsinghua University in Beijing, PRC for the 2005 - 2006 academic year under the BYU China Teachers Program.

John has taught a wide variety of planetary science and chemistry courses at both the undergraduate and graduate levels. His research interests are related to the application of chemistry to astronomical problems, including the origin of the Solar System, the evolution of planetary atmospheres, the origin of organic matter in planetary environments, the chemical structure and thermal history of icy satellites, the hazards of comet and asteroid bombardment of Earth, and the extraction, processing, and use of the energy and material resources of nearby space. He is the author of the standard cloud models of the Jovian planets, and the chemical interaction model for regulation of the atmospheric composition of Venus by surface minerals. He was first to propose the existence of a deep global ocean on Jupiter's moon Europa, and has made a number of specific proposals for the use of the natural material and energy resources of nearby space both to defray the costs of large-scale space exploration and to meet Earth's future needs. He served on the Board of Directors of American Rocket Company (AmRoc) during the development of hybrid rocket motors for the private launch business, a process that culminated in the use of an AmRoc-designed motor to propel SpaceShipOne to an altitude of over 100 km and win astronaut's wings for its pilots in 2004.

He has served as a member or Chairman of a wide variety of NASA and National Academy of Sciences (NAS) advisory committees and review panels on subjects ranging from exploration of Venus to outer planet entry probes to comet and asteroid missions to planetary quarantine. He has written, edited, or translated 17 books, including graduate and undergraduate texts and three popular science books (Rain of Iron and Ice, Mining the Sky, and Worlds without End), and has authored over 150 scientific publications.

He has been a guest commentator on Chinese Central Television in Beijing for the launches of the two-man Shenzhou 6 flight, the Chang'e lunar mission, and most recently for the three-man Shenzhou 7 flight and EVA in September, 2008, He has also made dozens of TV specials for the Discovery Channels in Canada, the US, and England, the History Channel, the Science Fiction Channel, and PBS, as well as for German and Japanese television.

AUTHORS' AFTERWORD

<u>2025 Space Command Mission</u>: The LTI-20 spacecraft must transport a 12 person crew to the far side of the planet, loiter in the atmosphere on-station for 2 weeks undetected, and return to the continental United States without "refueling" in the conventional sense – a "Mission Impossible" for all but beamed energy propulsion.

The astonishing vision conveyed by this book is just a small hint of what the future holds. The central concept in our story, Beamed Energy Propulsion (BEP), is far more than a dream or idea: it is a powerful enabling technology that will radically transform the future of air and space transportation. It is physics, not imagination. BEP permits us to build and fly hyper-energetic vehicles driven by remote sources of laser, microwave, and mm-wave power. Such vehicles provide unique performance that would be impossible to achieve with traditional, combustion-based engines. Vehicles driven by BEP will be "greener," safer, smaller, lighter, faster, and far more efficient than any currently existing means of flight transport. Beamed Energy Propulsion is inherently a clean technology. It uses electricity, which can be produced in an eco-friendly manner. It doesn't matter whether the electricity comes from Earth-based or space-based solar, wind, fission, fusion, hydroelectric (or other), so we can choose how to produce it – and we can choose from the start to produce it in an ecologically responsible way. We see here the emergence of nothing less than a sustainable energy infrastructure, one that permits us to relegate our crude oil to use as a chemical feedstock, a resource far too precious to burn.

Throughout the history of human transportation, two things have acted as the practical constraints on how fast, far, and high we can go: the power density of our engines and the energy density of our fuels. From walking to riding horses to the steam locomotive and onward, progress has depended on getting more power out of smaller packages and using fuels that pack more energy into smaller spaces. Steam engines gave way to piston engines, then to jet turbines, then to rocket engines, each surpassing the earlier engines in efficiency and performance. The range of these vehicles grew from a few miles to hundreds, then thousands of miles, reaching to the Moon, the planets, and into interstellar space. Speeds grew from feet per second to tens of miles per second.

Gasoline and diesel engines were a major breakthrough from the steam powerplants that came before. Instead of having to boil water and use bulky wood or coal for fuel, now the energy of very compact liquid fuels could be used more directly to move pistons up and down.

The next major breakthrough came with the invention of turbine engines. Again, a smaller, lighter powerplant using a fuel with a very high energy density allowed jets to soar to new heights and travel vast distances at high speeds.

All this progress in combustion-based propulsion has brought us to a plateau in performance and efficiency. The current pinnacle of transportation, the Space Shuttle, uses a propulsion technology (the hydrogen-oxygen rocket engine) that dates from the 1960s. But because it is powered by chemical propulsion, most of the power of the Shuttle's engines is "wasted" simply lifting the vehicle's fuel supply. Indeed, the Shuttle mission that carries 22 tons of payload into orbit weighs a staggering 2250 tons at liftoff, a payload ratio of 1%. It is impossible to improve upon this performance so long as the vehicle must carry its own propulsion system and propellant.

But what if we could leave the fuel behind – and exploit beamed energy propulsion instead? What if we could have the total output of a spaceship's powerplant devoted to lifting the payload and the structural shell that supports the engines and payload? Such a system would represent the next major breakthrough in aerospace propulsion, first enabling affordable access to space for launching constellations of nanosatellites, and finally transform humans from Earthlings into space-farers. We could not only lift 100 times as much payload with a given amount of energy – we could do so with cheap electric energy. By leaving the fuel behind, we also can make spaceflight much safer.

Cheap access to space is nothing short of revolutionary. Just as internal combustion, electricity, telephones, computers, the internet, and aviation have dramatically changed our lives, so will our lives be changed again by our gaining cheap access to nearby space. We will tap the vast energy resources of the Sun and material resources of the nearby asteroids while expanding human civilization into this unbounded new environment. We will enter a new era for humanity; a new Age of Exploration.

Is this some futuristic pipe dream? Interestingly, laser-powered lightcraft have already flown in miniature form. In a series of experiments nearly a decade ago, saucer-sized lightcraft were successfully launched using a laser (one non-optimized for propulsion) at White Sands Missile Range, New Mexico to 71 meters altitude. That altitude record still stands. In one generation, the science and technology needed to build and fly full-size Lightcraft has been developed to maturity, ripe for commercialization. All that's needed now is to actually build them. The problem has evolved from a scientific one to an engineering one.

With "wireless power transmission" technology in hand, other *fringe benefits* beyond access to space resources, space exploration, and environmental preservation will also appear. Energy beamed down from power stations in space can be used for electrical propulsion of cars, trucks, and trains, and for heating and cooling. Fossil fuel consumption and carbon dioxide generation will plummet. A global power system infrastructure with many parallel components, highly resistant to failure and sabotage, will emerge. The world will be a cleaner, safer place ... for *all life forms*.

The sky is no longer the limit.

ADDITIONAL PICTURE / ILLUSTRATION CREDITS

The following talented people from Rensselaer Polytechnic Institute (RPI) are responsible for the artwork in **Lightcraft Flight Handbook LTI-20** as indicated.

Russell Mohammed:
Figures. 6.3.3, 7.1.2, 7.5.2, 7.5.3, 7.5.5, 7.5.6, 13.1.2, 16.3.1, 16.3.2, 16.3.3, 16.3.4, 16.3.5, 16.3.6, 16.9.1, 16.9.2, 16.9.3, 16.9.4, 17.3.3, 17.3.4, 17.5.1, 17.5.3, 17.5.5, 20.3.1, 20.4.1, 20.4.2, 20.4.3, 20.4.4 and 20.6.1.

Chuck Lindgren:
Figures 3.5.1, 3.6.1, 10.2.1, 10.2.3, 10.2.7, 13.3.1, 14.9.1, 15.1.1, 15.2.1, 15.3.2, 15.3.1, 19.8.1, 19.8.2, 18.8.3, 19.8.6 through 19.8.12 and 20.8.4.

Scott Frazier:
Figure 19.9.10.

Tom Dickerson:
Figures 19.9.3 thru 19.9.11 and 19.9.12 through 19.9.14.

Joel Limmer:
Figures 5.7.2, 6.1.2, 6.1.3, 17.3.5 and 17.3.6.

Jeff Moder:
Figures 7.2.3, 7.2.5, 7.2.6 and 7.2.7

Doug Parker:
Figures 14.9.1 and 14.9.7.

Lightcraft CAD Team:
Figures 2.3.2, 2.3.3, 3.1.1, 5.2.2, 5.2.3, 5.2.4, 7.3.1, 8.3.1, 8.3.2, 9.5.1, 9.5.2, 10.2.2, 13.3.1, 14.3.1, 14.9.5, 14.9.8, 15.0.1, 15.1.2, 16.4.1, 16.5.1 and 17.2.3. The Lightcraft CAD Team members are: Joe Almeida, Tim Gallus, Dean Melone, Sam Raney, and Anette Strassberger.

Lightcraft Website Team:
Figures 3.5.1, 3.5.2, 3.6.1, 6.3.1, 6.3.2, 9.2.1, 10.2.3, 10.2.6, 10.2.8, 10.5.1, 12.2.2, 14.9.6, 15.0.2, 15.0.3, 16.3.7 and the Preface picture on page 5. The Lightcraft Website Team members are: Senter Reinhardt, Barry Kusumo, David Lewison, Hernan Orellana, and Chuck Lindgren.